Social Emergency Medicine

Harrison J. Alter • Preeti Dalawari
Kelly M. Doran • Maria C. Raven
Editors

Social Emergency Medicine

Principles and Practice

Springer

Editors
Harrison J. Alter
Department of Emergency Medicine
Highland Hospital
Alameda Health System
Oakland, CA
USA
Andrew Levitt Center for Social
Emergency Medicine
Berkeley, CA
USA

Kelly M. Doran
Departments of Emergency Medicine and
Population Health
NYU School of Medicine
New York, NY
USA

Preeti Dalawari
Division of Emergency Medicine
Saint Louis University School of Medicine
Saint Louis, MO
USA

Maria C. Raven
Department of Emergency Medicine
University of California
Philip R. Lee Institute for Health
Policy Studies
San Francisco, CA
USA

ISBN 978-3-030-65671-3 ISBN 978-3-030-65672-0 (eBook)
https://doi.org/10.1007/978-3-030-65672-0

This Springer imprint is published by the registered company Springer Nature Switzerland AG
The registered company address is: Gewerbestrasse 11, 6330 Cham, Switzerland

Foreword

> The tragedy of life is often not in our failure, but rather in our complacency;
> not in our doing too much, but rather in our doing too little;
> not in our living above our ability, but rather in our living below our capacities.
> Benjamin E. Mays, (1894–1984)

I offer a few statements by people whose words and actions helped me understand the meaning of social emergency medicine.

- I know what Rudolf Virchow (1821–1902) meant when he told his father: "I am no longer a partial man but a whole one in that my medical creed merges with my political and social creeds."
- I understood the lack of worker safety and food quality of the Chicago meat industry that Upton Sinclair (1878–1968) described in "The Jungle."
- My eyes were opened by Rachel Carson's (1907–1964) concerns for protecting the environment in the "Silent Spring." She taught us that whatever we do can affect every other human, animal, and plant through destruction of the quality of our air, water, and land.
- I appreciated the transformative thoughts of Gregory Pincus (1903–1967) as he discussed how his creation of the oral contraceptive would give women the right to control when they would become pregnant.
- I worked with Norman Pirie (1907–1997), a British biochemist who led an international team creating leaf protein for human consumption in an attempt to end kwashiorkor and marasmus.
- In the New Yorker, I read Berton Roueché's (1910–1994) monthly column "The Annals of Medicine" where he described people whose new diseases were treated by creative scientists and activist physicians.
- I read William Haddon's (1926–1985) papers on the role of an epidemiologist in searching for the factors that cause injury in the United States and the world. I learned to believe that his epidemiologic triad of the host, the agent, and the environment could be employed to investigate any problem I wished to address in emergency medicine.
- We all began to appreciate the remarkable civil rights advances led by Martin Luther King, Jr. (1929–1968) and the astounding health rights potential of the enactment of Medicaid (1965) and Medicare (1965) legislation.

When many of the earliest physicians in emergency medicine in the United States began caring for patients in "Emergency Rooms," there was little prior education in the field, little prehospital care, little or no graduate or postgraduate EM education, and very mixed opinions, if not outright rejection, of this work in emergency medicine by the leaders of organized academic medicine. I, for example, started my role at Bellevue Hospital with the support of New York City government and health leaders, but without support of the New York University School of Medicine. We worked to ensure that our doors would be open to everyone, under any circumstances, and as a right, independent of finances.

As we began this work, it became obvious that many individuals who were critically ill and injured came to our doors, receiving medical care never before available—often with remarkable results. Like those who arrived at Ellis Island, just a short distance from Bellevue Hospital, all of our patients were welcomed as they had been by Emma Lazarus.

> Give me your tired, your poor, your huddled masses yearning to breathe free,
> The wretched refuse of your teeming shore. Send these, the homeless, tempest-tost to me,
> I lift my lamp beside the golden door!
> Emma Lazarus (1849–1887)

In addition, members of our communities discussed in every chapter of this text—the neglected, discriminated against, abused, and needy—arrived. Those without food and shelter; those injured by domestic violence, industrial activities, traffic crashes, or child abuse; and those suffering from racism or misogyny and substance use or alcoholism came to our doors. We were ill prepared. We did not know enough social policy, public and population health, or human rights. It was obvious that our best efforts should have included writing prescriptions for food, clothing, housing, education, a job, and voter registration. Many hospitals were designed to serve communities that were more enfranchised and had fewer patients with overwhelming social determinant concerns: at the inception of emergency medicine, it had not been clear that addressing such concerns would become a hallmark of our field. It was the belief of some early leaders, particularly those in public hospitals who cared for the most disenfranchised, that emergency medicine might be more effective and better linked to a school of public health than a school of medicine. In the current climate, the bonds to medical centers, schools of medicine, and schools of public health are far stronger and vital, but still often representing complex, frequently incompatible interests.

The environment of the emergency department with our eyes on the community and our feet in the hospital has required us to be "doctors without intellectual or social borders." Emergency physicians must listen to our patients; we must look at them objectively and sympathetically and treat them to the best of our abilities in spite of our inadequacies and societal obstacles.

We must become Virchow's "natural advocates of the poor." We must do the essential scientific and humanistic work that restores public trust in science and medicine which will simultaneously prevent us from "clinician burnout." Our tasks in addressing the social determinants of our patients' health are enormous, but we have creative, purposeful investigators as demonstrated in this book who need collaborators. We must reimagine actions to address the social determinants of population health that have a strong social, ethical, and humanistic foundation and we must do so in our emergency departments. Precision medicine is the latest catchphrase meant to define the future of our field. We remain focused in this text on a program for creating a culture of precision *prevention* for population health in the ED, which arguably affects many more people on a deeper level. This approach to integrated, creative prevention will dramatically increase the focus on the social determinants as not only a medical but also a societal responsibility. In an ideal world, such an approach would diminish or even eliminate the need for the type of delayed rescue and inadequate stabilization that is often the norm in emergency care. These steps will be the only means of achieving the World Health Organization's (1986) definition of health as "a state of complete physical, mental and social well-being, and not merely the absence of disease or infirmity."

Our future will be developed by sensitive, humanistic, observant clinician-investigators who float intellectually between the community and the bedside. This book and these authors and editors have demonstrated the inadequacies in our society and our health education, and the critical deficits in systematically addressing the social forces faced by our patients. This book and the advances that many of the educators, clinicians, and investigators have described show us how we as emergency physicians and many others in society can play roles in improving the population's health and assuring the human rights of all individuals. This book enhances the foundation of social emergency medicine, demonstrates that we do see the injustices in our society, we know how to study these issues, and that we are finding pathways to implement essential changes necessary to overcome the social determinants that limit our patients' personal success and societal safety. We must address the social determinants that define and drive our patients' visits; we must create teams that cross all community, cultural, academic, political, and governmental borders to provide the research and evidence that will facilitate understanding and progress. This fine text demonstrates that precision medicine is an illusion for almost all of our society and how precisely we measure and successfully address the social determinants discussed in the text will determine how we live, the types and severity of illness we have, and how we die.

New York, NY, USA Lewis R. Goldfrank

Preface

Many people who choose to read this book may already be deeply invested in and knowledgeable about social emergency medicine. Others may be skeptical, wondering whether this is truly core content for emergency medicine or "part of our jobs"—especially as our jobs seem to become harder and more complex with each passing year. For anyone in the latter category, we are particularly glad that you have picked up this book. We hope that the chapters within will demonstrate clearly both *why* emergency medicine must concern itself with these issues as well as *how* we can, by incorporating social context, improve our practice of emergency medicine in small and large ways.

The practice of what has recently coalesced as social emergency medicine has been long underway, including at several safety-net institutions across the country. It also has a long historic precedent in fields outside emergency medicine and indeed outside medicine itself. Social emergency medicine has its roots in the concept of social determinants of health, described by *Healthy People 2020* as "conditions in the environments in which people are born, live, learn, work, play, worship, and age that affect a wide range of health, functioning, and quality-of-life outcomes and risks." [1] There has been an increasing public and scientific awareness that these conditions have a significant impact on health: experts estimate that while 10% of one's overall health is attributable to medical care and 30% to genetic predisposition (itself influenced by the environment as the growing field of epigenetics teaches us), 60% is related to social, economic, behavioral, and environmental influences [2].

As access to and legislation regarding our healthcare system changes over the years, emergency departments (EDs) consistently serve as our nation's safety net. While EDs cannot and should not be expected to solve all of society's failings, in our EDs we have a unique opportunity to bear witness to those failings. The willingness and skillset needed to address the social needs of our patients—and to understand the larger social and structural contexts in which they come to our doors—comprise a large part of our job, a part that is critical to our patients' well-being. Chapter authors describe how we can account for these factors in our individual patient interactions to provide better patient care. These experts also describe how we can productively collaborate with community organizations and advocate for policies that can more fundamentally remedy the inequities we witness daily in our EDs. We must do this work with humility, not as saviors but as partners and contributors.

The tent of social emergency medicine is wide, and the boundaries are not yet fully defined as the field continues to grow and mature. We struggled to decide what topics should constitute chapters in this inaugural textbook and apologize for any omissions. Some key concepts—such as public health practice and health inequities—are woven across multiple chapters rather than having their own specific chapter. We also acknowledge that we are biased by our own practice locations and therefore this book focuses most of its attention on the United States; we hope that the information it offers will be useful to readers elsewhere as well.

Each chapter follows the same general structure. Chapters begin with an abstract and key points. Next, a *Foundations* section includes background and a brief review of the evidence basis on the topic. The *Bedside and Beyond* section is organized according to the ecological model, with attention first to the level of the patient's bedside, then the hospital and healthcare system, and finally the societal level. We know that readers of this text will range from those practicing in hospitals that have perhaps never before considered addressing patient social needs or are poorly resourced, to those practicing at medical centers that are already well-versed in social emergency medicine. Therefore, in the *Recommendations for Emergency Medicine Practice* section, we asked chapter authors to give actionable recommendations at the basic, intermediate, and advanced levels. *Basic* recommendations are those that chapter authors felt every emergency provider and ED across the country should be doing now as part of providing quality emergency care. *Intermediate* recommendations are the next steps after an emergency provider or department has implemented the basics. *Advanced* recommendations often extend outside the ED to community involvement and advocacy including, for example, efforts that should be undertaken by emergency medicine specialty organizations, hospital groups, or others on a broader scale. Finally, each chapter ends with a *Teaching Case* including a clinical case, teaching points, and discussion questions. We asked authors to keep their chapters firmly grounded in the prior literature, so that chapters can serve as durable, evidence-based resources for readers. We aim for this text to be useful to a wide variety of emergency medicine practitioners: residents, attending physicians, nurses, physician assistants, nurse practitioners, social workers, administrators, and others. Its pages may also be useful for medical students, health policymakers, and others outside emergency medicine who are interested in a frontline view of social determinants of health and resultant social needs.

Our hope is that this text serves not only as a reference and educational resource, but as a guide for action. While some of the recommendations may currently seem aspirational for some ED settings, change begins with small steps made by each of us. Meaningful action could be as small as making a change in an element of one's own clinical practice. Or it could be as big as collaborating with local organizations on a program to better serve one's local community, advocating against health injustice, implementing new policies to address social needs within healthcare, or conducting groundbreaking research. For those new to social emergency medicine—and maybe even new to medicine itself—we would encourage you to dream big but not to fear starting small.

As we were putting the final touches on edits for this book, our world was besieged by two traumatic events: the COVID-19 pandemic and the murder of George Floyd. The pandemic wreaked havoc on many of our EDs, but even more pertinent to this book it put into sharp relief the profound health inequities in the United States. The inequities witnessed during the COVID-19 pandemic—borne of structural racism and many of the same social needs discussed in this book including financial insecurity and inadequate housing—strengthen our conviction that social emergency medicine is a vital part of emergency medicine. Similarly, the murder of George Floyd at the hands of police during a time when COVID-19 was already exposing—yet again—longstanding racial inequities has been a call to action to address racism within the many *structures* of American society, including healthcare. In viewing the social determinants of health through a structural lens, we can begin to understand the upstream social and economic policies that impact healthcare and outcomes. Most chapters of this book were already complete prior to the COVID-19 pandemic and George Floyd's murder and the resurgence of attention to structural racism that followed, and therefore do not discuss these events explicitly. Yet we hope that readers will be able to draw clear lines between the topics described in this book and these events—and will recommit themselves to fighting health injustice. We acknowledge that we are all learning. We hope that this book will help to foster dialogue within yourself, with your colleagues and in your health system, and beyond.

We would like to thank Springer Nature and the book editors Anila Vijayan and Sydney Keen. We would especially like to thank all of the chapter authors. We were blown away by your expertise and generosity with your time.

References

1. Healthy People 2020. Social determinants of health. Available at: https://www.healthypeople.gov/2020/topics-objectives/topic/social-determinants-of-health.
2. Schroeder SA. We can do better – improving the health of the American people. N Engl J Med. 2007; 357(12):1221–8. https://doi.org/10.1056/NEJMsa073350.

Oakland, CA, USA
Saint Louis, MO, USA
New York, NY, USA
San Francisco, CA, USA

Harrison J. Alter
Preeti Dalawari
Kelly M. Doran
Maria C. Raven

Contents

Contributors

Harrison J. Alter Department of Emergency Medicine Highland Hospital, Alameda Health System, Oakland, CA, USA

Andrew Levitt Center for Social Emergency Medicine, Berkeley, CA, USA

Erik S. Anderson Department of Emergency Medicine, Highland Hospital – Alameda Health System, Oakland, CA, USA

Jennifer Avegno Department of Emergency Medicine, University Medical Center New Orleans, New Orleans, LA, USA

Sukhveer K. Bains Departments of Emergency Medicine and Internal Medicine, University of Illinois at Chicago, Chicago, IL, USA

Edward Bernstein Department of Community Health Sciences, Boston University School of Public Health, Boston, MA, USA

Department of Emergency Medicine, Boston University School of Medicine, Boston, MA, USA

Marisa B. Brett-Fleegler Department of Pediatrics, Division of Emergency Medicine, Boston Children's Hospital, Boston, MA, USA

Stephen B. Brown Department of Emergency Medicine, University of Illinois Hospital & Health Sciences System, Chicago, IL, USA

Patrick M. Carter University of Michigan Injury Prevention Center, Department of Emergency Medicine, School of Medicine, Department of Health Behavior and Health Education, School of Public Health, University of Michigan, Ann Arbor, MI, USA

Makini Chisolm-Straker Department of Emergency Medicine, Icahn School of Medicine at Mount Sinai, Mount Sinai Queens, New York City, NY, USA

Christopher M. Colbert Department of Emergency Medicine, University of Illinois at Chicago, Chicago, IL, USA

Theodore Corbin Department of Emergency Medicine, St. Christopher's Hospital for Children/Drexel University College of Medicine, Philadelphia, PA, USA

Rebecca M. Cunningham University of Michigan Injury Prevention Center, Department of Emergency Medicine, School of Medicine, Department of Health Behavior and Health Education, School of Public Health, University of Michigan, Ann Arbor, MI, USA

Karen D'Angelo Department of Social Work, Southern Connecticut State University, New Haven, CT, USA

Adrian D. Daul Department of Emergency Medicine, Cooley Dickinson Hospital, Northampton, MA, USA

Marina Del Rios Department of Emergency Medicine, University of Illinois at Chicago, Chicago, IL, USA

Kelly M. Doran Departments of Emergency Medicine and Population Health, NYU School of Medicine, New York, NY, USA

Daniel A. Dworkis Department of Emergency Medicine, Keck School of Medicine of USC, LAC+USC, Los Angeles, CA, USA

Jahan Fahimi Department of Emergency Medicine, University of California, San Francisco, San Francisco, CA, USA

Eric W. Fleegler Department of Pediatrics, Division of Emergency Medicine, Boston Children's Hospital, Boston, MA, USA

Deborah A. Frank Department of Pediatrics, Boston Medical Center, Boston, MA, USA

Margaret B. Greenwood-Ericksen Department of Emergency Medicine, University of New Mexico Hospital, Albuquerque, NM, USA

Dennis Hsieh Contra Costa Health Plan, Harbor UCLA Medical Center, Martinez, CA, USA

Nathan Irvin Department of Emergency Medicine, Johns Hopkins University School of Medicine, Baltimore, MD, USA

Hemal K. Kanzaria Department of Emergency Medicine, University of California, San Francisco, San Francisco, CA, USA

Department of Care Coordination, Zuckerberg San Francisco General Hospital, San Francisco, CA, USA

Rebecca Karb Department of Emergency Medicine, Rhode Island Hospital/ Brown University, Providence, RI, USA

Wendy Macias-Konstantopoulos Department of Emergency Medicine, Massachusetts General Hospital, Boston, MA, USA

Shannon McNamara Department of Emergency Medicine, NYU School of Medicine, New York, NY, USA

Bryn Mumma Department of Emergency Medicine, University of California, Davis, Sacramento, CA, USA

Kian Preston-Suni David Geffen School of Medicine at UCLA, Greater Los Angeles VA Medical Center, Los Angeles, CA, USA

Megan L. Ranney Department of Emergency Medicine, Alpert Medical School, Brown University, Rhode Island Hospital, Providence, RI, USA

American Foundation for Firearm Injury Reduction in Medicine (AFFIRM Research), Providence, RI, USA

Brown-Lifespan Center for Digital Health, Providence, RI, USA

Maria C. Raven Department of Emergency Medicine, University of California, San Francisco, San Francisco, CA, USA

Karin Verlaine Rhodes, MD, MS Department of Emergency Medicine, Donald and Barbara Zucker School of Medicine at Hofstra Northwell, Manhasset, NY, USA

Bisan A. Salhi Department of Emergency Medicine, Emory University, Atlanta, GA, USA

Shamsher Samra Department of Emergency Medicine, Harbor-UCLA Medical Center, Torrance, CA, USA

Elizabeth A. Samuels Department of Emergency Medicine, Alpert Medical School of Brown University, Providence, RI, USA

Margaret E. Samuels-Kalow, MD, MPhil, MSHP Department of Emergency Medicine, Massachusetts General Hospital, Boston, MA, USA

Megan Sandel Department of Pediatrics, Boston Medical Center, Boston, MA, USA

Todd Schneberk Department of Emergency Medicine, Los Angeles County + University of Southern California Medical Center, Los Angeles, CA, USA

Amanda Stewart Division of Emergency Medicine, Boston Children's Hospital, Boston, MA, USA

Hanni Stoklosa Department of Emergency Medicine, Brigham and Women's Hospital, Boston, MA, USA

Breena R. Taira Department of Emergency Medicine, Olive View-UCLA Medical Center, Sylmar, CA, USA

Susi Vassallo Department of Emergency Medicine, Bellevue Hospital Center and NYU Langone Medical Center, New York City, NY, USA

Nancy Ewen Wang Department of Emergency Medicine/Pediatric Emergency Medicine, Stanford University School of Medicine, Palo Alto, CA, USA

Ziming Xuan Department of Community Health Sciences, Boston University School of Public Health, Boston, MA, USA

Part I

Underpinnings of Social Emergency Medicine

Social Emergency Medicine: History and Principles

1

Harrison J. Alter, Jahan Fahimi, and Nancy Ewen Wang

It is important for all of us to appreciate where we come from and how that history has really shaped us in ways that we might not understand [1].

Sonia Sotomayor

Key Points

- Social emergency medicine generally refers to the incorporation of social context into the structure and practice of emergency care.
- There are three main strands of history that intertwine to create the fabric of social emergency medicine. The first is the social medicine movement, rooted in the works of nineteenth century Rudolph Virchow, put into practice by the sociopolitical changes in Latin America in the mid twentieth century led by revolutionary physician Ernesto "Che" Guevara, and the vision of community clinics created by Jack Geiger in Mound Bayou, Mississippi in the 1960s.

H. J. Alter (✉)
Highland Hospital Department of Emergency Medicine, Alameda Health System, Oakland, CA, USA

Andrew Levitt Center for Social Emergency Medicine, Berkeley, CA, USA
e-mail: harrison_alter@levittcenter.org

J. Fahimi
Department of Emergency Medicine, University of California, San Francisco, San Francisco, CA, USA
e-mail: Jahan.fahimi@ucsf.edu

N. E. Wang
Department of Emergency Medicine/Pediatric Emergency Medicine, Stanford University School of Medicine, Palo Alto, CA, USA
e-mail: ewen@stanford.edu

H. J. Alter et al. (eds.), *Social Emergency Medicine*,
https://doi.org/10.1007/978-3-030-65672-0_1

- The second strand is the birth and growth of the specialty of emergency medicine. Emergency medicine was first officially recognized as a specialty in 1978 and imprinted with a social mission from the start.
- The third strand is the academic field of social epidemiology, most relevant for elaborating the social determinants of health. Research in this field has highlighted the fundamental and overwhelming contribution of "how we live, eat, work and play" to a person's health, well-being, and longevity, as compared to the contributions of medical care.

Social Medicine as a Political and Clinical Movement

Social medicine can be understood as the investigation of social, behavioral, and environmental factors influencing human disease and disability and the elucidation of methods of disease prevention and health promotion in individuals and communities [2]. Inherent throughout social medicine is its political mandate, to actively pursue change in social structures that suppress health and health equity.

Rudolf Virchow (1821–1902), commonly cited as the "Father of Pathology" is also one of the fathers of social medicine. Dr. Virchow was commissioned by the Prussian government to investigate a typhus outbreak in Upper Silesia (now in Poland) in 1848. His report laid clear blame for the outbreak on the miserable social conditions he found. He criticized government inaction, advocating for improved education, increased wages, and changes in agricultural policy [3]. Virchow's colleagues and students popularized the concept of medicine as a clinical social science in the interwar years. According to Porter's brief history of social medicine, "The interdisciplinary program between medicine and social science would provide medicine with the intellectual skills needed to analyze the social causes of health and illness in the same way as the alliance between medicine and the laboratory sciences had provided new insights into the chemical and physical bases of disease." [2]

The Latin American social medicine movement directly applied these principles to implement social change. So much so that they stated that social medicine policies should not be concerned with clinical medicine but rather with the conditions—the structures—that created the clinical situation. Thus Salvador Allende, a Chilean pathologist, as health minister and later as elected president of Chile, focused on social transformation—the alleviation of poverty, poor working conditions and lack of education—as fundamental to improving health. Dr. Ernesto "Che" Guevara's concept of revolutionary medicine similarly promoted teaching physicians about the social origins of illness and the need for social change to improve health. Overall, social medicine in Latin America focused on transforming the political and social structures underlying poverty, whereas public health worked within existing structures to create and implement public policy to benefit health [2].

In the US, during the 1960s, Drs. H. Jack Geiger and Count Gibson attempted to bridge the demand for structural change with the patient- and community-level effects of social inequality, establishing the first two community health centers in Bolivar County, Mississippi, (known as Mound Bayou) and the Columbia Point Public Housing Project in Boston, Massachusetts. The impetus to create these centers grew from the Medical Committee for Human Rights, a consortium of healthcare workers providing care to activists during the "Freedom Summer" volunteer movement for civil rights in Mississippi. Both Mound Bayou and Columbia Point provided much needed medical services in struggling communities [4]. They attempted to address the poverty, malnutrition, and unemployment as the roots of the poor health they observed. Geiger engaged local Black-owned grocers in Jim Crow Mississippi to honor food prescriptions written by clinicians for their malnourished patients at the Mound Bayou clinic. Geiger was famously quoted as saying, "The last time we looked in the book for specific therapy for malnutrition, it was food." [4] The community health center model, now codified in our Federally Qualified Health Centers (FQHCs), has spread widely—currently there are over 1000 centers throughout the US. This model relies on community engagement in a way that few other elements of the medical-industrial complex do, incorporating a community voice through advisory committees and patient advocacy panels. Geiger then went on to build the Social Medicine program at the City University of New York and Montefiore Hospital, which has trained generations of physician-activists.

The Specialty of Emergency Medicine and Its Social Mission

Emergency medicine is one of the youngest fields of medicine, not yet 50 years old in 2021. Unique among medical specialties, emergency medicine's specialty status is not based on an anatomic system, procedure, or specific patient population. Rather, emergency medicine is based on place and time. Emergency "rooms" are situated as the doorway to the hospital. As such, they are an entrance to social and medical services for the surrounding community. They also serve as a window into the community's health. Emergency care is predicated on a layperson's perception of an acute need and defined by access to care at any time of the day or night. By the definition endorsed by the American College of Emergency Physicians (ACEP), "The practice of emergency medicine includes the initial evaluation, diagnosis, treatment, coordination of care among multiple providers, and disposition of any patient requiring expeditious medical, surgical, or psychiatric care." [5] Or, put another way, we offer specialty care for "anyone" with "anything" at "anytime" [6].

Emergency medicine as a specialty arose out of the success of "curative" medicine and the development of modern hospitals housing diverse and increasingly effective diagnostic and treatment technology. After World War II, the US government put increased resources into building up the nation's health care infrastructure. The Hill-Burton Act of 1946 explicitly provided for hospital construction particularly in rural and small neighborhoods. Physicians' practices migrated from individual offices to hospitals, where they could provide efficient care and specialty access [7]. Although hospitals had emergency rooms, these had no designated medical staff. Private physicians or specialists would arrange to meet and care for their own patients in need and, if necessary, admit them to the hospital. Poor patients without a private physician would also go to the emergency room in search of help, often only to be seen by the least experienced personnel. Thus, the emergency room, though full of patients, had no specific personnel or expertise for evaluating and stabilizing patients with undifferentiated conditions.

The first known emergency medicine groups were formed in 1961 in Alexandria, Virginia, and Pontiac, Michigan. Brian Zink, emergency medicine's unofficial historian, points out that James D. Mills, the first emergency physician, was attracted to the practice in large part because of his realization that, "in serving as a full-time emergency physician … he could have more of an impact on improving health care for at least some of the poor and uninsured in his city" [6].

Demand for emergency medical care increased dramatically during this era. The Medicaid and Medicare programs implemented in 1963 gave recourse to the poor and elderly needing emergency care while providing financial incentives to physicians to care for them. Next, the Emergency Medical Treatment and Active Labor Act (EMTALA), passed in 1986, codified specific standards of care as a mandate: EMTALA required medical screening and stabilization for anyone who sought care within the grounds of a hospital. By law, though unfunded, no one, regardless of medical problem, ability to pay, or skin color, could be turned away from an emergency room.

While public policy was working to provide a solution to challenges arising from societal evolution, modernization, and changing demographics, the medical profession recognized the importance of structure, organization, standards, and a trained cadre of practitioners—the preconditions for establishment of a specialty. Thus, increasing demand for quality emergency care stimulated the creation of the American College of Emergency Physicians. In the early 1970s, the first emergency medicine residencies coalesced, followed quickly by the establishment of the American Board of Emergency Medicine, a formal examination and certification arm. The American Board of Medical Specialties approved emergency medicine as a specialty in 1979.

A new medical-social contract was forming from these developments. In the latter half of the twentieth century, those who were poor or disabled, who were immigrants, without primary care, or without the resources to prevent health complications or personal tragedies all now had a place to turn. Emergency rooms became emergency departments (EDs), equipped with the infrastructure, capability, workforce, and expertise to care for a larger segment of society. The principles of social medicine—as well as population and public health—were powerfully relevant to emergency medicine, which had been, in part, woven from "threads of egalitarianism, social justice, and compassion for the poor and underserved" [8].

The Horizon Expands: The Emergence of Social Epidemiology

We now understand the social conditions that Geiger and Gibson attempted to treat collectively as "social determinants of health." This concept began to materialize as the field of Social Epidemiology took shape in the early 1960s based in part on the work of Leonard Syme and Sir Michael Marmot. The concept of the social determinants of health emerged from early findings of the socioeconomic gradient in health, now recognized as one of the most robust relationships in biology [9]. For example, in studying the relationship between social mobility and coronary heart disease, Syme, like Virchow, found that social determinants largely predict health [10]. Syme's advantage was the tools of epidemiology, allowing him to demonstrate the concept more empirically.

The social determinants of health have since come to be defined by the World Health Organization (WHO) as the "conditions in which people are born, grow, live, work and age…shaped by the distribution of money, power and resources at global, national and local levels" [11]. Researchers and experts may expand these determinants to include income and income distribution; early life; education; housing; food security; employment and working conditions; unemployment and job security; social safety net; social inclusion/exclusion; and health services [12]. Increasingly, factors such as structural and community violence and racism are among the social forces included as social determinants of health [13].

As Social Epidemiology evolved, it took on some of the same characteristics that made emergency medicine unique. Whereas epidemiologists had been concerned with specific diseases—infectious disease outbreaks, injury, or cancer epidemiology—social epidemiology asserted itself in understanding the dynamics of the health of populations. This more holistic vantage meant that just as emergency physicians first saw patients with undifferentiated complaints and applied tools to make a definitive diagnosis, social epidemiologists studied the ubiquitous upstream drivers of health, applying them to a wide range of diseases.

Social epidemiology and emergency medicine share another conceptual framework: the care and study of populations. As one important arm of population health, EDs ensure that all persons have access to care, thereby somewhat reducing the impact of healthcare disparities. However, while social epidemiology studies social determinants of health, the practice of emergency medicine often addresses social needs, something that is best addressed at the bedside. Social needs may arise from social determinants of health, but these terms are not synonymous. For example, the relationship between an individual's hunger (the social need) and the structural determinants of the food landscape in that person's community (the social determinants of health) is complex. While a clinician interested in the relationship between social context and emergency care (i.e., social emergency medicine) may be interested in pushing both of these levers, action on the individual patient's hunger is often more direct and tangible in the ED. This is an illustration of the "upstream/downstream" dichotomy in social epidemiology [14].

In the current era, concepts relating to the social determinants of health are being rapidly refined. One way in which the dialogue is shifting is the sharpening focus on *structural* determinants of health, a concept which incorporates the way that social constructs such as racism, sexism, ablism, and other biases influence how society and institutions address health [15]. An example of such a focus is a study overlaying maps of acute asthma ED visits on historical "redlined" maps [16], which the federal government created for banks to exclude African-American and Latinx loan applicants from securing mortgages. The study's finding of increased ED visits within these neighborhoods supports the idea that structural racism, highlighted by the loan maps, has direct effects on health.

Social Emergency Medicine Comes Together

Any emergency clinician can glance at a list of social determinants and immediately understand how these and other social forces frequently complicate clinical encounters with their patients. These clinical experiences have long motivated clinician-scientists and health services researchers to investigate the distribution and impact of social determinants on the health of patients seeking care and help in EDs. Early examples of such inquiries include studies exploring the relationship between access to primary care and patterns of ED use [17, 18].

In 1994, Edward Bernstein led an authorship group on a paper entitled, "A Public Health Approach to Emergency Medicine: Preparing for the Twenty-First Century" [19]. This paper laid out an argument for a broader scope of practice in emergency medicine, an initial blueprint for what has become social emergency medicine. Their scope was somewhat limited, however, by the era; public health's incorporation into the medical model at that time meant essentially secondary prevention, identifying medical presentations whose recurrence could be prevented by social intervention, often taken to mean patient or public education.

Dr. Bernstein, an emergency physician, and Dr. Judith Bernstein, a public health and policy expert, then published *Case studies in emergency medicine and the health of the public,* a book which demonstrated opportunities for public health-style interventions in the ED through clinical cases [20]. The text introduces readers to cases about homelessness, partner violence, substance use disorder, and other social concerns, providing glimpses into practicing emergency medicine with a population health lens. In 1999, James Gordon published a paper in the *Annals of Emergency Medicine* further highlighting the interconnectedness of social and clinical care in EDs. Gordon's widely cited paper, "The Hospital Emergency Department as a Social Welfare Institution," deserves credit in many respects for launching the contemporary era of social emergency medicine.

Gordon lays out his vision for the twenty-first century ED:

> "How would a social triage system actually work? All patients presenting to the ED (or their proxy, when appropriate) would be screened by a short panel of questions built into the standard triage history or registration interview, designed to detect unmet social needs. The questions would reflect basic material, economic, social, and health factors important to maintain a minimum standard of well-being. Items would address such basic issues as: Can you pay your rent? Are your utilities working? Do you have enough food to eat? Can you get to the doctor? Can you afford medicines? Such simple questions are often never asked of the most disadvantaged and are usually absent from standard medical evaluations—yet the answers can profoundly reflect on overall well-being. If a major category of deprivation is identified, the patient would be referred to the social triage center for a more complete social evaluation, and a social care and referral plan established. This process would be designed not to interfere with the formal medical encounter, and could occur in the social triage area just before formal discharge" [21].

Gordon argues effectively that patients make a rational choice to seek care in the ED, and that as both a practical matter and a human one, EDs ought to be equipped to meet their needs.

For decades, the work of many clinicians and researchers from across the country has pointed towards this goal while building the field of social emergency medicine.

The label of social emergency medicine and its origins as a coordinated field began a few years after the publication of Gordon's roadmap, when EM physicians at Highland Hospital, in Oakland, California, partnered with the family of Andrew Levitt, a colleague who died unexpectedly, to honor his legacy by forming an independent non-profit research and advocacy institute to promote the concept of social emergency medicine. In 2008, they launched the Andrew Levitt Center for Social Emergency Medicine.

Meanwhile, the practice of social emergency medicine was not a new concept. Clinicians and leaders in emergency medicine from across the nation were training residents and building programs to think beyond the walls of the ED. For example, Lewis Goldfrank at NYU-Bellevue was shining a light on the importance of care for vulnerable populations and Stephen Hargarten at the Medical College of Wisconsin was studying violence and its impact on health. Clinician-investigators and socially oriented leaders worked together to bridge the gap from research to evidence-based implementation by addressing human trafficking, gun violence, homelessness, and a wide array of other issues affecting their patients.

Social EM: Current State and Future Aspirations

Soon after the creation of the Levitt Center, the idea of formalizing social emergency medicine began to take hold within academic and organized emergency medicine. Emergency medicine faculty at Stanford University and Highland Hospital simultaneously created the first training fellowships in social emergency medicine. In 2017, the Levitt Center, ACEP, and the Emergency Medicine Foundation organized a consensus conference in Dallas, Texas, funded by the Robert Wood Johnson Foundation. This event, titled "Inventing Social Emergency Medicine," drew a diverse array of investigators and innovators from across emergency medicine. Its proceedings, published as a supplemental issue of *Annals of Emergency Medicine* [22], constitute the most extensive collaboration of experts in the field. Shortly after the conference, a Social Emergency Medicine Section at ACEP and an Interest Group at the Society for Academic Emergency Medicine were created, to provide ongoing forums for collaboration among like-minded members of these specialty societies.

The range of initiatives proposed and undertaken by the members of these groups is vast. There are help desks for health-related social needs, such as the Highland Health Advocates [23]. There is a broad network of hospital-based violence intervention programs [24]. Numerous interventions recognize and address homelessness and unstable housing in ED patients. ED-based health coaches aid patients with chronic disease management [25]. After exploring the importance of the built environment, faculty and staff at the University of Pennsylvania ED have collaborated to "green" vacant lots, effectively reducing the community burden of medical emergencies [26]. Many of these innovations are documented in this textbook.

The basic precepts of the practice are emerging from the foundational and programmatic work. One of the recurrent themes is the notion of inreach; working with community partners to bring their social services into the ED. ED social workers, long the linchpin of addressing social needs, cannot do it all; between assessments, grief counselling and death notifications, family support, and so much more, there are limits on their capacity. For specialized services, such as bedside advocacy for violence survivors or housing needs, skilled community service providers with established relationships in the ED can meet patients in the ED. When services cannot be brought within the walls of the hospital, interprofessional teams have collaborated to develop "warm handoffs" for patients who need linkage to services to address their social needs [27].

Another theme arising as the historical precedent evolves into contemporary social emergency medicine is that the ED is a rational and potentially important location to address and assess patients' social needs. Though much focus of social medicine has centered on primary care, there is growing evidence that EDs have a unique role to play. For one, research has shown that—compared to patients in other settings—ED patients have uniquely high burdens of multiple social needs, including homelessness, food insecurity, exposure to violence, and others [28]. Relatedly, EDs accept patients at any hour and are mandated to serve all who seek care, therefore serving many—whether due to lack of access to other health care, patient preference, or other reasons—who do not receive regular outpatient care [29, 30]. Last, EDs serve as a social surveillance system, recognizing emerging individual and population social needs and creating capacity to address them at the bedside or within a larger system.

Parallel to the growth in social emergency medicine practice, there has been a surge in social emergency medicine research. Such inquiry is critical to push the field toward effective interventions and further solidify its standing as a rigorous, evidence-based part of emergency medicine. However much social emergency medicine has been about *doing*, it is crucial to also focus on *understanding*. As readers experience the breadth of topics in this text, attention should be paid to the underlying evidence to support the authors' conclusions, with an eye towards future high-quality research that will guide programs and interventions.

As this text highlights, a geographically and demographically diverse group of clinician-scientists and clinician-advocates have coalesced around a unifying movement [31]. Through sharing of insights, methods, and approaches, there now appears to be a collective voice advancing emergency care through incorporation of social context and social determinants of health.

References

1. Marinucci C. SCOTUS Justice Sonia Sotomayor in SF: on affirmative action, "terror" on the job — and witchcraft. In: PoliticsSource. Albany Times Union; 2013. Accessed 02/05 2020.
2. Porter D. How did social medicine evolve, and where is it heading? PLoS Med. 2006;3(10):e399. https://doi.org/10.1371/journal.pmed.0030399.

3. Virchow RC. Report on the typhus epidemic in Upper Silesia. Am J Public Health. 2006;96(12):2102–5. https://doi.org/10.2105/ajph.96.12.2102.
4. Ward TJ, Geiger HJ. Out in the rural: a Mississippi health center and its war on poverty. New York: Oxford University Press; 2017.
5. Definition of Emergency Medicine. "@acepnow". 2020. http://www.acep.org/patient-care/policy-statements/definition-of-emergency-medicine/.
6. Zink BJ. Anyone, anything, anytime: a history of emergency medicine. 2nd ed. Dallas: ACEP Bookstore; 2018.
7. Starr P. The social transformation of American medicine. Updated edition ed. New York: Basic Books; 2017.
8. Zink BJ. Social justice, egalitarianism, and the history of emergency medicine. AMA J Ethics. 2010;12(6):492–4.
9. Wilkinson RG, Pickett KE. Income inequality and socioeconomic gradients in mortality. Am J Public Health. 2008;98(4):699–704. https://doi.org/10.2105/AJPH.2007.109637.
10. Syme SL. Historical perspective: the social determinants of disease – some roots of the movement. Epidemiol Perspect Innov. 2005;2(1):2. https://doi.org/10.1186/1742-5573-2-2.
11. WHO | About social determinants of health. In: WHO. World Health Organization. 2017. https://www.who.int/social_determinants/sdh_definition/en/.
12. Raphael D. Addressing the social determinants of health in Canada: bridging the gap between research findings and public policy. Policy Options-Montreal. 2003;24(3):35–40.
13. Paradies Y, Ben J, Denson N, Elias A, Priest N, Pieterse A, et al. Racism as a determinant of health: a systematic review and meta-analysis. PLoS One. 2015;10(9):e0138511. https://doi.org/10.1371/journal.pone.0138511.
14. Braveman P, Egerter S, Williams DR. The social determinants of health: coming of age. Annu Rev Public Health. 2011;32:381–98. https://doi.org/10.1146/annurev-publhealth-031210-101218.
15. Shokoohi M, Bauer GR, Kaida A, Lacombe-Duncan A, Kazemi M, Gagnier B, et al. Social determinants of health and self-rated health status: a comparison between women with HIV and women without HIV from the general population in Canada. PLoS One. 2019;14(3):e0213901.
16. Nardone A, Casey JA, Morello-Frosch R, Mujahid M, Balmes JR, Thakur N. Associations between historical residential redlining and current age-adjusted rates of emergency department visits due to asthma across eight cities in California: an ecological study. Lancet Planetary Health. 2020;4(1):e24–31.
17. Baker DW. Patients who leave a public hospital emergency department without being seen by a physician. JAMA. 1991;266(8):1085. https://doi.org/10.1001/jama.1991.03470080055029.
18. Akin BV, Rucker L, Hubbell FA, Cygan RW, Waitzkin H. Access to medical care in a medically indigent population. J Gen Intern Med. 1989;4(3):216–20.
19. Bernstein E, Goldrank LR, Kellerman AL, Hargarten SW, Jui J, Fish SS, et al. A public health approach to emergency medicine: preparing for the twenty-first century. Acad Emerg Med. 1994;1(3):277–86. https://doi.org/10.1111/j.1553-2712.1994.tb02446.x.
20. Bernstein E, Bernstein J. Case studies in emergency medicine and the health of the public. Boston: Jones and Bartlett Publishers; 1996.
21. Gordon JA. The hospital emergency department as a social welfare institution. Ann Emerg Med. 1999;33(3):321–5.
22. Alter HJ. Foreword to conference proceedings, inventing social emergency medicine. Ann Emerg Med. 2019;74(5S):S1–2.
23. Losonczy LI, Hsieh D, Wang M, Hahn C, Trivedi T, Rodriguez M, et al. The Highland Health Advocates: a preliminary evaluation of a novel programme addressing the social needs of emergency department patients. Emerg Med J. 2017;34(9):599–605.
24. Bonne S, Dicker RA. Hospital-based violence intervention programs to address social determinants of health and violence. Curr Trauma Rep. 2020;6:23–28.
25. Capp R, Misky GJ, Lindrooth RC, Honigman B, Logan H, Hardy R, et al. Coordination program reduced acute care use and increased primary care visits among frequent emergency care users. Health Aff. 2017;36(10):1705–11.

26. Hohl BC, Kondo MC, Kajeepeta S, MacDonald JM, Theall KP, Zimmerman MA, et al. Creating safe and healthy neighborhoods with place-based violence interventions. Health Aff. 2019;38(10):1687–94.
27. Kelly T, Hoppe JA, Zuckerman M, Khoshnoud A, Sholl B, Heard K. A novel social work approach to emergency department buprenorphine induction and warm hand-off to community providers. Am J Emerg Med. 2020;38:1286.
28. Malecha PW, Williams JH, Kunzler NM, Goldfrank LR, Alter HJ, Doran KM. Material needs of emergency department patients: a systematic review. Acad Emerg Med. 2018;25(3):330–59. https://doi.org/10.1111/acem.13370.
29. Poon SJ, Schuur JD, Mehrotra A. Trends in visits to acute care venues for treatment of low-acuity conditions in the United States from 2008 to 2015. JAMA Intern Med. 2018;178(10):1342. https://doi.org/10.1001/jamainternmed.2018.3205.
30. Hwang SW, Chambers C, Chiu S, Katic M, Kiss A, Redelmeier DA, et al. A comprehensive assessment of health care utilization among homeless adults under a system of universal health insurance. Am J Public Health. 2013;103(S2):S294–301.
31. Directory — SocialEMpact. 2020. https://www.socialempact.com/directory.

Part II

Social Constructs, Structural Determinants, and Individual Identity

Race and Racism in Social Emergency Medicine

2

Sukhveer K. Bains, Christopher M. Colbert, and Marina Del Rios

Key Points

- Structural racism is defined as the macro-level systems, institutions, social forces, ideologies, and processes that generate and reinforce inequities among racial groups [1]. Emergency medicine physicians should be aware of how the history of structural racism has resulted in differential healthcare resource availability and health outcomes in the communities they serve.
- Implicit bias is an unconsciously held belief pertaining to a specific social group, related to the process that leads to stereotyping. Implicit bias helps explain how socialization can manifest in our unconscious and unintentional actions. It is a universal phenomenon, and awareness is key to control its negative effects on patient care.
- Emergency providers have a unique lens into health disparities as front-line healthcare workers. By actively working toward reducing implicit bias and advocating for systemic anti-racism strategies that dismantle structural racism, emergency providers are able to provide more equitable care at the bedside.

Foundations

Background

Race and Structural Racism

> Race is not a biological category that naturally produces health disparities because of genetic differences. Race is a social category that has staggering biological consequences because of its impact of social inequality on people's health
>
> – Dorothy E. Roberts, J.D [2].

S. K. Bains
Departments of Emergency Medicine and Internal Medicine, University of Illinois at Chicago, Chicago, IL, USA

C. M. Colbert · M. Del Rios (✉)
Department of Emergency Medicine, University of Illinois at Chicago, Chicago, IL, USA

H. J. Alter et al. (eds.), *Social Emergency Medicine*,
https://doi.org/10.1007/978-3-030-65672-0_2

The definition of race rests on external characteristics of color and other phenotypic attributes we categorize socially [3]. In the literature, and in society, race is often confounded with ethnicity [4], which refers to elements such as culture, language, heritage, history, shared geography, and the practices and norms that individuals come to share through their socialization. For example, the term African American is often used interchangeably with Black when describing the race of a population. This verbiage negates the heterogeneity of both terms, as there are many individuals who are categorized as Black and trace their ancestry to the Caribbean, Asia, or South America. Race and ethnicity are important axes of social stratification in the US [5]. Given the conflation of race and ethnicity in common language and medical literature, there will be some overlap of these terms within this chapter. We have used the original verbiage of the research studies in the citations.

Racism is when the "presumed superiority of one or more racial groups is used to justify the inferior social position or treatment of other racial groups" [6]. Structural racism is defined as the "ways in which historical and contemporary racial inequities are perpetuated by social, economic and political systems... It results in systemic variation in opportunity according to race" [7].

The history of the US as a slaveholding republic and a colonial settler nation cannot be minimized when discussing how race impacts health in the present day. The modern concept of "racism" emerged as early European settlers sought to preserve an economy largely on the basis of the labor of enslaved people [8]. Colonists established legal categories based on the premise that Black and indigenous individuals were different, less than human, and innately, intellectually, and morally inferior—and therefore subordinate—to White individuals [9]. These ideologies were foundational to the creation of systems and institutions that led to the formation of the US. In the post emancipation era, the US government remained complicit in the promotion of racial discrimination right into the civil rights movement of the 1960s and 1970s; and this history continues to manifest today. While interpersonal racism, bias, and discrimination in healthcare settings can directly affect health through poor health care, it is essential to recognize the broader context within which healthcare systems operate. Over 100 years of exclusionary housing policies resulting in segregated neighborhoods [10, 11] and segregated hospitals [12, 13]; voter suppression of racial minorities [14]; discriminatory criminal justice practices and incarceration [15]; and barriers to financial assistance [16], all of which have significant repercussions on the health of racial minorities today [17]. These manifestations of structural racism are often overlooked as root causes of health inequities [1].

One example of government sanctioned discrimination with longstanding health repercussions is the Home Owners' Loan Corporation (HOLC) established in 1933. Formed under the New Deal initiative as a depression-era emergency agency, the HOLC was a measure to refinance defaulted home mortgages and prevent foreclosures. However, the agency systematically graded neighborhoods that were predominantly inner-city, Black, and immigrant as dangerous, and outlined these neighborhoods in red on maps, creating the term "redlining." Neighborhoods with higher property values, better housing quality, and fewer individuals who were people of color and "foreign-born" were considered lower risk. This practice helped institutionalize and perpetuate racial segregation by driving divestment from redlined communities and in turn, decreasing educational and employment opportunities [11], diminishing accumulation of wealth, and decreasing appreciation of home values [18]. Residential segregation results in dramatic variations in factors conducive to the practice of healthy or unhealthy behaviors, such as the availability of open spaces like parks and playgrounds [19] and of healthful products in grocery stores [20, 21]. In addition, redlining and divestment have also resulted in inequitable distribution of healthcare infrastructure and services by neighborhood, thereby exposing racial minorities to unequal health services [22–25].

Implicit Bias and Interpersonal Racism

Implicit biases are defined as unintentional or habitual preferences and behaviors that are relatively inaccessible to conscious awareness or control; they are "habits of mind" [26]. Implicit bias is not problematic in and of itself; it is simply one of the many well-established factors that influence human behavior. The implicit biases we hold may be unconscious manifestations of stereotypes we have for certain groups that result in unintentional preferences. Interpersonal racism can arise when these biases manifest in behaviors that are racially preferential and consequential in their outcomes, regardless of intent [27]. Socialization does not occur in a vacuum, and implicit biases are acquired through our societal ideologies, social interactions, and institutions; all of which are informed by our history, which includes a legacy of racism.

Given the necessity of heuristic clinical assessments in emergency medicine (EM), emergency care providers are at high risk for exhibiting implicit bias. Although the intent is to administer evidence-based, objective clinical care, the larger environment within which we practice can influence and impact our actions. In order to eliminate racial disparities in emergency care and outcomes, it is important to discern why these disparities exist and how our actions, consciously or unconsciously, perpetuate them. It is through these lenses of structural racism and implicit bias that we can understand the effect and impact of race and racism in emergency care.

Evidence Basis

The last three decades have witnessed a growing body of research on the topics of implicit bias and racism in EM [28–31]. Wide disparities in prehospital [32, 33], triage [34, 35], and emergency department (ED) assessment [36] and treatment have been identified and are associated with worse outcomes among patients who are categorized as racial minorities. Most evidence comes from large surveillance studies, prospective and retrospective observational studies, and some systematic reviews. After controlling for geography, hospital size or type where care was received, insurance status, and multiple patient variables including age, sex, and comorbidity, the vast majority of research concerning emergency care indicates that racial minorities are less likely than White people to receive needed services, including clinically necessary and potentially life-saving procedures [37].

Regardless of clinical setting (i.e., community-based or academic, urban or rural) the indirect application of racism is apparent as evidenced by the significant disparities in life expectancy when comparing people from racial and ethnic minority groups to non-Hispanic White people. Black men have a life expectancy of 7 years less than the aggregate population [38]. Hispanic people in the US have higher mortality rates than non-Hispanic White people for cancers of the stomach, liver, and cervix; diabetes mellitus; and liver disease [39, 40]. African American people have higher rates than White people for all-cause mortality in all groups aged less than 65 years. Compared with White people, Black people in age groups under 65 years have higher levels of some self-reported risk factors and chronic diseases, and mortality from cardiovascular diseases (CVD) and cancer; diseases that are most common among persons aged 65 years and older [41]. Native American people with CVD have a 20% higher mortality rate compared to other races [42].

Overall, Black, Latinx, and Native American patients seeking care in the ED have longer wait times to be seen compared to non-Hispanic White patients [43–45] with ED wait time disparities most pronounced as illness severity, as measured by triage acuity, decreases [46, 47]. Black, Latinx, and Native American patients seeking care in the ED are more often assigned less acute triage severity scores than their non-Hispanic White counterparts, even after adjusting for age, comorbidity, vital signs, and time and day of presentation [47–50]. In the case of potentially life-threatening complaints, such as chest pain, African American and Latinx patients are less likely to be triaged emergently [51] or to have a cardiac monitor or pulse oximetry ordered upon arrival compared to their non-Hispanic White counterparts [52]. Disparities in triage assessment have also been recognized when children present to the ED with potentially high acuity complaints such as fever, abdominal pain, and/or difficulty breathing. Black, Latinx, and Native American children are more likely to be assigned lower acuity scores compared to White children for similar presenting complaints [34]. Lower triage acuity score designation is, in turn, also associated with delayed analgesia for back and abdominal pain in racial minority patients compared to White patients with the same complaints [53–56].

Prehospital emergency medical systems (EMS) management literature has several examples of implicit bias potentially impacting the management of patients. Among patients picked up and transported by EMS with blunt traumatic injury, Black and Latinx patients are less likely to receive prehospital opioid analgesia compared to their non-Hispanic White counterparts [57–59]. When examining the evaluation of stroke-like symptoms, EMS hospital pre-notification – a factor associated with improved evaluation, timelier diagnosis, and treatment with thrombolytics – is less likely to occur when transporting Black or Latinx patients with subsequent diagnosis of stroke when compared to non-Hispanic White patients [55, 60–62]. Failure to recognize life-threatening emergencies on the part of EMS providers, combined with assignment of lower acuity scores during triage assignment, results in avoidable delays in therapeutic interventions in patients who are racial minorities and can result in poor clinical outcomes [63, 64].

A declaration of the existence of these disparities and recommendations to move toward ending them were highlighted by the 2002 Institute of Medicine (IOM) Report "Unequal Treatment: Confronting Racial and Ethnic Disparities in Health Care" [37, 63]. These recommendations include: collecting and reporting health care access and utilization data by patients' race/ethnicity, encouraging the use of evidence-based guidelines and quality improvement, supporting the use of language interpretation services in the clinical setting, increasing awareness of racial/ethnic disparities in health care, increasing the proportion of underrepresented minorities in the health care workforce, integrating cross-cultural education into the training of all health care professionals, and conducting further research to identify sources of disparities and promising interventions. Evidence of substandard care based on group membership raises concern that provision of care is inconsistently and subjectively administered, therefore exposing a threat to the quality of care of all Americans [65]. Despite the IOM's call to action to end disparities in health care, few randomized or even prospective studies have focused on interventions aimed at decreasing these disparities.

Emergency Department and Beyond

Bedside

Because emergency care providers are constantly confronted with situations requiring quick decisions, as is the case in overcrowded emergency departments or when balancing the care of life-threatening emergencies with less urgent conditions, we are at high risk for acting based on implicit bias. Often we must make quick decisions with incomplete or missing objective indicators of health such as past medical histories and/or laboratory values. In such an information vacuum, healthcare providers rely more on heuristics; thus, diagnostic and treatment decisions, as well as feelings about patients, can be influenced by patients' race and ethnicity [37].

Disparities in emergency pain management of minority patients are one way in which implicit bias manifests in EM. Racial minorities are systematically undertreated for pain [66, 67]. Black and Latinx patients are less likely to receive opioid analgesia in EDs for long bone fractures, back pain, and abdominal pain, even when controlling for pain scores, compared to their White patient counterparts [53, 68–70]. They are also less likely to receive opioid prescriptions at discharge for similar complaints [54]. Racial disparities are also reflected in provider decisions when evaluating patients. In the case of acute headache, Black patients are less likely to undergo advanced diagnostic imaging (CT/MRI) compared to White patients independent of clinical or demographic factors [71].

When providers do have access to information about their patients' medical and social histories, there is a risk that the presence of some medical conditions or social needs that may be disproportionately prevalent in certain races will activate implicit bias. For example, patients with pain due to sickle cell disease (SCD) experience 25–50% longer times for evaluation as measured by door to provider time when compared with patients who present with other painful conditions [72]. Provider bias serves as a significant barrier to delivery of high quality care to persons with SCD [73]. Having negative impressions of SCD patients, greater frustration in caring for SCD patients, and assuming that there is a high prevalence of opioid use disorder among the SCD patients are associated with decreased adherence by ED providers to recommended ED pain management strategies [74, 75]. Provider bias is ameliorated in settings where there is doctor–patient race concordance [76]. African American providers are more likely than providers of other races to have more positive feelings of affiliation with SCD patients, and to be more aware of the role that race plays in the delivery of quality care to this population [74, 77].

The unequal treatment in racial minority children is especially concerning. When looking at "potentially pain-related conditions," Black children are more likely to receive non-opioid analgesics compared to White children [78]. Analgesia is largely underutilized in the pediatric population in the case of acute appendicitis. However, Black children are less likely to receive any pain medication for moderate pain and are less likely to receive opioids for severe pain due to appendicitis, compared to non-Hispanic White children while in the ED [79]. A 2016 study by Hoffman et al. demonstrated that false beliefs in regard to biological differences in pain tolerance between Black and White patients continue to exist among White laypersons, medical students, and residents, suggesting that these implicit biases may continue to perpetuate racial disparities in the evaluation and treatment of pain [66].

Emergency care providers can play a role in reducing the impact of structural racism in their individual patient interactions by engaging in open dialogues within their institutions (with peers, coworkers, and trainees) on how implicit biases and societal forces may be impacting their practice. Project Implicit, a nonprofit international research collaboration, has resulted in a substantial body of literature that provides insight into the pervasive nature of implicit bias. There is consistent evidence of racial preference toward White people across multiple contexts in the U.S [27, 80]. Taking an Implicit Association Test (IAT) can help us recognize our own unconscious biases so we can then train ourselves to overcome them (https://implicit.harvard.edu) [80]. One example of implicit bias training is the Bias Reduction in Internal Medicine (BRIM) curriculum created by a group of researchers at the University of Madison-Wisconsin. BRIM offers a three hour evidence-based workshop to teach how implicit bias is a habit, how to become bias literate, and evidence-based strategies on how to break implicit biases (https://brim.medicine.wisc.edu/) [81].

Structural competency contextualizes social determinants of health in the broader structural, historical, and ideological drivers that lead to health inequities [82]. Structural competency is the capacity for health professionals to recognize and respond to health and illness as the downstream effects of broad social, political, and economic structures [83]. Integrating implicit bias and structural competency training in curricular content can help practitioners understand the impact of race and racism in clinical practice. It is time to go beyond describing health disparities and shift attention to forces that influence health outcomes beyond individual interactions [82–84]. Having a structural understanding may help combat implicit bias and facilitate intentional anti-racism practice by shifting focus and blame away from individual or cultural factors of a patient's care to the larger forces affecting a patient's health [82–84].

Hospital/Healthcare System

Racial disparities in ED care and outcomes are heavily influenced by the fact that racial minorities more frequently receive care in lower performing hospitals [85]. Access to high-quality care varies considerably by area—by state, between rural and urban areas, as well as across smaller communities [86]. Structural racism manifests as historical patterns of segregation and discrimination affecting the geographic availability of healthcare institutions. High levels of residential segregation, in combination with a high percentage of poor residents, confers a higher likelihood of hospital closure [25, 87]. Because racial minorities are more likely to live near and access hospitals with fewer resources, including financial, infrastructure and technical resources, and human capital, they on average have unequal health outcomes compared to non-Hispanic White people [88].

Limited resources in minority-serving healthcare systems results in ED overcrowding and unequal implementation of evidence-based care [89–92]. ED overcrowding leads to ambulance diversion, which occurs more frequently in hospitals treating a high share of patients who are racial minorities [93, 94]. Ambulance diversion can have a negative impact on patients who have to be diverted elsewhere, as it may delay time-sensitive interventions. Moreover, diversion in hospitals serving a large proportion of minorities may indicate a fundamental mismatch in supply and demand of emergency department services [93, 94]. Systemic issues related to ED overcrowding can also result in delayed delivery of life-saving therapeutic interventions [93, 94].

An old adage attributed to Lord Kelvin says "To measure is to know; If you cannot measure it, you cannot improve it" [95]. A growing body of literature supports the implementation of surveillance programs tied to quality improvement initiatives such as improving access to quality primary and secondary preventive care and social services, protocol driven care, and clinical decision support tools as a path to bridge the racial gap in clinical processes and outcomes. Quality improvement programs that assess adherence to recommended processes of care have led to decreased disparities in outcomes by race. The Center for Medicare and Medicaid Services (CMS) Hospital Inpatient Quality Reporting program authorized CMS to pay hospitals that successfully report designated quality measures a higher annual update to their payment rates, and led to decreased disparities between 2005–2010 in acute myocardial infarction, heart failure, and pneumonia [96, 97]. However, it is important to note that not all quality improvement programs have the same effect. There is concern that the Hospital Readmission Reduction Program, a Medicare value-based purchasing program that reduces payments to hospitals with excess readmissions, disproportionately penalizes hospital systems taking care of disadvantaged individuals with unaccounted social needs [98–100]; and there is conflicting evidence on whether or not it reduces racial health disparities [101–103].

The implementation of heavily protocol-driven processes has also been identified as a potential solution to reducing racial disparities in clinical outcomes. In one randomized trial, a protocol-driven care model for patients with chest pain (including placement in an observation unit for serial cardiac markers with an expectation for stress imaging) reduced previously observed disparities with regards to diagnostic testing, revascularization, and clinical outcomes by race compared to standard inpatient care, which was left to the discretion of the clinicians [104]. Another study found that implementation of computerized triage order sets, education for the medical provider team, and requiring that team members follow quality guidelines was associated with decreased time to first dose pain medications, improved patient satisfaction, and decreased length of stay for patients with sickle cell disease presenting with pain due to vasocclusive crisis [105]. In another example, a computerized clinical decision support tool involving completion of checklists to review venous thromboembolism (VTE) risk factors and contraindications to pharmacologic prophylaxis followed by recommendation of the most appropriate form of VTE prophylaxis was able to reduce disparities in VTE prophylaxis by race and sex [106]. Enhanced recovery after surgery (ERAS) protocol implementation, which spans the continuum of surgical care and includes processes such as patient education, multimodal analgesia, and early mobility, has been shown to ameliorate racial disparities in postoperative length of stay in patients undergoing colorectal surgery [107]. By tying surveillance programs to quality improvement initiatives, protocol driven care, and clinical decision support tools, it is possible to lessen the impact of race and racism on health outcomes. Therefore, one strategy for hospitals to decrease disparities in clinical outcomes is through the implementation of rigorous quality improvement programs and by providing incentives to providers who adhere to evidence-based clinical care guidelines and protocol-driven care.

Another opportunity to address race and racism in healthcare is to deliberately implement interventions to diversify the healthcare workforce. Race, ethnicity, and language concordance are thought to foster trust, communication, and better patient–provider interaction [108–110]. While there is no conclusive evidence to support that patient–provider race concordance is associated with better health outcomes for minorities [111], microaggressions and implicit bias against underrepresented minorities lead to less satisfaction with their care [112]. Compared to patients whose primary physicians are of a different race/ethnicity, patients who have concordant race/ethnicity as their physician are more likely to use needed health services and are less likely to postpone or delay seeking care [113, 114].

The long- and short-term advantages of diversity in the healthcare professions should not be underemphasized. Ensuring workplace diversity enhances critical thinking, problem-solving, and employee professional skills [115, 116]. Additionally, it enables health professionals to attract a wider patient base, improve attractiveness, and exhibit commitment to the community. While diversity in organizations is increasingly respected as a fundamental characteristic, it will reap no benefit if the vantage points of diverse people are not valued. Rather, a diverse and inclusive workforce is the goal. Inclusion enhances an organization's ability to achieve better business results by engaging people from diverse backgrounds and perspectives through participatory decision-making [117, 118].

A strategy applied by many training institutions is to establish programs that seek to increase representation of underrepresented groups in health care by partnering with student associations (e.g., Student National Medical Association), developing pipeline programs, and innovative curricula (e.g., bias literacy training and University of Chicago Pritzker School of Medicine's health disparity course) [81, 119]. Additionally, curricular interventions are being developed to directly educate future health care professionals about the impact of structural racism and implicit bias on health. A multidisciplinary and multiracial group at the University of Minnesota has come together to develop and pilot an intervention for first year medical students informed by Public Health Critical Race Praxis (PHCRP). This methodology seeks to "dismantle group power relations and to systematically promote and sustain inter-racial dialogue in a professional setting by encouraging participants to systematically assess and address racism-related factors that may influence research and practice" [120]. The intent of this pilot is to promote and facilitate open dialogue about race and racism by having students learn about lived experiences from the perspective of racially marginalized groups. The examples described above support the importance for healthcare providers to learn about the effects of systemic racism on minority health, as well as the implementation of initiatives to increase the number of underrepresented minorities in medical fields, as ways to improve the quality of care delivered.

Another strategy to increase the number of medical students and faculty from racial and ethnic minority backgrounds is by increasing federal funding to support their recruitment and retention and provide incentives to practice in medically underserved urban and rural areas [121]. Federal research and training grants, workforce programs, scholarships, and loan forgiveness programs are just some examples that help reduce the financial burden faced by underrepresented minorities in medicine and increase the physician workforce in underserved communities. Financial incentives paired with purposeful outreach, mentoring, and tutoring may encourage more students from racial and ethnic minority backgrounds to pursue careers in science and health.

Leadership at the hospital level can also play a significant role in bridging the healthcare gap and improving the quality of care provided to patients of all racial backgrounds. One survey study found substantial differences between the executive boards of Black-serving hospitals (defined as nonprofit hospitals in the top decile based on proportion of discharged elderly Black patients) and non-Black-serving hospitals in their engagement with quality of care issues. Boards of Black-serving hospitals were less likely to report having expertise with quality of care issues, being knowledgeable about specific quality programs, or identifying quality as a top priority for board oversight or the evaluation of CEO performance [122]. Many hospital chairpersons do not recognize the prevalence and persistence of racial disparities in quality of care. Nearly 90% of respondents admitted that they do not examine quality of care data stratified by race and ethnicity [122]. One opportunity for reducing disparities and improving quality of care for racial minorities is by engaging hospital boards in quality improvement activities. By regularly comparing hospital performance with trends and national benchmarks, and routinely reviewing key quality measures disaggregated by race and ethnicity, hospitals can provide more effective and higher quality care [123].

Societal Level

Racial and ethnic disparities in healthcare occur in the context of broader historic and contemporary social and economic inequality and persistent racial and ethnic discrimination in many sectors of American life. Conceptually, structural racism incorporates the totality of ways in which societies foster racial discrimination through mutually reinforcing systems of housing, education, employment, earnings, benefits, credit, media, health care, and criminal justice. These patterns and practices in turn reinforce discriminatory beliefs, values, and distribution of resources [124].

Throughout US history, the belief that Black people are fundamentally and biologically different was championed by physicians, scientists and slave owners alike to justify inhumane treatment in medical research [125]. The 1932 US Public Health Service study known as the Tuskegee experiment, wherein Black men were used to examine how untreated syphilis affected the body, is one of the most widely recognized examples of how structural racism has manifested across our social institutions [126]. Other examples of unethical experimentation in minority populations include the disproportionate sterilization rates of Latinas under California's eugenics law [127] and of Native American women as recently as the 1970s [128], and the whole or partial body irradiation of African American cancer patients for over a decade by Dr. Eugene Sanger at the University of Cincinnati [129]. These experiments have had lasting impacts for racial minorities on their trust of the healthcare system and are evidence of how medical institutions operationalize implicit bias and structural racism [130, 131].

Some of the most notable consequences of structural racism are evidenced by the wide disparities in ED utilization associated with poor outpatient management of chronic conditions [132, 133]. For example, lack of preventive care leads to disparities in the presentation of acute exacerbation of chronic obstructive pulmonary disease (COPD) in the ED. Latinx and Black patients are less likely to have a primary care physician and have more frequent and lower acuity ED visits for COPD compared to non-Hispanic White patients [134]. Among patients with a history of seizure presenting for emergency care, Black patients are less likely to have regular ambulatory care and are more likely to have missed or ran out of antiepileptic medications [135]. These disparities are reflective of the important role that the ED plays as a safety net for many racial minority patients. However, this downstream effect of structural racism and lack of access to preventative care may also reinforce the stereotype or perception that certain minority groups do not take preventative measures regarding their health, and unnecessarily over-utilize the ED [136].

Racial and ethnic disparities are also evidenced as a disproportionate burden of infectious disease in Black, Latino, and Native American communities. African Americans account for about 13% of the US population, yet represent almost half of new AIDS diagnoses [137]. Native Americans experience higher rates of meningitis and invasive bacterial disease from *Hemophilus influenzae* type B (Hib) than do other groups [138]. The H1N1 pandemic of 2008 saw a higher risk of hospitalization and death among Black, Latinx, and Native American patients compared to non-Hispanic Whites [139]. More recently, African American, Latinx, and Native American populations represented a disproportionately higher incidence of COVID-19 infections compared to the White population; and higher morbidity and mortality based on population per capita [140]. The disproportionate impacts of infectious disease in racial minorities are the legacy of structural racism and result from living in crowded conditions, lower income, susceptibility to complications caused by chronic disease, and (for the COVID-19 pandemic) differential exposure due to the inability to self-isolate and continuing to work in the midst of stay at home orders [141].

Another consequence of structural racism is maladaptive patterns of interactions with the healthcare system. Specific to emergency care, multiple studies have cited delayed EMS activation by minorities as a contributing factor to poor outcomes. Reasons for delays in calling 911 for life-threatening conditions are multifactorial and include disparities in health literacy [142] and concerns about costs associated with ambulance transport [143, 144]. Other consequences of structural racism cited by Black and Latinx patients as a reason for delaying calls to 911 include distrust and fear of law enforcement [145, 146]. Disparities in EMS utilization for stroke by race were measured in the Get with the Guidelines stroke registry [147]. Latinx and Asian men and women had lower adjusted odds of using EMS versus their White counterparts; Black women were less likely than White women to use EMS [148]. Delayed EMS activation may explain why Black patients are more likely to present later after symptom onset for acute MI compared to White people (180 min versus 120 min) [149] and why African American and Latinx males have lower odds of achieving a door to balloon time for cardiac catheterization under 90 min compared to White males [150].

The systematic and pervasive disparities in the treatment and outcomes of patients who are racial minorities operate within the context of larger systemic inequities including unequal access to health care infrastructure and services. Emergency medicine is uniquely positioned and can function as a barometer for how well the health care system is performing and how social movements influence it. As patient volumes in our EDs continue to increase and income gaps continue to widen, our specialty is confronted with adopting an increasingly disproportionate role in serving patients with unmet social needs. Emergency medicine training prepares us with the expertise to care for critically ill patients with complicated diseases, but the success of many emergency care interventions is entirely dependent on the reality of our patients outside of our clinical encounters. Therefore, it is imperative to develop new interdisciplinary care models to bridge the racial gap in healthcare outcomes.

One example of a novel interdisciplinary care model involving the emergency care system is the Coordinated Healthcare Interventions for Childhood Asthma Gaps in Outcomes (CHICAGO) Plan. Black and Latinx children with asthma are twice as likely to visit an ED and are less likely to be prescribed or effectively use asthma medication compared to non-Hispanic White children [151, 152]. The CHICAGO Plan aims to test evidence-based strategies to improve the care and outcomes of Black and Latinx children with uncontrolled asthma presenting to EDs in Chicago. Innovative features of the CHICAGO Plan include early and continuous engagement of children, caregivers, the Chicago Department of Public Health, and other stakeholders to inform the design and implementation of the study [153]. Children are randomized to receive a patient-centered ED discharge tool, a patient-centered ED discharge tool plus community health worker (CHW) home visitation, or usual care. The study's two intervention arms, a patient-centered ED discharge tool and assignment of a community health worker (CHW), aim to improve care coordination post-discharge. The CHW arm also addresses home environmental remediation to reduce indoor allergens and asthma symptoms and improve quality of life, a topic that is not usually an area of focus in the ED encounter. While the results of this study have not yet been published, the study shows promise in addressing some of the downstream effects of structural racism, such as barriers related to care coordination post-discharge and home environment [154].

Multidisciplinary healthcare system-wide quality improvement projects can also help reduce disparities resulting from structural racism and unequal access. One example is the HeartRescue Project, a multistate initiative to develop regional cardiac resuscitation systems of care for out-of-hospital cardiac arrest (OHCA) by implementing guideline-based best practices for bystander, pre-hospital, and hospital care and standardized data reporting (including disaggregated outcomes by race) [155]. The Illinois HeartRescue Project directed in partnership with the Chicago Fire Department implemented evidence-based quality improvement initiatives, focused community engagement, and surveillance to address disparities in OHCA treatment and survival. Implementation of dispatch assisted CPR training coupled with bystander CPR training in predominantly minority neighborhoods with low OHCA survival resulted in a greater than 50% increase in bystander CPR rates [156–158]. These improved bystander CPR rates coupled with new resuscitation and post-resuscitation care, destination protocols for EMS, and case review resulted not only in narrowed survival disparities in predominantly minority neighborhoods [158], but also improved Chicago's overall OHCA survival rates [159].

Recommendations for Emergency Medicine Practice

> Stories can name a type of discrimination, once named it can be combated. If Race is not real or objective but constructed, racism and prejudice should be capable of deconstruction.
>
> – Critical Race Theory, pg. 49 [160]

Basic

- *Name racism when you see it.* Do not ignore or minimize the impact of race and racism in your professional interactions in your immediate clinical environment and the surrounding healthcare system; or in your personal interactions. Not being racist is not the same as being anti-racist. We must be allies and amplifiers for our colleagues, patients, and communities to dismantle structural racism and implicit biases.
- *Recognize your own implicit biases and how structural racism affects your practice.* Introspection is key as we must look inward and ask ourselves why we make the decisions we make when treating racial minority patients. Taking a moment to reflect on the care you provide to someone with a discordant race from your own or a marginalized race presents an opportunity to reflect on our implicit biases in our decision making. To do this work we must be open to confronting our unconscious biases and actively combat their consequences.
- *Become bias literate.* Take an implicit bias test at Project Implicit (https://implicit.harvard.edu) [81]. Bias is implicit and ubiquitous and without recognizing biases in ourselves, we cannot move to combat their effects.

Intermediate

- *Seek opportunities to educate yourself about the impact of race and racism in emergency care and on our patients.* The American College of Emergency Physicians (ACEP), through the establishment of the Diversity, Inclusion, and Health Equity (DIHE) Section, has made a significant effort to emphasize the need for this type of education in our practice [161]. The Society of Academic Emergency Medicine (SAEM) also supports the Academy for Diversity and Inclusion in Emergency Medicine (ADIEM) where there are many opportunities for education and advocacy [162].
- *Engage in efforts within your own practice setting to include education for medical students, residents, and hospital staff on issues of racial disparities in health care and how to reduce these inequities.* Two excellent resources for curricular content and research can be found from the Structural Competency Working Group [163] and the Anti-Racism in Medicine Collection from the AAMC MedEdPortal [164].
- *Undertake quality improvement initiatives within your department that include metrics of race and their impact on health care delivery mechanisms and clinical outcomes.* Deliberately measuring the impact of race in patient outcomes facilitates transparency in our practice environments to compare institutional data with national trends in racial health disparities. We must be willing to find possible disparities in order to engage in anti-racism work toward health equity.

Advanced

- *Advocate for racial diversity and inclusion in the workplace*. Participate in search committees within your hospital, and if you are in academics, volunteer your time to interview residency candidates and applicants to medical school.
- *Engage in advocacy through emergency medicine professional organizations*. Participate in advocacy efforts and become members of sections focused on reducing racial disparities in health; several emergency medicine professional organizations have such sections, including SAEM (ADIEM), ACEP (DIHE), AAEM (Diversity, Equity and Inclusion Committee), and the National Medical Association. Participating in national efforts can help shape how we bring awareness of issues of race and racism in our specialty and actively encourage the work of anti-racism.
- *Collaborate with other governmental, private, and health care organizations (e.g., American Medical Association, American Hospital Association) that are working actively in the sphere of combating racism and racial disparities*. This includes coordinating efforts between medical societies with a focus on minority health and larger professional organizations of healthcare providers that have power and visibility to amplify this issue. Use our voices to lobby within hospital systems to coordinate efforts: to call attention to the impact of race and racism on the health of our patients; to call for greater transparency of practice patterns on the impact of race and racism on community health; to call out unequal care when it occurs; and to share best practices and resources.
- *Advocate for funding opportunities that prioritize implementation and demonstration projects related to reducing racial disparities in emergency medicine*. A strong body of literature has repeatedly demonstrated the existence of health disparities that can be attributed to racism. It is now time to prioritize supporting researchers and clinicians who seek to implement projects and interventions that will reduce known disparities and help us move to more justly appropriate resources to care for our most marginalized populations, including racial minorities.

Teaching Case

Clinical Case

The emergency department is overflowing with patients after the closure of yet another urban hospital. You just called to ask to be put on diversion because you are boarding patients and there are no inpatient beds left. You are in the middle of getting a patient with a subarachnoid hemorrhage transferred to a facility with neurocritical care, when the charge nurse approaches you. There is a patient identified at triage as a potential COVID-19 case and was provided a surgical mask and placed in an isolation room. The charge nurse requests that you see this patient before the nurse that is assigned to the room, in order to minimize staff exposure as your hospital is running low on personal protective equipment. The charge nurse tells you that the patient has stable vital signs and appears in no distress, so the patient waits in a room alone for two hours while you manage other unstable patients.

When you finally make it into the room, you find a 48-year-old non-toxic African American female with complaints of progressive shortness of breath and fever. The patient is anxiously rocking in her bed. She shares with you that she recently visited her grandmother in the nursing facility that was just on the news due to an outbreak of cases of COVID-19. As the patient speaks, you note that she sometimes stops in between sentences to take a deep breath. She shares that she is having some difficulty climbing stairs and walking long distances, but as long as she is at rest, she does not feel short of breath. She currently does not have a primary care provider, having been recently laid off from her job as a seamstress.

Her past medical history includes diabetes, hypertension, and hypercholesterolemia. Pertinent findings on her physical exam include a fever of 102 degrees F, but otherwise normal vital signs and she is noted to have slightly diminished breath sounds. Her laboratory testing returns with normal results CBC, BMP, with no noted elevation in troponin nor b-type natriuretic peptide. There is a slightly elevated d-dimer and ferritin level, but her chest x-ray demonstrates no obvious infiltrates, although her lungs are hypoinflated. Your hospital does not possess point of care respiratory virus testing capacity, so you obtain a sample to send out for COVID-19 testing.

This patient has multiple comorbidities putting her at high risk for negative outcomes due to COVID-19, but she currently does not require immediate hospitalization. You would like to discharge this patient home but are preoccupied about the challenges this patient may face in securing close follow-up. The patient does not have an active mobile phone plan. She provides the cell phone number of her younger sister who lives with her, so she can be informed of her respiratory panel results. You decide that you will make a note to yourself to call the patient with her respiratory panel results and to check on her symptom development, but you worry about the patient having reliable access to the number she provided.

Teaching Points

1. Lack of resources in hospitals serving predominantly minority neighborhoods significantly impacts the reliability of epidemiologic data during pandemics.
2. Overcrowding and ambulance diversion are one effect of hospital closures in many minority neighborhoods.
3. There are significant disparities in medical resources, such as personal protective equipment for medical providers, critical care space, and availability of specialty services that result in delayed delivery of life-saving therapeutic interventions.

Discussion Questions

1. What disease management options are available to hospitals with limited resources during a pandemic?
2. In what ways can you facilitate follow-up care in patient populations without insurance, with limited transportation options, limited communication resources, and no designated primary care provider?
3. What improvement opportunities exist in your emergency department that will enhance healthcare for a socially disenfranchised patient population?

References

1. Gee GC, Ford CL. Structural racism and health inequities: old issues, new directions. Du Bois Rev. 2011;8(1):115.
2. Roberts D. The problem with race-based medicine. In: Obasogie OK, editor. Beyond bioethics: toward a new biopolitics. Berkeley: University of California Press; 2019. p. 410–4.
3. Hardeman RR, Medina EM, Kozhimannil KB. Structural racism and supporting black lives – the role of health professionals. N Engl J Med. 2016;375(22):2113–5.
4. Clair M, Denis JS. Sociology of racism. The International Encyclopedia of the Social and Behavioral sciences. 2015;19:857–63. [Publisher's version accessed at: https://www.sciencedirect.com/science/article/pii/B9780080970868321225].
5. Ford CL, Harawa NT. A new conceptualization of ethnicity for social epidemiologic and health equity research. Soc Sci Med. 2010;71(2):251–8.
6. Clair M, Denis JS. Racism, sociology of. International encyclopedia of the social & behavioral sciences. 2015;858.
7. Pallok K, De Maio F, Ansell DA. Structural racism – a 60-year-old black woman with breast cancer. N Engl J Med. 2019;380(16):1489–93.
8. Omi M, Winant H (2014). Racial formation in the United States. New York, New York: Routledge Taylor and Francis Group.
9. Hammonds E. The nature of difference: sciences of race in the United States from Jefferson to genomics. 2009.
10. Williams DR, Collins C. Racial residential segregation: a fundamental cause of racial disparities in health. Public Health Rep. 2016;116:404.
11. Anderson KF, Fullerton AS. Residential segregation, health, and health care: answering the Latino question. Race Soc Probl. 2014;6(3):262–79.
12. Smith DB. The racial segregation of hospital care revisited: Medicare discharge patterns and their implications. Am J Public Health. 1998;88(3):461–3.
13. Largent EA. Public health, racism, and the lasting impact of hospital segregation. Public Health Rep. 2018;133(6):715–20.
14. Hing AK. The right to vote, the right to health: voter suppression as a determinant of racial health disparities. J Health Disparities Res Pract. 2018;12(6):5.
15. Binswanger IA, Redmond N, Steiner JF, Hicks LS. Health disparities and the criminal justice system: an agenda for further research and action. J Urban Health. 2012;89(1):98–107.
16. Faber J, Friedline T. The racialized costs of banking (2018). New America; Washington, DC.
17. Bailey ZD, Krieger N, Agénor M, Graves J, Linos N, Bassett MT. Structural racism and health inequities in the USA: evidence and interventions. Lancet. 2017;389(10077):1453–63.
18. Nardone A, Casey JA, Morello-Frosch R, Mujahid M, Balmes JR, Thakur N. Associations between historical residential redlining and current age-adjusted rates of emergency department visits due to asthma across eight cities in California: an ecological study. Lancet Planetary Health. 2020;4(1):e24–31.
19. Hamstead ZA, Fisher D, Ilieva RT, Wood SA, McPhearson T, Kremer P. Geolocated social media as a rapid indicator of park visitation and equitable park access. Comput Environ Urban Syst. 2018;72:38–50.
20. Chen D, Jaenicke EC, Volpe RJ. Food environments and obesity: household diet expenditure versus food deserts. Am J Public Health. 2016;106(5):881–8.
21. Lee A, Cardel M, Donahoo WT. Social and environmental factors influencing obesity. Endotext [Internet]. South Dartmouth: MDText.com, Inc.; 2019.
22. Hussein M, Roux AVD, Field RI. Neighborhood socioeconomic status and primary health care: usual points of access and temporal trends in a major US urban area. J Urban Health. 2016;93(6):1027–45.
23. Edward J, Biddle DJ. Using geographic information systems (GIS) to examine barriers to healthcare access for Hispanic and Latino immigrants in the US south. J Racial Ethn Health Disparities. 2017;4(2):297–307.

24. Marley TL. Ambiguous jurisdiction: governmental relationships that affect American Indian health care access. J Health Care Poor Underserved. 2019;30(2):431–41.
25. Ko M, Needleman J, Derose KP, Laugesen MJ, Ponce NA. Residential segregation and the survival of US urban public hospitals. Med Care Res Rev. 2014;71(3):243–60.
26. Carnes M, Devine PG, Manwell LB, Byars-Winston A, Fine E, Ford CE, et al. Effect of an intervention to break the gender bias habit for faculty at one institution: a cluster randomized, controlled trial. Acad Med: J Assoc Am Med Coll. 2015;90(2):221.
27. Banaji MR, Greenwald AG (2016). Blindspot: Hidden biases of good people: New York, NY; Bantam Books.
28. Dehon E, Weiss N, Jones J, Faulconer W, Hinton E, Sterling S. A systematic review of the impact of physician implicit racial bias on clinical decision making. Acad Emerg Med. 2017;24(8):895–904.
29. Maina IW, Belton TD, Ginzberg S, Singh A, Johnson TJ. A decade of studying implicit racial/ethnic bias in healthcare providers using the implicit association test. Soc Sci Med. 2018;199:219–29.
30. Ben J, Cormack D, Harris R, Paradies Y. Racism and health service utilisation: a systematic review and meta-analysis. PLoS One. 2017;12(12):e0189900.
31. Yearby R. Racial disparities in health status and access to healthcare: the continuation of inequality in the United States due to structural racism. Am J Econ Sociol. 2018;77(3–4):1113–52.
32. York Cornwell E, Currit A. Racial and social disparities in bystander support during medical emergencies on US streets. Am J Public Health. 2016;106(6):1049–51.
33. Hewes HA, Dai M, Mann NC, Baca T, Taillac P. Prehospital pain management: disparity by age and race. Prehosp Emerg Care. 2018;22(2):189–97.
34. Zook HG, Kharbanda AB, Flood A, Harmon B, Puumala SE, Payne NR. Racial differences in pediatric emergency department triage scores. J Emerg Med. 2016;50(5):720–7.
35. Zipp SA, Krause T, Craig SD, editors. The impact of user biases toward a virtual human's skin tone on triage errors within a virtual world for emergency management training. Proceedings of the Human Factors and Ergonomics Society Annual Meeting. Los Angeles: SAGE Publications; 2017.
36. Johnson TJ, Hickey RW, Switzer GE, Miller E, Winger DG, Nguyen M, et al. The impact of cognitive stressors in the emergency department on physician implicit racial bias. Acad Emerg Med. 2016;23(3):297–305.
37. Nelson A. Unequal treatment: confronting racial and ethnic disparities in health care. J Natl Med Assoc. 2002;94(8):666–8.
38. Statistics NCFH. Life expectancy at birth, by sex, race, and Hispanic origin: United states, 2006–2016 2017 [cited 2019 April 1, 2019]. Available from: https://www.cdc.gov/nchs/data/hus/2017/fig01.pdf.
39. Prevention CfDCa. Hispanic health: preventing type 2 diabetes 2017 [Accessed 1 Apr 2019]. Available from: https://www.cdc.gov/features/hispanichealth/index.html.
40. McDonald JA, Paulozzi LJ. Parsing the paradox: hispanic mortality in the US by detailed cause of death. J Immigr Minor Health. 2019;21(2):237–45.
41. Cunningham TJ, Croft JB, Liu Y, Lu H, Eke PI, Giles WH. Vital signs: racial disparities in age-specific mortality among blacks or African Americans—United States, 1999–2015. MMWR Morb Mortal Wkly Rep. 2017;66(17):444.
42. Veazie M, Ayala C, Schieb L, Dai S, Henderson JA, Cho P. Trends and disparities in heart disease mortality among American Indians/Alaska Natives, 1990–2009. Am J Public Health. 2014;104(S3):S359–S67.
43. Wu BU, Banks PA, Conwell DL. Disparities in emergency department wait times for acute gastrointestinal illnesses: results from the National Hospital Ambulatory Medical Care Survey, 1997-2006. Am J Gastroenterol. 2009;104(7):1668–73.
44. Opoku ST, Apenteng BA, Akowuah EA, Bhuyan S. Disparities in emergency department wait time among patients with mental health and substance-related disorders. J Behav Health Serv Res. 2018;45(2):204–18.

45. James CA, Bourgeois FT, Shannon MW. Association of race/ethnicity with emergency department wait times. Pediatrics. 2005;115(3):e310–e5.
46. Qiao WP, Powell ES, Witte MP, Zelder MR. Relationship between racial disparities in ED wait times and illness severity. Am J Emerg Med. 2016;34(1):10–5.
47. Schrader CD, Lewis LM. Racial disparity in emergency department triage. J Emerg Med. 2013;44(2):511–8.
48. Vigil JM, Alcock J, Coulombe P, McPherson L, Parshall M, Murata A, et al. Ethnic disparities in emergency severity index scores among US Veteran's affairs emergency department patients. PLoS One. 2015;10(5):e0126792.
49. Pollock C. A preliminary study exploring racial differences in triage, hospitalization status, and discharge medication in an emergency department in Graniteville, SC. 2013.
50. Okunseri C, Okunseri E, Chilmaza CA, Harunani S, Xiang Q, Szabo A. Racial and ethnic variations in waiting times for emergency department visits related to nontraumatic dental conditions in the United States. J Am Dent Assoc. 2013;144(7):828–36.
51. Davey K, Olivieri P, Saul T, Atmar S, Rabrich JS, Egan DJ. Impact of patient race and primary language on ED triage in a system that relies on chief complaint and general appearance. Am J Emerg Med. 2017;35(7):1013–5.
52. Lopez L, Wilper AP, Cervantes MC, Betancourt JR, Green AR. Racial and sex differences in emergency department triage assessment and test ordering for chest pain, 1997-2006. Acad Emerg Med. 2010;17(8):801–8.
53. Mills AM, Shofer FS, Boulis AK, Holena DN, Abbuhl SB. Racial disparity in analgesic treatment for ED patients with abdominal or back pain. Am J Emerg Med. 2011;29(7):752–6.
54. Tamayo-Sarver JH, Hinze SW, Cydulka RK, Baker DW. Racial and ethnic disparities in emergency department analgesic prescription. Am J Public Health. 2003;93(12):2067–73.
55. Lin CB, Peterson ED, Smith EE, Saver JL, Liang L, Xian Y, et al. Emergency medical service hospital prenotification is associated with improved evaluation and treatment of acute ischemic stroke. Circ Cardiovasc Qual Outcomes. 2012;5(4):514–22.
56. Singhal A, Tien Y-Y, Hsia RY. Racial-ethnic disparities in opioid prescriptions at emergency department visits for conditions commonly associated with prescription drug abuse. PLoS One. 2016;11(8):e0159224.
57. Young MF, Hern HG, Alter HJ, Barger J, Vahidnia F. Racial differences in receiving morphine among prehospital patients with blunt trauma. J Emerg Med. 2013;45(1):46–52.
58. Kennel J, Withers E, Parsons N, Woo H. Racial/ethnic disparities in pain treatment: evidence from Oregon emergency medical services agencies. Med Care. 2019;57(12):924–9.
59. Lord B, Khalsa S. Influence of patient race on administration of analgesia by student paramedics. BMC Emerg Med. 2019;19(1):32.
60. Govindarajan P, Friedman BT, Delgadillo JQ, Ghilarducci D, Cook LJ, Grimes B, et al. Race and sex disparities in prehospital recognition of acute stroke. Acad Emerg Med. 2015;22(3):264–72.
61. Vaughan Sarrazin M, Limaye K, Samaniego EA, Al Kasab S, Sheharyar A, Dandapat S, et al. Disparities in inter-hospital helicopter transportation for Hispanics by geographic region: a threat to fairness in the era of thrombectomy. J Stroke Cerebrovasc Dis: Off J Nat Stroke Assoc. 2019;28(3):550–6.
62. Kimball MM, Neal D, Waters MF, Hoh BL. Race and income disparity in ischemic stroke care: nationwide inpatient sample database, 2002 to 2008. J Stroke Cerebrovasc Dis: Off J Nat Stroke Assoc. 2014;23(1):17–24.
63. Smedley B, Stith A, Nelson A. Unequal treatment: what healthcare providers need to know about racial and ethnic disparities in healthcare. Washington, DC: National Academy Press. Retrieved February. 2002;12:2004.
64. Hanchate AD, Paasche-Orlow MK, Baker WE, Lin M-Y, Banerjee S, Feldman J. Association of race/ethnicity with emergency department destination of Emergency Medical Services transport. JAMA Netw Open. 2019;2(9):e1910816-e.

65. Institute of Medicine (US) Committee on Quality of Health Care in America. Crossing the Quality Chasm: A New Health System for the 21st Century. Washington (DC): National Academies Press (US); 2001.
66. Hoffman KM, Trawalter S, Axt JR, Oliver MN. Racial bias in pain assessment and treatment recommendations, and false beliefs about biological differences between blacks and whites. Proc Natl Acad Sci U S A. 2016;113(16):4296–301.
67. Green CR, Anderson KO, Baker TA, Campbell LC, Decker S, Fillingim RB, et al. The unequal burden of pain: confronting racial and ethnic disparities in pain. Pain Med. 2003;4(3):277–94.
68. Todd KH, Samaroo N, Hoffman JR. Ethnicity as a risk factor for inadequate emergency department analgesia. JAMA. 1993;269(12):1537–9.
69. Mossey JM. Defining racial and ethnic disparities in pain management. Clin Orthop Relat Res®. 2011;469(7):1859–70.
70. Dickason RM, Chauhan V, Mor A, Ibler E, Kuehnle S, Mahoney D, et al. Racial differences in opiate administration for pain relief at an academic emergency department. West J Emerg Med. 2015;16(3):372–80.
71. Harris B, Hwang U, Lee WS, Richardson LD. Disparities in use of computed tomography for patients presenting with headache. Am J Emerg Med. 2009;27(3):333–6.
72. Haywood C Jr, Tanabe P, Naik R, Beach MC, Lanzkron S. The impact of race and disease on sickle cell patient wait times in the emergency department. Am J Emerg Med. 2013;31(4):651–6.
73. Haywood C Jr, Lanzkron S, Hughes M, Brown R, Saha S, Beach MC. The association of clinician characteristics with their attitudes toward patients with sickle cell disease: secondary analyses of a randomized controlled trial. J Natl Med Assoc. 2015;107(2):89–96.
74. Freiermuth CE, Haywood C Jr, Silva S, Cline DM, Kayle M, Sullivan D, et al. Attitudes toward patients with sickle cell disease in a multicenter sample of emergency department providers. Adv Emerg Nurs J. 2014;36(4):335–47.
75. Glassberg JA, Tanabe P, Chow A, Harper K, Haywood C Jr, DeBaun MR, et al. Emergency provider analgesic practices and attitudes toward patients with sickle cell disease. Ann Emerg Med. 2013;62(4):293–302 e10.
76. Poma PA. Race/ethnicity concordance between patients and physicians. J Natl Med Assoc. 2017;109(1):6–8.
77. Telfair J, Myers J, Drezner S. Does race influence the provision of care to persons with sickle cell disease?: perceptions of multidisciplinary providers. J Health Care Poor Underserved. 1998;9(2):184–95.
78. Rasooly IR, Mullins PM, Mazer-Amirshahi M, van den Anker J, Pines JM. The impact of race on analgesia use among pediatric emergency department patients. J Pediatr. 2014;165(3):618–21.
79. Goyal MK, Kuppermann N, Cleary SD, Teach SJ, Chamberlain JM. Racial disparities in pain management of children with appendicitis in emergency departments. JAMA Pediatr. 2015;169(11):996–1002.
80. Xu K, Nosek B, Greenwald AG. Psychology data from the Race Implicit Association Test on the Project Implicit Demo website. J. Open Psychol. Data. 2014;2(1):e3.
81. Carnes M, Devine PG, Isaac C, Manwell LB, Ford CE, Byars-Winston A, et al. Promoting institutional change through bias literacy. J Divers High Educ. 2012;5(2):63.
82. Neff J, Holmes SM, Knight KR, Strong S, Thompson-Lastad A, McGuinness C, Duncan L, Saxena N, Harvey MJ, Langford A, Carey-Simms KL. Structural Competency: Curriculum for Medical Students, Residents, and Interprofessional Teams on the Structural Factors That Produce Health Disparities. MedEdPORTAL: the journal of teaching and learning resources. 2020;16:10888.
83. Metzl JM, Hansen H. Structural competency: theorizing a new medical engagement with stigma and inequality. Soc Sci Med. 2014;103:126–33.
84. Neff J, Knight KR, Satterwhite S, Nelson N, Matthews J, Holmes SM. Teaching structure: a qualitative evaluation of a structural competency training for resident physicians. J Gen Intern Med. 2017;32(4):430–3.

85. Hasnain-Wynia R, Baker DW, Nerenz D, Feinglass J, Beal AC, Landrum MB, et al. Disparities in health care are driven by where minority patients seek care: examination of the hospital quality alliance measures. Arch Intern Med. 2007;167(12):1233–9.
86. Bulatao RA, Anderson NB. Understanding racial and ethnic differences in health in late life: a research agenda. Washington (DC): National Academies Press; 2004.
87. Yearby R. Sick and tired of being sick and tired: putting an end to separate and unequal health care in the United States 50 years after the Civil Rights Act of 1964. Health Matrix. 2015;25:1.
88. Lewis VA, Fraze T, Fisher ES, Shortell SM, Colla CH. ACOs serving high proportions of racial and ethnic minorities lag in quality performance. Health Aff. 2017;36(1):57–66.
89. Mathison DJ, Chamberlain JM, Cowan NM, Engstrom RN, Fu LY, Shoo A, et al. Primary care spatial density and nonurgent emergency department utilization: a new methodology for evaluating access to care. Acad Pediatr. 2013;13(3):278–85.
90. Luo W, Wang F. Measures of spatial accessibility to health care in a GIS environment: synthesis and a case study in the Chicago region. Environ Plann B: Plann Des. 2003;30(6):865–84.
91. Macinko J, Starfield B, Shi L. Quantifying the health benefits of primary care physician supply in the United States. Int J Health Serv: Plann Adm Eval. 2007;37(1):111–26.
92. Brown EJ, Polsky D, Barbu CM, Seymour JW, Grande D. Racial disparities in geographic access to primary care in Philadelphia. Health Aff. 2016;35(8):1374–81.
93. Shen Y-C, Hsia RY. Do patients hospitalised in high-minority hospitals experience more diversion and poorer outcomes? A retrospective multivariate analysis of Medicare patients in California. BMJ Open. 2016;6(3):e010263.
94. Hsia RY, Sarkar N, Shen Y-C. Impact of ambulance diversion: black patients with acute myocardial infarction had higher mortality than whites. Health Aff. 2017;36(6):1070–7.
95. Merton RK, Sills DL, Stigler SM. The Kelvin dictum and social science: an excursion into the history of an idea. J Hist Behav Sci. 1984;20(4):319–31.
96. Edmund Anstey D, Li S, Thomas L, Wang TY, Wiviott SD. Race and sex differences in management and outcomes of patients after ST-elevation and non-ST-elevation myocardial infarct: results from the NCDR. Clin Cardiol. 2016;39(10):585–95.
97. Trivedi AN, Nsa W, Hausmann LR, Lee JS, Ma A, Bratzler DW, et al. Quality and equity of care in U.S. hospitals. N Engl J Med. 2014;371(24):2298–308.
98. Sjoding MW, Cooke CR. Readmission penalties for chronic obstructive pulmonary disease will further stress hospitals caring for vulnerable patient populations. Am J Respir Crit Care Med. 2014;190(9):1072–4.
99. Caracciolo CM, Parker DM, Marshall E, Brown JR. Excess readmission vs excess penalties: readmission penalties as a function of socioeconomics and geography. J Hosp Med. 2017;12(8):610.
100. Wadhera RK, Maddox KEJ, Wasfy JH, Haneuse S, Shen C, Yeh RW. Association of the Hospital Readmissions Reduction Program with mortality among Medicare beneficiaries hospitalized for heart failure, acute myocardial infarction, and pneumonia. JAMA. 2018;320(24):2542–52.
101. Figueroa JF, Zheng J, Orav EJ, Epstein AM, Jha AK. Medicare program associated with narrowing hospital readmission disparities between black and white patients. Health Aff. 2018;37(4):654–61.
102. Nastars DR, Rojas JD, Ottenbacher KJ, Graham JE. Race/ethnicity and 30-day readmission rates in medicare beneficiaries with COPD. Respir Care. 2019;64(8):931–6.
103. Chaiyachati KH, Qi M, Werner RM. Changes to racial disparities in readmission rates after Medicare's Hospital Readmissions Reduction Program within safety-net and non–safety-net hospitals. JAMA Netw Open. 2018;1(7):e184154-e.
104. Miller CD, Stopyra JP, Mahler SA, Case LD, Vasu S, Bell RA, et al. ACES (accelerated chest pain evaluation with stress imaging) protocols eliminate testing disparities in patients with chest pain. Crit Pathw Cardiol. 2019;18(1):5–9.
105. Kim S, Brathwaite R, Kim O. Evidence-based practice standard care for acute pain management in adults with sickle cell disease in an urgent care center. Qual Manag Health Care. 2017;26(2):108–15.

106. Lau BD, Haider AH, Streiff MB, Lehmann CU, Kraus PS, Hobson DB, et al. Eliminating healthcare disparities via mandatory clinical decision support: the venous thromboembolism (VTE) example. Med Care. 2015;53(1):18.
107. Wahl TS, Goss LE, Morris MS, Gullick AA, Richman JS, Kennedy GD, et al. Enhanced recovery after surgery (ERAS) eliminates racial disparities in postoperative length of stay after colorectal surgery. Ann Surg. 2018;268(6):1026–35.
108. Cooper LA, Roter DL, Johnson RL, Ford DE, Steinwachs DM, Powe NR. Patient-centered communication, ratings of care, and concordance of patient and physician race. Ann Intern Med. 2003;139(11):907–15.
109. Blanchard J, Nayar S, Lurie N. Patient–provider and patient–staff racial concordance and perceptions of mistreatment in the health care setting. J Gen Intern Med. 2007;22(8):1184–9.
110. Kumar D, Schlundt DG, Wallston KA. Patient-physician race concordance and its relationship to perceived health outcomes. Ethn Dis. 2009;19(3):345.
111. Meghani SH, Brooks JM, Gipson-Jones T, Waite R, Whitfield-Harris L, Deatrick JA. Patient–provider race-concordance: does it matter in improving minority patients' health outcomes? Ethn Health. 2009;14(1):107–30.
112. Spevick J. The case for racial concordance between patients and physicians. AMA J Ethics. 2003;5(6):163–5.
113. Saha S, Komaromy M, Koepsell TD, Bindman AB. Patient-physician racial concordance and the perceived quality and use of health care. Arch Intern Med. 1999;159(9):997–1004.
114. LaVeist TA, Nuru-Jeter A, Jones KE. The association of doctor-patient race concordance with health services utilization. J Public Health Policy. 2003;24(3–4):312–23.
115. Okoro EA, Washington MC. Workforce diversity and organizational communication: analysis of human capital performance and productivity. J Divers Manag (JDM). 2012;7(1):57–62.
116. Saxena A. Workforce diversity: a key to improve productivity. Procedia Econ Finance. 2014;11(1):76–85.
117. Jordan TH. Moving from diversity to inclusion. Profiles in Diversity Journal. March 22, 2011.
118. Stevens FG, Plaut VC, Sanchez-Burks J. Unlocking the benefits of diversity: all-inclusive multiculturalism and positive organizational change. J Appl Behav Sci. 2008;44(1):116–33.
119. Figueroa O. The significance of recruiting underrepresented minorities in medicine: an examination of the need for effective approaches used in admissions by higher education institutions. Med Educ Online. 2014;19:24891.
120. Hardeman RR, Burgess D, Murphy K, Satin DJ, Nielsen J, Potter TM, et al. Developing a medical school curriculum on racism: multidisciplinary, multiracial conversations informed by public health critical race praxis (PHCRP). Ethn Dis. 2018;28(Suppl 1):271–8.
121. Williams DR, Cooper LA. Reducing racial inequities in health: using what we already know to take action. Int J Environ Res Public Health. 2019;16(4):606.
122. Jha AK, Epstein AM. Governance around quality of care at hospitals that disproportionately care for black patients. J Gen Intern Med. 2012;27(3):297–303.
123. Millar R, Mannion R, Freeman T, Davies HT. Hospital board oversight of quality and patient safety: a narrative review and synthesis of recent empirical research. Milbank Q. 2013;91(4):738–70.
124. Bailey ZD, Krieger N, Agenor M, Graves J, Linos N, Bassett MT. Structural racism and health inequities in the USA: evidence and interventions. Lancet. 2017;389(10077):1453–63.
125. Trawalter S, Hoffman KM. Got pain? Racial bias in perceptions of pain. Soc Personal Psychol Compass. 2015;9(3):146–57.
126. Feagin J, Bennefield Z. Systemic racism and U.S. health care. Soc Sci Med. 2014;103:7–14.
127. Novak NL, Lira N, O'Connor KE, Harlow SD, Kardia SL, Stern AM. Disproportionate sterilization of Latinos under California's eugenic sterilization program, 1920–1945. Am J Public Health. 2018;108(5):611–3.
128. Ralstin-Lewis DM. The continuing struggle against Genocide: Indigenous women's Reproductive Rights. Wicazo Sa Review. 2005 Apr 1:71-95.
129. Goodwin M. Vulnerable subjects: why does informed consent matter? J Law Med Ethics. 2016;44(3):371–80.

130. Baker R. Minority distrust of medicine: a historical perspective. Mount Sinai J Med, New York. 1999;66(4):212.
131. Brandon DT, Isaac LA, LaVeist TA. The legacy of Tuskegee and trust in medical care: is Tuskegee responsible for race differences in mistrust of medical care? J Natl Med Assoc. 2005;97(7):951.
132. Johnson PJ, Ghildayal N, Ward AC, Westgard BC, Boland LL, Hokanson JS. Disparities in potentially avoidable emergency department (ED) care: ED visits for ambulatory care sensitive conditions. Medical Care. 2012 Dec 1:1020–8.
133. Biello KB, Rawlings J, Carroll-Scott A, Browne R, Ickovics JR. Racial disparities in age at preventable hospitalization among US adults. Am J Prev Med. 2010;38(1):54–60.
134. Tsai CL, Camargo CA Jr. Racial and ethnic differences in emergency care for acute exacerbation of chronic obstructive pulmonary disease. Acad Emerg Med. 2009;16(2):108–15.
135. Fantaneanu TA, Hurwitz S, van Meurs K, Llewellyn N, O'Laughlin KN, Dworetzky BA. Racial differences in emergency department visits for seizures. Seizure. 2016;40:52–6.
136. Smedley BD. The lived experience of race and its health consequences. Am J Public Health. 2012;102(5):933–5.
137. Control CfD, Prevention. Diagnoses of HIV infection in the United States and dependent areas, 2018 (Updated). HIV Surveill Rep. 2020;31:1–119.
138. Singleton R, Hammitt L, Hennessy T, Bulkow L, DeByle C, Parkinson A, et al. The Alaska Haemophilus influenzae type b experience: lessons in controlling a vaccine-preventable disease. Pediatrics. 2006;118(2):e421–9.
139. Dee DL, Bensyl DM, Gindler J, Truman BI, Allen BG, D'Mello T, et al. Racial and ethnic disparities in hospitalizations and deaths associated with 2009 pandemic influenza A (H1N1) virus infections in the United States. Ann Epidemiol. 2011;21(8):623–30.
140. Webb Hooper M, Nápoles AM, Pérez-Stable EJ. COVID-19 and Racial/Ethnic Disparities. JAMA. 2020;323(24):2466–2467.
141. van Dorn A, Cooney RE, Sabin ML. COVID-19 exacerbating inequalities in the US. Lancet (London, England). 2020;395(10232):1243.
142. McGruder HE, Greenlund KJ, Malarcher AM, Antoine TL, Croft JB, Zheng ZJ. Racial and ethnic disparities associated with knowledge of symptoms of heart attack and use of 911: National Health Interview Survey, 2001. Ethn Dis. 2008;18(2):192–7.
143. Skolarus LE, Murphy JB, Zimmerman MA, Bailey S, Fowlkes S, Brown DL, et al. Individual and community determinants of calling 911 for stroke among African Americans in an urban community. Circ Cardiovasc Qual Outcomes. 2013;6(3):278–83.
144. Quinones C, Shah MI, Cruz AT, Graf JM, Mondragon JA, Camp EA, et al. Determinants of pediatric EMS utilization in children with high-acuity conditions. Prehosp Emerg Care: Off J Nat Assoc EMS Phys Nat Assoc State EMS Dir. 2018;22(6):676–90.
145. Chenane JL, Wright EM, Gibson CL. Traffic stops, race, and perceptions of fairness. Policing and society. 2020 Jul 2;30(6):720–37.
146. Sasson C, Haukoos JS, Ben-Youssef L, Ramirez L, Bull S, Eigel B, et al. Barriers to calling 911 and learning and performing cardiopulmonary resuscitation for residents of primarily Latino, high-risk neighborhoods in Denver, Colorado. Ann Emerg Med. 2015;65(5):545–52. e2
147. Ekundayo OJ, Saver JL, Fonarow GC, Schwamm LH, Xian Y, Zhao X, et al. Patterns of emergency medical services use and its association with timely stroke treatment: findings from get with the guidelines-stroke. Circ Cardiovasc Qual Outcomes. 2013;6(3):262–9.
148. Mochari-Greenberger H, Xian Y, Hellkamp AS, Schulte PJ, Bhatt DL, Fonarow GC, et al. Racial/ethnic and sex differences in emergency medical services transport among hospitalized US stroke patients: analysis of the National Get With The Guidelines–Stroke Registry. J Am Heart Assoc. 2015;4(8):e002099.
149. Bolorunduro O, Smith B, Chumpia M, Valasareddy P, Heckle MR, Khouzam RN, Reed GL, Ibebuogu UN. Racial difference in symptom onset to door time in ST elevation myocardial infarction. J. Am. Heart Assoc. 2016;5(10):e003804.
150. Cavender MA, Rassi AN, Fonarow GC, Cannon CP, Peacock WF, Laskey WK, et al. Relationship of race/ethnicity with door-to-balloon time and mortality in patients undergoing

primary percutaneous coronary intervention for ST-elevation myocardial infarction: findings from get with the guidelines–coronary artery disease. Clin Cardiol. 2013;36(12):749–56.
151. Urquhart A, Clarke P. US racial/ethnic disparities in childhood asthma emergent health care use: National Health Interview Survey. J Asthma. 2013–2015;2019:1–11.
152. Crocker D, Brown C, Moolenaar R, Moorman J, Bailey C, Mannino D, et al. Racial and ethnic disparities in asthma medication usage and health-care utilization: data from the National Asthma Survey. Chest. 2009;136(4):1063–71.
153. Krishnan JA, Martin MA, Lohff C, Mosnaim GS, Margellos-Anast H, DeLisa JA, et al. Design of a pragmatic trial in minority children presenting to the emergency department with uncontrolled asthma: the CHICAGO plan. Contemp Clin Trials. 2017;57:10–22.
154. Martin MA, Press VG, Nyenhuis SM, Krishnan JA, Erwin K, Mosnaim G, et al. Care transition interventions for children with asthma in the emergency department. J Allergy Clin Immunol. 2016;138(6):1518–25.
155. van Diepen S, Abella BS, Bobrow BJ, Nichol G, Jollis JG, Mellor J, et al. Multistate implementation of guideline-based cardiac resuscitation systems of care: description of the HeartRescue project. Am Heart J. 2013;166(4):647–53. e2
156. Beck EH, Richards C, Kiely M, Weber JM, Markul E, Stein-Spencer L, et al. Identifying hot spots for EMS-treated acute cardiovascular events-stroke, STEMI, and cardiac arrest-in a large metropolitan city. Circulation. 2013;128:A16267.
157. Del Rios M, Han J, Cano A, Ramirez V, Morales G, Campbell TL, et al. Pay it forward: high school video-based instruction can disseminate CPR knowledge in priority neighborhoods. West J Emerg Med. 2018;19(2):423.
158. Del Rios MWJ, Campbell T, Nyguen H, Pugach O, Markul E, Gerber B, Sharp L, Vanden Hoek T, Hoek TV. Dispatch-assisted cardiopulmonary resuscitation coupled with bystander cardiopulmonary resuscitation training in priority neighborhoods improves cardiac arrest survival. Acad Emerg Med. 2018;25(S1):A636.
159. Del Rios M, Weber J, Pugach O, Nguyen H, Campbell T, Islam S, et al. Large urban center improves out-of-hospital cardiac arrest survival. Resuscitation. 2019;139:234–40.
160. Delgado R, Stefancic J (2017). Critical race theory: an introduction. New York, NY; New York University Press.
161. American College of Emergency Physicians. Why does diversity matter? [updated 2020]. Available from: https://www.acep.org/life-as-a-physician/why-does-diversity-matter/.
162. Academy for Diversity and Inclusion in Emergency Medicine [updated 2020]. Available from: https://www.saem.org/adiem.
163. Structural Competency Working Group [updated 2020]. Available from: https://www.struct-comp.org/.
164. Anti-racism in Medicine Collection [updated 2020]. Available from: https://www.mededportal.org/anti-racism.

Immigration as a Social and Structural Determinant of Health

3

Todd Schneberk and Shamsher Samra

Key Points

- Immigration and especially undocumented status can create barriers to health for immigrant populations. Barriers should be viewed through a structural lens and framework, as opposed to being viewed as purely behavioral or cultural issues.
- Undocumented immigrants' disadvantaged health status and barriers to care increase the likelihood that the ED will be their most likely touch point in the healthcare system.
- The ED visit represents a potent opportunity to address acute and upstream causes of poor health in immigrant populations.
- Healthcare systems can be optimized to provide immigration-informed care. This can be done through knowledge of local access barriers and development of referral systems to help address health related and other structural barriers immigrants can face (e.g., access to care through insurance or primary care programs, legal aid resources, sanctuary status of health settings).

Foundations

Background

In the year 2018, approximately 44 million people, or 13.7% of the entire US population, were thought to be foreign born, the highest proportion since 1910 [1]. Among US children, 19.6 million were foreign born or had at least one

T. Schneberk (✉)
Department of Emergency Medicine, Los Angeles County + University of Southern California Medical Center, Los Angeles, CA, USA

S. Samra
Department of Emergency Medicine, Harbor-UCLA Medical Center, Torrance, CA, USA

H. J. Alter et al. (eds.), *Social Emergency Medicine*,
https://doi.org/10.1007/978-3-030-65672-0_3

parent who was foreign born [2]. Far from being homogenous, the US immigrant community is diverse in culture, history, and beliefs. Legal classification of immigrant groups has significant bearing on their social stability, access to health resources, and consequently, health [3]. The US Census Bureau divides foreign-born US residents into four primary categories: naturalized US citizens (those who have attained US citizenship), lawful permanent residents (LPR) ("green card holders"), humanitarian migrants (refugees and asylees), and unauthorized migrants ("undocumented"). It is important to recognize that terms "illegal immigrants" and "illegal aliens" are sometimes used to refer to undocumented persons. The use of illegal implies that the illegality is inherent to the person rather than an externally applied legal categorization that is malleable over time. This phrase risks dehumanizing the individual and making professional obligations to the patient and the right to health subservient to politicized categories. These terms have been shown to engender negative attitudes towards these patients and thus we discourage use of these terms by providers [4].

Almost half of foreign-born US residents are naturalized citizens [5]. As citizens, this population faces no immigration-based exclusion from healthcare or social services, though they may still face stigma and prejudice that hinders health care access [6]. Another 30% of foreign-born US residents are LPRs, with most being eligible for naturalization over time. LPRs are generally not eligible for Medicaid or Children's Health Insurance Program (CHIP) coverage unless they have maintained their LPR status for at least 5 years. Twenty-three percent of LPRs are uninsured compared to 8% of US citizens [3].

A humanitarian immigrant (refugee and asylee) is a "person(s) who is unable or unwilling to return to his or her country of nationality because of persecution or a well-founded fear of persecution on account of race, religion, nationality, membership in a particular social group, or political opinion" [7]. Humanitarian immigrants are eligible for LPR status after 1 year and naturalization after 5 years. While refugees apply for status outside of the US, asylees apply for status either within the US or at a port of entry. In 2017, 146,003 refugees and asylees adjusted their status to lawful permanent residents in the US, of whom 120,356 (82%) were refugees and the remainder, 25,647 (18%), were asylees, making up a very small percentage of all foreign-born persons living in the US [8]. Immigrants who are granted humanitarian status are generally eligible for Medicaid, CHIP, and other public benefits.

Undocumented US residents refer to residents who lack legal standing in the US and are at risk for deportation. More than 11.3 million undocumented people currently reside throughout the US [9, 10]. Among this population, 47% are women, and approximately 9% are minors. The majority of undocumented individuals are from Mexico (56%), followed by Central America (15%) and Asia (14%) [11]. Although children make up a small proportion of the entire undocumented US population, four million US citizen children have at least one undocumented parent [7].

Evidence Basis

It is important for emergency medicine practitioners to recognize that immigration is a social determinant of health in its own right, in addition to being highly correlated with other social determinants. A large body of research highlights distinct behavioral and cultural characteristics of minority and immigrant subcommunities in influencing lifestyle practices and perceptions of health, healthcare, and illness [12, 13]. Such findings have led to an emphasis on cross-cultural understanding of individual patients in an attempt to decrease health inequities of marginalized populations [12]. In regards to immigration, this simplistic view of cultural competence overlooks the structural forces and structural violence that drive migration, impart physical and mental trauma, limit access to healthcare and services, and constrain healthy behavior [12, 14]. Therefore, structural competence—the ability to discern forces that influence health outcomes at levels above individual interaction—is imperative to emergency practitioners' understanding and promotion of health among immigrant communities [13]. Structural competency consists of: (1) recognizing the sociopolitical structures that shape clinical interactions; (2) developing a language of structure outside of the medical lens; (3) rearticulating "cultural" formulations in structural terms; (4) observing and imagining structural interventions; and (5) developing structural humility [12]. This process includes, but is not limited to, recognition of domestic policies that promote displacement and migration of foreign populations, including an analysis of historical and contemporary US military and neoliberal economic policies that undermine sovereignty (i.e., those policies that serve to destabilize foreign governments to extract and export wealth and natural resources). Relevant examples to the US context include US interventions in Central American conflicts at the end of the twentieth century and free trade policies such as the North American Free Trade Agreement (NAFTA) and the Dominican Republic-Central America Free Trade Agreement (CAFTA-DR) which have displaced millions of individuals from Mexico and Central America, and drive migration to the US [15, 16]. Research from a structural framework has focused mostly on immigration status affected by limited access to healthcare and health-protective resources, agnostic to these larger forces at the root of US migration [13].

Limited Access to Healthcare

Rates of emergency department (ED) utilization are lower for noncitizens than for citizens; annual ED use rates are 12.2% vs. 15.4% and 19.3%, respectively, for undocumented individuals, naturalized citizens and US-born citizens [17]. Despite that, undocumented individuals remain uniquely dependent on the ED for care due to insurance barriers to outpatient care [18, 19]. Undocumented populations, including children, are explicitly excluded from expansion of eligibility for Medicaid and Medicare coverage under the Patient Protection and Affordable Care Act (PPACA) [20]. Participants in the Deferred Action for Childhood Arrivals (DACA) are similarly excluded from eligibility in most states [21]. More than 45% of non-elderly undocumented immigrants are uninsured [3]. As the overall percentage of uninsured Americans decreases, the percentage of the uninsured population that is undocumented is predicted to increase to 25% [3, 17]. Though children of undocumented immigrants are eligible for services through CHIP, research has shown these children have both significantly fewer medical appointments and ED visits compared to children of US citizens [22, 23].

Limited Access to Health Protective Resources

The undocumented immigrant is the most vulnerable when compared to documented immigrants, facing compounding layers of structural barriers that ultimately have negative impacts on health [24]. Multiple social, economic, and political factors framed by local or national policies affect immigrant health. Undocumented immigrants have fewer employment opportunities and are susceptible to extortion and workplace abuse as a result of working in the informal economy, which is exacerbated by a reluctance to report crimes to authorities due to fears of immigration enforcement [25]. They additionally have less access to educational opportunities and the social safety net, which includes assistance with food, wages, housing, health insurance, and healthcare systems in general [26]. Overall, undocumented persons have fewer opportunities for upward mobility compared to documented persons, leading to feelings of reduced agency and empowerment [18, 26, 27].

Emergency Department and Beyond

Immigration status is both a structural and a behavioral barrier that permeates and disadvantages the immigrant globally by exacerbating all other social risks [13]. There are significant barriers to care before an immigrant becomes an ED patient. The ED visit represents a limited window of opportunity to direct patients to appropriate care and resources.

Bedside

Emergency provider understanding and awareness is the linchpin in the care of the immigrant patient. Knowledge of the patient's risks, as well as structural competency, language competency, and the wherewithal to deliver care in a compassionate way reinforce patient-centeredness. Emergency providers can establish and reinforce the sanctuary state of the hospital and healthcare setting in the patient encounter. Sanctuary status of health care settings designates a safe space for care, with policies and a culture reducing cooperation with immigration enforcement [28]. Emergency physicians should realize that there might be a lack of trust in the healthcare setting. Undocumented immigrants cite a fear of discovery and deportation even in use of the ED, which worsened after the rhetoric and immigration policies following the 2016 US presidential election [29, 30]. Patients need to feel that their provider is concerned with their health and safety regardless of background, country of origin, or immigration status [31]. Asking basic questions regarding insurance status and empanelment in primary care can be helpful. Specific follow-up care questions can also be useful surrogates for asking about documentation status, which requires the practitioner to develop local knowledge of the population (e.g., an undocumented person in California could be someone born outside of the country who is enrolled in limited scope Medicaid). Depending on state and local policies, there may be populations that are at high likelihood of being undocumented, and thus the ED visit presents an opportunity to offer both primary care enrollment resources together with immigration resources [32].

In addition, providers should recognize that while the medical record is protected by HIPAA, there are limitations to these protections. Notation of the patient's undocumented status in the medical record could imperil patients if accessed by immigration agencies, and also subject patients to stigma. Recognizing this risk, providers must be thoughtful about the purpose of including citizenship status in medical records, if it is to be included at all [31]. Unlike other social determinants of health where documenting or using ICD-10 codes can help determine the scope of the issue, immigration status is more delicate and nuanced. Proxies using insurance status and knowledge of local populations as discussed above need to be developed and validated based on local infrastructure and resources.

Specifically, providers should recognize that undocumented populations are particularly vulnerable to labor and sex trafficking in addition to other abuses. It is estimated that the majority (67%) of labor trafficking victims and a large (17%) percentage of sex trafficking victims are non-US citizens [33]. Their tenuous legal status creates barriers to leaving a dangerous social and work dynamic. Similarly, in nontrafficked undocumented patients, lack of legal status risks abuse including but not limited to domestic violence, wage theft, and unsafe labor conditions [25]. ED providers must maintain a high degree of suspicion regarding exploitation. It's essential to promote confidentiality by separating the patient from employers or domestic partners when obtaining a history [34]. Patients should be advised regarding their rights and offered referrals to appropriate local support services. Providers should be aware that undocumented victims of trafficking, domestic violence, torture, and other crimes may be eligible for adjustment to legal citizenship, which may support escape from exploitative conditions. While the complexity of immigration status adjustment falls beyond the scope of ED practice, providers should make referrals to legal partners, and emergency departments can establish medical–legal partnerships to improve identification of eligible cases [35].

Apart from the clinical setting, emergency providers can also be involved in advocacy for refugee and asylum seekers by performing forensic evaluations in conjunction with immigration attorneys to substantiate legal cases such as asylum claims. Medical asylum clinics can be an important site of medical–legal partnerships, where physicians can actively contribute to an asylum seeker's legal case. Working within the infrastructure of a local asylum clinic, an emergency provider can obtain training from organizations such as Physicians for Human Rights and can volunteer to perform these evaluations.

Hospital/Healthcare System

At the hospital level, immigration-informed care starts with effective communication, which means providing adequate resources for those patients who have limited English proficiency (LEP) [36]. Despite demonstrating LEP in the clinical setting, immigrants are often treated in English or another language inadequately [37]. It is up to hospital systems to understand their demographics and provide appropriate resources for their patient populations (i.e., ensuring languages spoken are available from interpreter services). Intertwined with, and dependent on LEP, is health literacy [38]. Inadequate measures to accommodate LEP and reduced health literacy impede a hospital's ability to provide effective treatment to immigrant populations.

The hospital system must also mitigate barriers to immigrant patients entering the health care setting by creating a supportive and welcoming environment for this population. It is important to understand that the culture of fear has been layered on top of a baseline vulnerability, as demonstrated in a study in 2013 which showed that one in every eight undocumented patients reported fear of discovery and subsequent deportation during an ED visit [29]. Patients who are most vulnerable may be accessing the healthcare system as their only touch-point to any social or governmental services due to this culture of fear. The unique opportunity to deliver resources to an undocumented person or asylum seeker is rare and requires a cohesive system that is capable of addressing the needs of this special population without introducing stigma or reinforcing fear.

Overall, undocumented patients are more likely to be unfamiliar with the complicated US health system and to experience difficulty in navigating care. Patient navigator interventions have been successful in improving outcomes and overcoming this barrier [39]. Additionally, there is minimal literature documenting outcomes of efforts to reduce fear in healthcare settings, but there are multiple case examples of best practices. Making hospitals "sanctuary sites" may improve use of healthcare and decrease fear among immigrant populations [28, 40]. New York City has pioneered methods of communicating with immigrant communities, with signs declaring "You are safe here" and "We care about your health not your documentation status" in healthcare settings, as well as publishing an open letter to immigrants explaining the importance of healthcare and ensuring their safety from immigration enforcement in health settings [41]. New York City also offers free or low-cost health coverage to all residents regardless of ability to pay or documentation status [41]. Some health systems have also issued statements noting that they will not cooperate with Immigration and Customs Enforcement (ICE), and others have prepared trainings to help providers respond to protect patients if ICE officers attempt to use health facilities for enforcement activities [28]. Patient-centered programs like these can be developed in conjunction with community immigrant advocacy organizations that can hone the messaging, and aid with receptivity among patients.

Emergency department and healthcare systems can develop an immigration-informed social referral pathway to intervene upon the structural barriers that these patients face. This can be done through medical–legal partnerships or in conjunction with local legal and community advocates [32, 42]. A sensitive and discreet screening process for undocumented status and other structural barriers should be combined with effective referrals to community-based immigrant rights organizations, immigration legal advocacy and other forms of community-based accompaniment and care navigation. This type of referral system should not be dependent upon the emergency provider alone but upon the emergency department system of care, including social workers, case managers, community health workers, and financial services.

The ED at Los Angeles County + USC Medical Center uses this model to offer undocumented patients immigration legal services [32]. The patient who remains uninsured, as in "residually uninsured" after expansion of the Affordable Care Act, represents a patient with a high likelihood of being undocumented. This categorization is used as a proxy for undocumented status, and enables providers to focus on referral to a co-located community resource center for insurance and primary care enrollment, as is standard practice in this ED. Patients are met by structurally and linguistically competent staff at the co-located resource center, where they are presented with options for insurance enrollment. If they are only eligible for the county level primary care access plan (a program for undocumented persons), they are offered immigration legal services referral. Similar programs are necessary for EDs to address upstream factors of disease and ultimately practice more immigration-informed care, but they start with research of the local immigrant access infrastructure in order to discreetly direct and refer undocumented patients to needed resources.

Societal Level

At this time, the rights and vulnerabilities of immigrant populations in the US are closely tied to their documentation status [43]. The various levels of documentation from undocumented immigrant to naturalized US citizen have corresponding levels of opportunity within our society. While federal policies largely define the scope of public benefits available to immigrant populations, state and regional institutions can mediate the impact on their constituents. In the case of healthcare, the Affordable Care Act largely excludes health insurance access to undocumented populations [20]. States and local municipalities sometimes find ways to fund health care for this population (i.e., emergency and hospital-based care for acute health events can still be covered by Emergency Medicaid in some instances) [44]. Structural barriers to health insurance, preventative care, and routine care promotes use of hospital and emergency services for catastrophic care [26, 27]. This is exemplified in undocumented hemodialysis dependent patients. Those living in municipalities that only provided emergency hemodialysis suffered a 14-fold increase in mortality compared to those living in a municipality that funded standard regular hemodialysis [45]. This highlights not only the health impact of constraining services to immigrant populations but also the possibility of state and regional bodies in mediating the impact.

Federal anti-immigrant political rhetoric has been tied to a perceived lack of safety amongst both documented and undocumented ED patients [30]. This sense of societal prejudice and insecurity permeates communities and has been linked to increased anxiety and depression, and higher mortality in both documented and undocumented Latinx immigrant populations [6, 46]. Increased ICE enforcement and presence in the news also portends detrimental mental and physical health outcomes among undocumented immigrants [47–51]. Fear and perceptions of discrimination undermine trust in social institutions such as health care, social services, and law enforcement, leaving immigrant populations vulnerable to both crime and poor health outcomes [30, 49].

The impact of anti-immigrant political rhetoric on health seeking behavior has been well documented. In 1994, California passed Proposition 187 which barred undocumented immigrants from using nonemergency services. In response, Latinx populations responded by seeking fewer low acuity and preventative mental health care visits, but increasing amounts of high acuity visits [52, 53]. In Arizona, Senate Bill 1070 increased leniency for traffic stops by law enforcement for immigration purposes. The passage of the bill was associated with decreased prenatal and well-child visits, and interval reductions in birth weights in local Latinx populations [54, 55]. In Georgia, the passage of House Bill 87 which granted local law enforcement the authority to enforce immigration law resulted in a fewer number of Latinx pediatric ED visits, but increased visit acuity and hospitalization rates [56]. These studies exemplify the risk of anti-immigrant rhetoric and legislation in exacerbating inequities in care by deterring care seeking behavior. Consequently sociopolitical conditions may drive patients to defer preventative and routine care until disease progression demands higher acuity ED care [57].

Minimal healthcare utilization as a result of structural barriers and behavioral de-incentivization through anti-immigrant rhetoric have recently been exacerbated by the February 2020 expansion of the *public charge* rule, which introduces immigration enforcement consequences for use of health-related services [58]. The proposed rule creates immigration consequences for use of foundational health assistance programs including: housing assistance such as housing support (Section 8) vouchers, cash assistance programs, food stamps, and long-term care facility use through payment programs including Medicaid [59]. These changes are likely to discourage patients from seeking safety net resources that are both high value from a health standpoint and necessary for ensuring a baseline level of subsistence, especially among needy families and children [60–62]. Not only undocumented immigrants, but mixed documentation status families are likely to be discouraged from using resources because of the fear of enforcement against family members [23, 62]. This includes patients who are citizens or legal permanent residents, who may reduce their own use of vital and high value services in an effort to indemnify their less documented family members against immigration enforcement [62]. Public charge compounds an already difficult pattern of access for these patients, underscoring the importance for ED providers to make the most of the emergency department presentations.

Of special consideration especially in recent times is the population of refugees and asylees. This group has experienced a high rate of violence and trauma in their home countries, which has significant effects on mental and physical wellbeing [63, 64]. While processes exist to provide some protection and access to services once their cases are approved, the system is currently overwhelmed with a backlog of cases and more frequent denial of status, compounded by increasing impediments to approval. Witnessing or experiencing violence is more common now in migrants who seek safety in the US than in migrants presenting in previous years [65]. Emergency providers should recognize that this special vulnerability to victimization and violence does not belong only to those officially recognized as refugees and asylees, but to a large percentage of foreign-born persons in the US.

Our role as emergency providers begins with using an equity lens to approach each patient as deserving of care and resources and not respond to external hierarchies of deservedness in our society [66]. Educational resources can then be used to inform ourselves about immigrant health and how current policies may affect those barriers [55]. It is important for immigrants and undocumented populations themselves to be included in informing health practices and policy. This may include the creation of a hospital community advisory board, appointment of immigrants to leadership positions, partnership with local health advocacy organizations, and using community-based participatory research methods to study ongoing care [67]. Grounded in these relationships, medical providers can intervene upon barriers, improve messaging, and create welcoming health-centered language to reassure patients about our therapeutic alliance [68, 69]. Efforts to improve the health of these populations without the involvement of immigrant community voices will not only perform poorly but also violate the equity premise in which the effort is rooted [70]. Relationships between community-based organizations and healthcare providers can provide local-, state-, and national-level opportunities for policy advocacy and activism, as well as training future healthcare providers [71].

Recommendations for Emergency Medicine Practice

Basic

- Create a welcoming and supportive environment for immigrants that extends from the bedside throughout the hospital. Include adding signage and messaging throughout the hospital campus that assures equitable treatment and confidentiality.
- Ensure language justice and appropriate translation services.
- Understand your local context: who are the immigrant populations at risk, what are their health care utilization patterns and what are the specific barriers to health they face, including health insurance access barriers? This will allow individualized responses to structural barriers relevant to the local immigration context
- Connect with local immigrant advocacy groups that can inform the ED and provide patient perspective to move care upstream for these patients. Ideally, this connection would provide the foundation for a larger community partnered

relationship, but at minimum, it can be a vital source of information about how to tailor care in the ED to immigrant patients' needs.
- Recognize that certain undocumented patients may be eligible for legal status change due to their presenting trauma and warrant referral to legal service providers.

Intermediate

- Advocate for healthcare settings to be free of anti-immigrant enforcement and anything else that might discourage health utilization. Lobby local health municipalities to make healthcare settings sanctuary sites [40].
- Create referral pathways to remove barriers to healthcare access. Determine which community support organizations serve immigrant populations and create direct conduits from the ED to those places of care.
- Develop relationships with federally qualified health centers or analogous clinics that are hubs for the care of undocumented patients and streamline referrals to them [72].
- Advocate for your municipality and hospital system to provide specialized pathways of care coverage for undocumented immigrant patients, and to have transparent insurance or fee systems to support immigrants' use of needed health care. Arguments can be made to the county level that preventative and primary care access can be cost saving over time. In counties or states without coverage programs for certain immigrant groups, hospitals may have to set up their own systems of charity care [73].
- Advocate for adoption of a patient advisory board model where patients can provide feedback on hospital decisions and advise implementation of programming across the hospital. Lobby hospital administration to have undocumented or immigrant community representation.
- Ensure that your ED and hospital administrators are aware of the expansion of the public charge rule and its implications for patients' access to care. Advocate that frontline staff who may interact with services subject to public charge such as patient financial services, registration and social work personnel are aware of the expanded rule and can avoid imperiling patients' immigration status. For example, social work staff need to know that if they offer Section 8 housing assistance to a patient, they should explain the public charge risk if the patient is undocumented.

Advanced

- Build a system of screening and referral of patients to medicolegal partnerships from the ED [74]. Use the knowledge of barriers and local environment to target resources to immigrant patients in a culturally appropriate and unintimidating way. The LA County + USC medical legal partnership is one such example.

- Seek or provide training for individual physicians in forensic medical evaluations through organizations like Physicians for Human Rights or Healthright International and work with lawyers to substantiate asylum legal cases.
  - Form an asylum clinic with other trained physicians to receive referrals from local lawyers to perform forensic asylum evaluations [75].
  - Develop systems of detention advocacy for asylum seekers and other detainees, which can be analogous to asylum clinics but focused on identifying and advocating for those who require release from detention on health-related grounds.
- Organize as healthcare providers and advocates that immigration be treated as a social determinant of health and that we should be concerned about access within this population based on our duty to advocate for population and public health [76]. Organize healthcare providers to engage with domestic policies that drive migration and imperil migrants (i.e., military interventions, neoliberal trade policies, and border militarization).
- Push national organizations and advocacy groups to support expansion of health insurance to undocumented immigrant populations and denounce anti-immigrant rhetoric and legislation such as public charge [44]. Create spaces and positions of power for immigrant communities to inform health practices and policy.
- Develop community-based participatory research in conjunction with relevant community-based organizations to evaluate how effectively community conduits and care management programs improve patient outcomes, reduce ED recidivism, and encourage high value care.

Teaching Case

Clinical Case

A 38-year-old female presents to the ED after a six-foot fall from a ladder. She could not ambulate at the scene. Because she is distraught when providers attempt to examine her and will not communicate what happened or where she is most hurt, she receives CT scans of her head, c-spine, thorax, abdomen, and pelvis. She is signed out to you as the oncoming doctor pending the radiology reading of her CT scans. You enter the room to find a patient with a vacant stare. You attempt to engage her and she screams, "¡No me toca!" Her sister is at bedside and with the help of the interpreter you ask her about the patient and what could be going on.

She reveals the patient's backstory. She is originally from El Salvador where a predominant gang was extorting her. They came every week asking for a higher amount, until she was unable to make enough money to pay the fee. She feared for her life and in an attempt to escape their extortion and threats, she paid a coyote (smuggler) to smuggle her and her children into the US. Midway through the journey, the coyote sold her to a drug cartel that held her in captivity for 10 months. She was repeatedly sexually and physically assaulted while her family in the US collected enough money to pay a ransom. She was then freed and brought to Los Angeles. Her sister describes how she hasn't been the same since this experience and she has significant residual mental health issues, including intermittent episodes where she seems to go blank like in today's presentation that resemble flashbacks. These episodes are increasingly affecting the patient's ability to function and especially to parent, so other family members often watch her children. The family has been trying to convince her to seek medical help and especially mental health care, but she is too afraid of deportation and the thought of returning to the nightmare she escaped.

She works cleaning houses to help support her family. Today's fall happened at work. She was asked by her employer to clean the windows on the outside of the house. Despite pleading that she didn't know how to use a ladder, the boss insisted. She complied for fear of losing her job, as work has been hard for her to find due to her lack of documentation.

The patient's CT scans are read as negative. After time, reassurance, and anxiolytic medications, the patient is able to engage in conversation and returns to her baseline mentation. After offering her several community resources, you are able to have the social worker connect her with a local federally qualified clinic that specializes in care for this population and with a local immigrant rights organization. Through these organizations, she receives treatment of her psychiatric disease with medication and therapy. She is also connected to immigration legal services where, with their help, she submits a trafficking visa application to acquire legal status.

Teaching Points

1. Immigration is an important and often under-recognized social determinant of health.
2. Undocumented patients are largely excluded from public services, including, but not limited to, health insurance.
3. The ED has become a primary touch-point for this population, as a social and health care safety net.
4. In certain cases, undocumented individuals may be eligible for asylum or other pathways to legal permanent residence and naturalization if connected to appropriate legal partners.

Discussion Questions

1. In this scenario, what are the barriers to health and health care encountered by the patient? What are some barriers in your ED and health system when taking care of the immigrant population?
2. Please compare a cultural competence and structural competence lens for reviewing this case. Do the resulting interventions differ?
3. Blueprint what an emergency department and health system might look like to best meet the needs of this patient. What avenues may be available in your ED, health system, and local community for collaborative work to help the undocumented immigrant population?

References

1. U.S. Census Bureau. (2012). 2009-2011 American Community Survey 3-year Public Use Microdata Samples [SAS Data file]. Retrieved from https://factfinder.census.gov/faces/nav/jsf/pages/searchresults.xhtml?refresh=t.
2. Child Trends. 2019. Immigrant children – child trends. [online] Available at: https://www.childtrends.org/indicators/immigrant-children [Accessed 2 May 2019].
3. The Henry J. Kaiser Family Foundation. 2019. Health Coverage of Immigrants. [online] Available at: https://www.kff.org/disparities-policy/fact-sheet/health-coverage-of-immigrants/ [Accessed 2 May 2019].
4. Ommundsen R, Van der Veer K, Larsen KS, Eilertsen DE. Framing unauthorized immigrants: the effects of labels on evaluations. Psychol Rep. 2014;114(2):461–78.
5. Jie Zong J. Naturalization Trends in the United States. [online] migrationpolicy.org. 2019. Available at: https://www.migrationpolicy.org/article/naturalization-trends-united-states [Accessed 2 May 2019].
6. Morey BN, Gee GC, Muennig P, Hatzenbuehler ML. Community-level prejudice and mortality among immigrant groups. Soc Sci Med. 2018;199:56–66.
7. Dhs.gov. 2019. Annual flow report. [online] Available at: https://www.dhs.gov/sites/default/files/publications/Refugees_Asylees_2017.pdf [Accessed 2 May 2019].
8. Pimienti M, Polkey C. Snapshot of US immigration 2019. March 29, 2019; Available from: https://www.ncsl.org/research/immigration/snapshot-of-u-s-immigration-2017.aspx.
9. Baker B, Rytina N. Estimates of the unauthorized immigrant population residing in the United States: January 2012: Office of Immigration Statistics; 2012.
10. Fazel-Zarandi MM, Feinstein JS, Kaplan EH. The number of undocumented immigrants in the United States: estimates based on demographic modeling with data from 1990 to 2016. PLoS One. 2018;13(9):e0201193.
11. Rosenblum M, Ruiz Soto AG. An analysis of unauthorized immigrants in the United States by country and region of birth. Washington, DC: Migration Policy Institute; 2015.
12. Metzl JM, Hansen H. Structural competency: theorizing a new medical engagement with stigma and inequality. Soc Sci Med. 2014;103:126–33.
13. Castaneda H, Holmes SM, Madrigal DS, Young ME, Beyeler N, Quesada J. Immigration as a social determinant of health. Annu Rev Public Health. 2015;36:375–92.
14. Bourgois P, Holmes SM, Sue K, Quesada J. Structural vulnerability: operationalizing the concept to address health disparities in clinical care. Acad Med. 2017;92(3):299–307.
15. Cone J, Bosch-Bonacasa M. Invisible war: Central America's forgotten humanitarian crisis. Brown J World Aff. 2017;24:225.
16. Miller T. Empire of Borders: The Expansion of the US Border Around the World. United Kingdom: Verso Books. 2019.
17. Wallace SP, Torres J, Sadegh-Nobari T, Pourat N, Brown ER. Undocumented immigrants and health care reform: UCLA Center for Health Policy Research; 2013. p. 26. http://healthpolicy.ucla.edu/publications/Documents/PDF/undocumentedreport-aug2013.pdf.
18. Pourat N, Wallace SP, Hadler MW, Ponce N. Assessing health care services used by California's undocumented immigrant population in 2010. Health Aff (Millwood). 2014;33(5):840–7.
19. Nandi A, Galea S, Lopez G, Nandi V, Strongarone S, Ompad DC. Access to and use of health services among undocumented Mexican immigrants in a US urban area. Am J Public Health. 2008;98(11):2011–20.
20. Wallace S, Tones J, Sadegh-Nobari T, Pourat N, Brown R. Undocumented Immigrants and Health Care Reform. UCLA: Center for Health Policy Research. Retrieved from https://escholarship.org/uc/item/8sv4w4m4. 2013.

21. Linton JM, Ameenuddin N, Falusi O. Pediatricians awakened: addressing family immigration status as a critical and intersectional social determinant of health. Am J Bioeth. 2019;19(4):69–72.
22. Callaghan T, Washburn DJ, Nimmons K, Duchicela D, Gurram A, Burdine. Immigrant health access in Texas: policy, rhetoric, and fear in the Trump era. BMC Health Serv Res. 2019;19(1):342.
23. Cohen MS, Schpero WL. Household immigration status had differential impact on medicaid enrollment in expansion and nonexpansion states. Health Aff (Millwood). 2018;37(3):394–402.
24. Hacker K, Anies M, Folb BL, Zallman L. Barriers to health care for undocumented immigrants: a literature review. Risk Manag Healthc Policy. 2015;8:175–83.
25. Flynn MA, Eggerth DE, Jacobson CJ Jr. Undocumented status as a social determinant of occupational safety and health: the workers' perspective. Am J Ind Med. 2015;58(11):1127–37.
26. Vargas Bustamante A, Fang H, Garza J, Carter-Pokras O, Wallace SP, Rizzo JA, et al. Variations in healthcare access and utilization among Mexican immigrants: the role of documentation status. J Immigr Minor Health. 2012;14(1):146–55.
27. Khullar D, Chokshi DA. Immigrant health, value-based care, and emergency medicaid reform. JAMA. 2019;321(10):928–9.
28. Saadi A, McKee M. Hospitals as places of sanctuary. BMJ. 2018;361:k2178.
29. Maldonado CZ, Rodriguez RM, Torres JR, Flores YS, Lovato LM. Fear of discovery among Latino immigrants presenting to the emergency department. Acad Emerg Med. 2013;20(2):155–61.
30. Rodriguez RM, Torres JR, Sun J, Alter H, Ornelas C, Cruz M, et al. Declared impact of the US President's statements and campaign statements on Latino populations' perceptions of safety and emergency care access. PLoS One. 2019;14(10):e0222837.
31. Kim G, Molina US, Saadi A. Should immigration status information be included in a patient's health record? AMA J Ethics. 2019;21(1):E8–16.
32. Saadi A, Cheffers ML, Taira B, Trotzky-Sirr R, Parmar P, Samra S, et al. Building immigration-informed, cross-sector coalitions: findings from the Los Angeles County Health Equity for Immigrants Summit. Health Equity. 2019;3(1):431–5.
33. Macias-Konstantopoulos W. Human trafficking: the role of medicine in interrupting the cycle of abuse and violence. Ann Intern Med. 2016;165(8):582–8.
34. Shandro J, Chisolm-Straker M, Duber HC, Findlay SL, Munoz J, Schmitz G, et al. Human trafficking: a guide to identification and approach for the emergency physician. Ann Emerg Med. 2016;68(4):501–508.e1.
35. Samra S, Taira BR, Pinheiro E, Trotzky-Sirr R, Schneberk T. Undocumented patients in the emergency department: challenges and opportunities. West J Emerg Med. 2019;20(5):791–8.
36. Ngai KM, Grudzen CR, Lee R, Tong VY, Richardson LD, Fernandez A. The association between limited English proficiency and unplanned emergency department revisit within 72 hours. Ann Emerg Med. 2016;68(2):213–21.
37. Taira BR. Improving communication with patients with limited English proficiency. JAMA Intern Med. 2018;178(5):605–6.
38. Becerra BJ, Arias D, Becerra MB. Low health literacy among immigrant Hispanics. J Racial Ethn Health Disparities. 2017;4(3):480–3.
39. Shommu NS, Ahmed S, Rumana N, Barron GR, McBrien KA, Turin TC. What is the scope of improving immigrant and ethnic minority healthcare using community navigators: a systematic scoping review. Int J Equity Health. 2016;15:6.
40. Saadi A, Ahmed S, Katz MH. Making a case for sanctuary hospitals. JAMA. 2017;318(21):2079–80.
41. New York City Health and Hospitals Corporation. Seek care without fear. 2018. https://www.nychealthandhospitals.org/immigrant/.
42. Tobin Tyler E. Medical-legal partnership in primary care: moving upstream in the clinic. Am J Lifestyle Med. 2019;13(3):282–91.

43. Martinez O, Wu E, Sandfort T, Dodge B, Carballo-Dieguez A, Pinto R, et al. Evaluating the impact of immigration policies on health status among undocumented immigrants: a systematic review. J Immigr Minor Health. 2015;17(3):947–70.
44. Kelley AT, Tipirneni R. Care for undocumented immigrants – rethinking state flexibility in medicaid waivers. N Engl J Med. 2018;378(18):1661–3.
45. Cervantes L, Tuot D, Raghavan R, Linas S, Zoucha J, Sweeney L, et al. Association of emergency-only vs standard hemodialysis with mortality and health care use among undocumented immigrants with end-stage renal disease. JAMA Intern Med. 2018;178(2):188–95.
46. Krieger N, Huynh M, Li W, Waterman PD, Van Wye G. Severe sociopolitical stressors and preterm births in New York City: 1 September 2015 to 31 August 2017. J Epidemiol Community Health. 2018;72(12):1147–52.
47. Hacker K, Chu J, Arsenault L, Marlin RP. Provider's perspectives on the impact of Immigration and Customs Enforcement (ICE) activity on immigrant health. J Health Care Poor Underserved. 2012;23(2):651–65.
48. McLeigh JD. How do Immigration and Customs Enforcement (ICE) practices affect the mental health of children? Am J Orthopsychiatry. 2010;80(1):96–100.
49. Hacker K, Chu J, Leung C, Marra R, Pirie A, Brahimi M, et al. The impact of immigration and customs enforcement on immigrant health: perceptions of immigrants in Everett, Massachusetts, USA. Soc Sci Med. 2011;73(4):586–94.
50. Vargas ED, Sanchez GR, Juarez M. Fear by association: perceptions of anti-immigrant policy and health outcomes. J Health Polit Policy Law. 2017;42(3):459–83.
51. Lopez WD, Kruger DJ, Delva J, Llanes M, Ledón C, Waller A, et al. Health implications of an immigration raid: findings from a Latino community in the midwestern United States. J Immigr Minor Health. 2017;19(3):702–8.
52. Fenton JJ, Catalano R, Hargreaves WA. Effect of Proposition 187 on mental health service use in California: a case study. Health Aff (Millwood). 1996;15(1):182–90.
53. Berk ML, Schur CL. The effect of fear on access to care among undocumented Latino immigrants. J Immigr Health. 2001;3(3):151–6.
54. Torche F, Sirois C. Restrictive immigration law and birth outcomes of immigrant women. Am J Epidemiol. 2019;188(1):24–33.
55. Toomey RB, Umaña-Taylor AJ, Williams DR, Harvey-Mendoza E, Jahromi LB, Updegraff KA. Impact of Arizona's SB 1070 immigration law on utilization of health care and public assistance among Mexican-origin adolescent mothers and their mother figures. Am J Public Health. 2014;104(Suppl 1):S28–34.
56. Beniflah JD, Little WK, Simon HK, Sturm J. Effects of immigration enforcement legislation on Hispanic pediatric patient visits to the pediatric emergency department. Clin Pediatr (Phila). 2013;52(12):1122–6.
57. Rodríguez MA, Bustamante AV, Ang. A. Perceived quality of care, receipt of preventive care, and usual source of health care among undocumented and other Latinos. J Gen Intern Med 2016. 2009;24(Suppl 3):508–13.
58. The Lancet. US public charge rule: pushing the door closed. Lancet. 2019 Jun 1;393(10187):2176. https://doi.org/10.1016/S0140-6736(19)31233-4. PMID: 31162066.
59. Services, U.S.C.a.I. Public charge rule. 2020; Available from: https://www.uscis.gov/greencard/public-charge.
60. Zallman L, Finnegan KE, Himmelstein DU, Touw S, Woolhandler S. Implications of changing public charge immigration rules for children who need medical care. JAMA Pediatr. 2019;173:e191744.
61. Bleich SN, Fleischhacker S. Hunger or deportation: implications of the Trump Administration's Proposed Public Charge Rule. J Nutr Educ Behav. 2019;51(4):505–9.
62. Perreira KM, Yoshikawa H, Oberlander J. A new threat to immigrants' health – the public-charge rule. N Engl J Med. 2018;379(10):901–3.
63. Cleary SD, Snead R, Dietz-Chavez D, Rivera I, Edberg MC. Immigrant trauma and mental Health outcomes among Latino Youth. J Immigr Minor Health. 2018;20(5):1053–9.

64. Steel JL, Dunlavy AC, Harding CE, Theorell T. The psychological consequences of pre-emigration trauma and post-migration stress in refugees and immigrants from Africa. J Immigr Minor Health. 2017;19(3):523–32.
65. Eisenman DP, Gelberg L, Liu H, Shapiro MF. Mental health and health-related quality of life among adult Latino primary care patients living in the United States with previous exposure to political violence. JAMA. 2003;290(5):627–34.
66. Perreira KM, Pedroza JM. Policies of exclusion: implications for the health of immigrants and their children. Annu Rev Public Health. 2019;40:147–66.
67. Berlinger N, Zacharias RL. Resources for teaching and learning about immigrant health care in health professions education. AMA J Ethics. 2019;21(1):E50–7.
68. Acosta DA, Aguilar-Gaxiola S. Academic health centers and care of undocumented immigrants in the United States: servant leaders or uncourageous followers? Acad Med. 2014;89(4):540–3.
69. Moreno G, Rodríguez MA, Lopez GA, Bholat MA, Dowling PT. Eight years of building community partnerships and trust: the UCLA family medicine community-based participatory research experience. Acad Med. 2009;84(10):1426–33.
70. Casale CR, Clancy CM. Commentary: not about us without us. Acad Med. 2009;84(10):1333–5.
71. Polk S, DeCamp LR, Guerrero Vázquez M, Kline K, Andrade A, Cook B, et al. Centro SOL: a community-academic partnership to care for undocumented immigrants in an emerging Latino area. Acad Med. 2019;94(4):538–43.
72. Losonczy LI, Hsieh D, Wang M, Hahn C, Trivedi T, Rodriguez M, et al. The Highland Health Advocates: a preliminary evaluation of a novel programme addressing the social needs of emergency department patients. Emerg Med J. 2017;34(9):599–605.
73. Katz MH, Brigham TM. Transforming a traditional safety net into a coordinated care system: lessons from healthy San Francisco. Health Aff (Millwood). 2011;30(2):237–45.
74. Regenstein M, Trott J, Williamson A, Theiss J. Addressing social determinants of Health through medical-legal partnerships. Health Aff (Millwood). 2018;37(3):378–85.
75. Bernhardt LJ, Lin S, Swegman C, Sellke R, Vu A, Solomon BS, et al. The Refugee Health Partnership: a longitudinal experiential medical student curriculum in refugee/asylee health. Acad Med. 2019;94(4):544–9.
76. Eisenstein L. To fight burnout, organize. N Engl J Med. 2018;379(6):509–11.

Language and Literacy

4

Kian Preston-Suni and Breena R. Taira

Key Points

- Patients with limited English proficiency have a right to receive their medical care with the assistance of a certified healthcare interpreter.
- Universal health literacy precautions and teach-backs are simple tools to improve patient understanding.
- Cultural humility means exploring your patient's motivations, context, and perspectives. It can help you and your patient come to a shared understanding of the disease and treatment plan.
- Self-reflection and introspection are critical elements of the lifelong learning that cultural humility requires.

Foundations

Background

Communication is the core of the doctor–patient relationship. Clear, bidirectional communication facilitates the transfer of information required to make a diagnosis, determine the treatment plan, and explain its details. When effective, it can positively impact care, improving the likelihood of medication adherence and follow-up. Conversely, poor communication impedes care, and can leave

K. Preston-Suni (✉)
David Geffen School of Medicine at UCLA, Greater Los Angeles VA Medical Center, Los Angeles, CA, USA
e-mail: kpsuni@ucla.edu

B. R. Taira
Department of Emergency Medicine, Olive View-UCLA Medical Center, Sylmar, CA, USA
e-mail: btaira@g.ucla.edu

H. J. Alter et al. (eds.), *Social Emergency Medicine*,
https://doi.org/10.1007/978-3-030-65672-0_4

patients confused about instructions, increasing the risk of poor outcomes. Medical communication is multifaceted with three central domains: language, health literacy, and culture [1]. Patient-centered communication requires attention to each.

Linguistic diversity in the US is rapidly increasing. Since 1980, the proportion of people who speak a language other than English at home has doubled. As of 2011, over 20% of the population spoke a non-English language at home, and more than 25 million people spoke English less than "very well" [2]. Despite this, the most commonly spoken non-English languages remain underrepresented among physicians entering the workforce [3]. Accordingly, patients with limited English proficiency (LEP) experience frequent communication barriers with their healthcare providers and are at risk for receiving poor quality care in a wide range of healthcare settings [4, 5].

Health literacy is defined by the National Academy of Medicine as the degree to which individuals have the capacity to obtain, process, and understand basic health information and services needed to make appropriate health decisions [6]. Low health literacy is widespread, with 36% of adults having basic or below basic health literacy [7]. People from historically disenfranchised groups: those with LEP, low socioeconomic status, the elderly and the deaf have even higher rates of low health literacy [7–9]. Conceptually, health literacy encompasses three realms: oral literacy, print literacy, and numeracy. *Oral literacy*, comprised of listening and speaking, is needed to interact with physicians, to ask questions, and to understand verbal explanations of medical conditions. *Print literacy*, composed of reading and writing, allows for the comprehension of medication bottles, discharge instructions and written consent forms. *Numeracy* is the ability to understand quantitative concepts including medication dosage and the concept of risk. Low health literacy is associated with less knowledge of medical conditions, difficulty taking medications appropriately, worse understanding of medication labels, and increased mortality among the elderly [10, 11].

All communication is filtered through the perspective of culture. The shared beliefs, attitudes, practices, and values of a particular group frame how events, information, and illness are understood and incorporated into one's worldview. Cultural differences between patients and their healthcare providers have the potential to create misunderstanding and conflict. Cultural humility on the part of the health care provider incorporates long term commitment to self-evaluation, awareness of power imbalances and development of a mutually beneficial and nonpaternalistic partnership with the patient in an effort to improve health [12]. When acknowledged and explored with humility, cultural differences can be incorporated into a greater understanding of the individual and the context in which they live. In this chapter, we review evidence and best practices for the three domains of communication: language, health literacy, and culture. When each is addressed by employing language assistance, usable health information and cultural humility, clear bidirectional communication can be achieved (see Fig. 4.1).

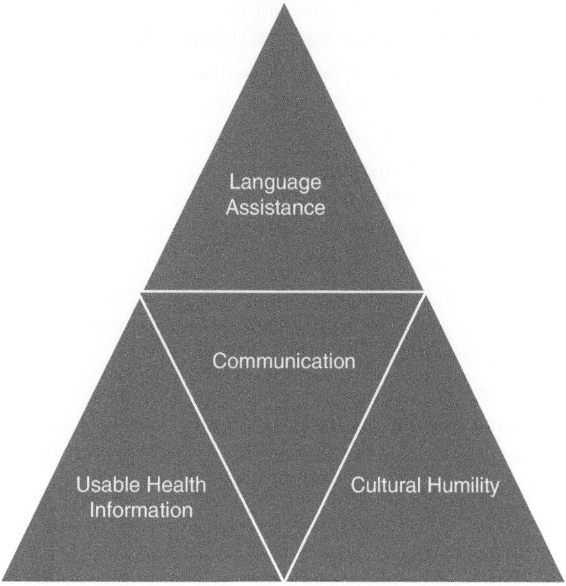

Fig. 4.1 The Communication Triad for Optimal Clinical Care. This figure demonstrates how the tools of language assistance, cultural humility, and usable health information together facilitate clear bidirectional communication

Evidence Basis

Language

In the emergency department (ED), LEP is associated with reduced satisfaction with care and worse understanding of discharge instructions [13–16]. Patients with LEP receive a larger burden of diagnostic testing during their ED visits [4, 17, 18]. LEP also predicts an increased risk of unexpected ED return visits and decreased probability of receiving a follow-up appointment [5, 19–21]. The mainstay of patient-centered communication for patients with LEP is interpretation provided by a certified healthcare interpreter. Patients who receive care with the assistance of an interpreter have increased satisfaction compared with those who do not receive professional language services [15, 22–24]. Hospitalized patients with LEP who receive phone interpretation experience a reduced rate of hospital readmission compared to those who do not receive language assistance [25]. The use of a professional interpreter is associated with a lower rate of unexpected ED return visits, reduced disparity in diagnostic test utilization, and fewer significant communication errors [26, 27].

Federal statute requires the provision of language services for patients with LEP. Title VI of the Civil Rights Act of 1964, which prohibits discrimination on the basis of national origin, requires entities receiving federal funds to provide interpreter services to patients with LEP at no charge [28]. Section 1557 of the Patient Protection and Affordable Care Act of 2010 expands on these protections. The law requires certification for healthcare interpreters, sets minimum standards for video interpretation, and explicitly bans the use of family, friends and minor children as ad hoc interpreters [29].

Bilingual providers also influence the quality of care for LEP patients. Patients with LEP who receive their care from physicians who speak their language experience higher quality of care. Language concordance between physicians and patients with LEP is associated with less reliance on diagnostic resources, higher likelihood of receiving follow-up appointments and increased satisfaction with care [4, 21, 23]. In the clinic setting, diabetic patients with LEP have better glycemic control and lower ED utilization when cared for by physicians who speak their language [30, 31]. Admitted patients with LEP cared for by language concordant physicians have lower rates of ED visits after discharge [32]. Patients with LEP whose physicians speak their language ask more questions and receive more health education, which may mediate the relationship between language concordance and quality of care [33, 34].

These improvements in the quality of care for LEP patients, however, occur only when the provider is truly bilingual. The majority of physicians with secondary language skills are not native speakers [35]. Moreover, physicians do not accurately estimate their own non-English language proficiency, and tend to overestimate their patients' English proficiency [30, 36]. Such misjudgments contribute to physicians choosing to "get by" with their own limited language skills rather than seeking the assistance of an interpreter [37]. In one study, over 20% of patients with LEP seen in a public ED reported that their provider spoke in Spanish but should have used an interpreter [38]. Despite the evidence of benefit and regulatory requirement to obtain language assistance for patients with LEP, emergency physicians' use of inadequate secondary language skills remains high [39]. ED providers should thus only use their own non-English language skills for patient care if they have been tested and certified proficient by the health system.

Health Literacy

Patients with low health literacy experience worse health outcomes including a higher rate of admission and more medication errors than those with adequate health literacy [40, 41]. Low health literacy is also associated with increased ED utilization and a higher rate of 2-week return visits [42]. Physicians are unreliable at assessing their patients' health literacy and routinely overestimate their ability to understand the information presented during a clinical encounter [43–45]. Patients may be hesitant to disclose lack of understanding due to the associated stigma and shame, and low health literacy may not be readily apparent among those with high verbal fluency [46, 47]. Accordingly, communication is likely to be ineffective if the clinician depends on their judgment of a patient's medical literacy and relies on the patient to overtly express confusion or request clarification.

Culture

The patient–provider dyad may consist of a near limitless combination of sociocultural differences based on gender, socioeconomic status, ethnicity, religion, and language, to name only a few. Biomedicine exists within a predominantly European-American framework valuing autonomy and individuality, which may be at odds with the perspectives of the patient.

Culturally competent communication requires a specific skillset with four components. The first is the communication repertoire, which includes the skills that allow for relationship building, information gathering, and management of the patient's problems [48]; specific associated behaviors include showing empathy, active listening, seeking the patient's perspective, and taking adequate time [49, 50]. The second component is situational awareness, which allows the culturally competent physician to notice the subtle cues indicating unspoken disagreement or confusion [48]. The adaptable physician, using the third component of the skillset, responds to these cues and adjusts communication during the encounter to address any perceived miscommunication. Lastly, knowledge of core cultural issues that lead commonly to misunderstanding may help avoid cultural dissonance. These core issues include beliefs about gender roles in traditional societies, the tendency to defer to physician authority among some cultures, and mores related to physical space and touching.

Emergency Department and Beyond

Bedside

Our health care system is experiencing an epidemic of under-communication, with the impact felt most acutely at the bedside [51]. Oftentimes, language and literacy create barriers to care between patients and providers [52]. Communication between physicians and patients is often physician-centric, rooted in the dominant biomedical culture, using advanced medical terminology and often the physician's own limited secondary language skills [39, 53].

Emergency departments throughout the country continue to report low rates of interpreter use despite the legal mandate and evidence of improved outcomes with certified healthcare interpreter utilization [38, 39, 54]. This may be related to variability in language assistance availability and difficulties associated with individual modalities of interpretation [55]. Under-communication is an attribute of poor-quality care associated with worse comprehension of discharge instructions and increased risk of ED return visits [19, 53], as well as ineffective consultations that result in poor adherence to treatment regimens [56]. Language discordance between a patient and physician, a subset of under-communication, leads to worse understanding by both the physician and patient, poorer explanations by the physician, less patient involvement in decision-making and reduced trust in the physician [57].

Often intended to address communication barriers experienced by Spanish-speaking patients, instruction in medical Spanish is widely offered in US medical schools [58]. Although these courses may increase comfort and familiarity with medical vocabulary and improve the results of language testing [59], concerningly, medical Spanish instruction is also associated with decreased interpreter use and an increased rate of major language errors [60, 61]. Medical Spanish instruction may cause harm through overconfidence, and is unlikely to provide students or physicians with the proficiency necessary to handle the nuanced discussions required to provide high quality care [62]. Thus, unless a provider's non-English language skills have been tested and certified, it is not recommended that providers use their own non-English language skills in the clinical setting.

A consideration of deafness in the clinical encounter is adjacent to that of language but has somewhat broader implications. Deafness has been traditionally viewed through the biomedical lens according to its underlying pathology. However it is better understood by its age of onset relative to language acquisition, with an appreciation of the sociocultural effects this entails [63]. Hearing loss of adulthood is characterized by associated stigma and the social isolation that comes with inability to communicate effectively. In contrast, sign language allows for unimpeded communication in those with pre-linguistic and childhood onset deafness. The Deaf community exists as a sociocultural entity with a unique language, customs, and beliefs. Deaf people function without difficulty within this setting. Only on interacting with the hearing community does deafness become a "disability." In the clinic setting, deaf patients cared for by physicians who do not sign receive less preventive care than those with providers fluent in American Sign Language (ASL) or assisted by an ASL interpreter [64, 65].

It is important to recognize that English is a second language for many people with pre-linguistic and childhood onset deafness and that ASL may be their first language. Relying on lip-reading or communicating in written English with deaf patients is no more appropriate than it would be for a primary Korean-speaker [66]. A deaf patient whose preferred language is ASL must receive sign language interpretation during the medical encounter, whether in-person or video-assisted, to ensure accurate bidirectional communication as required by the Americans with Disabilities Act [67]. Further, it is important to ascertain whether the patient is fluent in ASL or uses a different form of sign language, especially if the patient was born outside the US. The hospital ASL interpreter may be able to help the clinician ascertain a deaf patient's best form of communication if it is in question.

Given the prevalence of low health literacy in the ED setting, we recommend universal health literacy precautions in all patient communication [68]. Universal health literacy precautions, composed of a range of simple techniques, help all patients, not only those with low health literacy, and can be incorporated into all areas of clinical care. These precautions include jargon-free explanations and clear, simple instructions, both of which have been shown to increase understanding and disease management for patients at all literacy levels [69, 70]. Additional techniques include making eye contact, listening carefully, using simple sentence construction, and providing concrete instructions with no more than 3–5 pieces of information. Simple explanatory drawings, where appropriate, can also be used to help communicate medical concepts [68].

The teach-back technique, in which the physician asks the patient to explain (or teach back) the topic that was just discussed, helps ensure the patient's understanding and highlights areas needing clarification. In the ED, teach-back improves comprehension and retention of after-care instructions for patients of all ages and education levels [71, 72]. Teach-back can also be used to mitigate the impact of low numeracy, where calculations or ambiguity can cause confusion. Commonly seen in the ED, low numeracy impedes the comprehension of medication instructions and risk quantification, among other topics [73]. Saying, "take 1 pill in the morning, afternoon and evening," is preferable to "take 1 pill 3 times daily" [74]. In addition, when explaining risk, whole numbers should be used rather than fractions or decimals. For example, "1 in 1000" may be easier to grasp than "0.1%."

Self-reflection on one's identity and cultural conception of health and healthcare is a critical step in understanding the expectations we place on patients [75]. Culturally competent, patient-centered communication requires understanding the individual patient's context, perspectives and motivations in order to achieve a mutual understanding of illness and the treatment plan [76, 77]. Familiarity with diverse cultures is important, but the emphasis of the clinical encounter should be on the individual. Memorization of cultural attributes and culture-bound syndromes risks emphasizing stereotypes; instead one should seek the mutually respectful partnerships that define cultural humility [12]. This approach considers the patient's communication style including nonverbal behaviors and preferences for family involvement in decision making. The communities to which we belong play a fundamental role in the unconscious assignment of values to others. These automatic associations are a core feature of the cognitive system that allow us to make split-second decisions in the ED, but they are also responsible for the propagation of healthcare disparities [78]. Recognition of our own implicit biases is essential to understanding their influence on our interactions, and is required to mitigate their effects on clinical care [79]. Individuation, whereby one focuses on individual unique attributes of the patient rather than their group membership, is one effective strategy to address such implicit biases [80]. An understanding of implicit bias is inseparable from providing culturally competent care.

Hospital/Healthcare System

At the hospital and health system level, the detriments of communication barriers are often neglected [81]. Failure of the health system to connect and interact with communities leads to misunderstandings. Lack of community understanding of available resources for interpretation also exists. The majority of people with LEP are unaware of federal laws requiring provision of healthcare interpreters [82]. Moreover, simply being aware of language laws is inadequate as patient awareness of language regulations is not associated with increased interpreter utilization [82]. Further, providers may not be educated in language regulations well enough themselves to address this lack of awareness. In the absence of clear directives regarding language services and their seamless integration into workflows, physicians experiencing time constraints will choose the communication method they find least burdensome despite recognition of the potential harm [37]. Physicians may also rely on the uncertified language skills of staff, which increases the risk of interpretation error [83]. Despite widespread prevalence of limited English proficiency, hospitals are not adequately recruiting physicians with proficiency in the most commonly needed languages [3].

Universal assessment of language preference is recommended in all clinical settings [84]. It is incumbent on healthcare facilities to systematically capture their patients' need for language assistance at the first point of contact, and to ensure this need is met effectively [85]. For those respondents who state they speak another language at home, the US Census Bureau asks, "How well do you speak English?" and defines LEP as speaking English less than "very well." This question has high sensitivity for identifying patients likely to benefit from language assistance in a healthcare setting [86] and can be incorporated into the electronic health record. Addition of a question assessing the patient's language preference for receiving medical care increases the specificity without sacrificing significant sensitivity, thereby excluding patients who may be able to complete their care using English.

At the organizational level, a number of strategies are recommended to ensure patient-centered communication. Health literate health care organizations make genuine patient understanding a priority, with leaders who ensure literacy is a core attribute of care processes throughout the system [87]. A literacy-focused organization incorporates measures of patient understanding into implementation, quality improvement, and ongoing evaluation, while confirming patient understanding at all transitions of care and avoiding stigmatizing patients with low health literacy [87].

In addition to clear verbal communication, hospitals must ensure that reading materials given to patients provide clear instructions and are easily understood. The average adult in the US reads between an 8th and 9th grade reading level, with 1 in 5 reading below the 6th grade level. The reading level of a text is calculated based on a combination of the average sentence and word length, and is easily measured using widely available computer software. To ensure readability, written materials should be no higher than a 6th grade reading level [68].

Societal Level

Failure to surmount communication barriers and achieve clear bidirectional communication in the medical sector is a contributor to population-level health disparities. Widespread neglect of the components of communication has resulted in a health system that perpetuates disadvantages experienced by minority groups [88]. Systemic contributors to poor health among disadvantaged groups are exacerbated by underrepresentation of healthcare providers within communities of color and among ethnic minoritie groups.

A lack of uniformity characterizes national language policy. There are no national standards for bilingual certification of providers nor a national certification for healthcare interpreters. There is significant variation in determination of proficiency for physicians' secondary language skills between health systems [89, 90]. No standards exist for healthcare interpreters, with various healthcare systems, states and organizations employing different methods of training and certification [29]. Such variability leaves open the possibility of inadequate language services perpetuating health disparities for patients with LEP. Further, the lack of enforcement of legislation on language access may lead institutions to de-prioritize improvements in communication in favor of other metrics that are more heavily scrutinized. The Office of Civil Rights investigates complaints of healthcare discrimination covered under Section 1557 of the Affordable Care Act, but this requires a proactive complaint submission. The Joint Commission enforces only the collection of the patient's preferred language and the requirement of certification of bilingual staff who perform interpretation, not the provision of interpreter services [91]. There is currently no systematic enforcement of the requirement that patients with LEP be provided certified healthcare interpretation.

Changes in federal regulations can strengthen or weaken existing laws protecting patient communication. Physician advocacy in the regulatory and legislative arenas has the potential to meaningfully impact care received at the bedside. In addition to federal policy, state regulations can impact patient–provider communication barriers. States providing Medicaid reimbursement for healthcare interpreter assistance have higher rates of utilization [92].

Recommendations for Emergency Medicine Practice

Basic

- *Know how your hospital identifies LEP*. Recording the patient's preferred language is required, but determination of LEP varies. Know what questions are asked regarding patient language to assess LEP at your institution and where that information is recorded in the electronic health record [86].
- *Understand the resources available for interpretation and know best practices for working with an interpreter, using your own non-English language skills only if you have been tested and certified.* Interpreter services may be provided over the phone, using a video interface, or in-person. Know the techniques for effectively interviewing a patient with assistance of an interpreter. These include making eye contact with the patient, appropriate positioning, speaking directly to the patient and outlining the purpose of the encounter beforehand with the interpreter [93]. Using a secondary language in which you have limited proficiency increases the likelihood of a significant misunderstanding, potentially harming your patient. Identify and follow your hospital policy regarding certification, ideally using a validated language test [89].
- *Use universal precautions for health literacy*. Speak clearly using simple terminology, make eye contact, limit your take-home points, verify understanding and use clear drawings when appropriate. These techniques benefit all patients, regardless of literacy level [69]. Medication instructions often implicitly require calculations. Rather than leaving the calculations up to your patient, be explicit in your instructions [68]. For example, telling your patient to take "no more than 3000 mg of acetaminophen in one day" may be misunderstood. It is better to instruct your patient to take acetaminophen as needed "in the morning, at lunch, in the evening and before bed"; and to "count the number of acetaminophen

tablets taken; take no more than six tablets per day." Do not leave your instructions open to interpretation. Rather than saying "drink plenty of fluids," instruct your patient to drink a tall glass of water at breakfast, in the morning, at lunch, in the afternoon and at dinner.

- *Invite the patient's perspective and attend to subtle cues of lack of understanding*. Ask your patient what she thinks. Cultural humility means understanding her perspective, motivations, and circumstances. This is critical to formulating a collaborative treatment plan. Patient-centered communication is associated with increased trust in the treating physician, and patient trust is in turn associated with higher rates of medication adherence [94, 95]. Acknowledgement of differences in opinion between physicians and patients is also associated with higher levels of trust [56]. Moreover, your patient may not be comfortable telling you he doesn't understand. Cultural competence requires that you are alert to nonverbal signs of disagreement and confusion.
- *Use your interpreters as a cultural resource when communication seems hindered*. If you find there remains a communication barrier despite appropriate language assistance and approaching the encounter with cultural humility, ask your interpreter for clarification. Interpreters can provide valuable cultural context [96].

Intermediate

- *Advocate for LEP patients by providing feedback to trainees on appropriate use of language assistance and demonstrate this behavior yourself*. Model the ideal behavior and evaluate patients with LEP with the assistance of an interpreter. Require your residents, advanced practice providers and medical students to do the same. Patients with LEP may lack a voice at your hospital, make sure their needs are represented in committees and understood by decision-makers.
- *Implement teach-backs in the discharge process in your ED*. Discharge instructions are hard to remember and easy to misunderstand. Using the teach-back technique is a simple and effective way to improve understanding and retention of discharge information [72].
- *Ask who helps at home with medical information and invite the patient to call that person to listen in on the discharge*. Patients with low health literacy often do not disclose their difficulty understanding, and may depend on others for help in medical settings [46]. Involving this trusted person is a strategy to improve retention and comprehension.
- *Reflect on wording choices and eliminate those that could be perceived as judgmental*. Cultural humility means respecting the autonomy of your patient. The blame implied with judgmental statements is counterproductive and hinders effective bidirectional communication. For example, the statement "You should have seen a doctor sooner" may make the patient feel defensive and looked down upon. In this case, it would be better to ask questions that explore barriers to care, without making assumptions.

- *Review and update emergency department policies on language service use and bilingual certification and provide education to colleagues on policies and legal requirements.* Many emergency physicians may not be aware of the laws relating to healthcare language assistance and may use uncertified secondary language skills in their evaluation of patients with LEP. Know your local policies and update them to ensure they reflect the available literature and are compliant with federal and state laws.

Advanced

- *Review hospital and health system policies and provide feedback to the administration if outdated.* Ensure your hospital's policies reflect communication best practices and are compliant with legislation requiring certified healthcare interpreters for patients with LEP. Evidence-based interventions at the level of the hospital and health system include improving the accessibility of interpreter services, incorporation of language skills into the hiring process, educational initiatives, and tracking appropriate interpreter utilization [81].
- *Engage in personal reflection after a challenging case where communication barriers play a role, to advance one's own cultural humility.* Cultivating your self-awareness is a key component of cultural humility which requires a commitment to lifelong learning and introspection [12]. Increasing your knowledge about a patient group is important but can lead to unintentional stereotyping. Instead, thoughtful reflection can influence your future attitude and behavior in cross-cultural interactions, improving your adaptability.
- *Advocate for your patients who experience communication barriers.* Advocate for changes in policy that will benefit your patients who experience communication barriers as result of LEP, low health literacy, or cultural membership.
- *Implement teaching modules for trainees on communication and language justice in educational curricula.* Improve resident and medical student knowledge of communication barriers and how to address them by implementing structured education for trainees. This may be incorporated into the residency curriculum or on-shift bedside teaching [97, 98].

Teaching Case

Clinical Case

You are evaluating a woman in her mid-fifties for abdominal pain who has no significant comorbidities. She came to the emergency department with her daughter, expecting that her daughter would interpret between English and Arabic, her native language. Knowing federal laws and the negative impact ad hoc interpreters have on communication, you call an Arabic interpreter using the hospital's video interpretation equipment. With the assistance of the interpreter you perform a history and physical exam, which are notable for recent nausea, vomiting, dysuria and left flank tenderness with signs of dehydration. The urinalysis is consistent with your clinical diagnosis of pyelonephritis. Her symptoms improve with intravenous fluids and antiemetics.

While explaining home treatment, you notice her body language becomes more guarded and she stops asking questions. Attuned to these signals, you ask if she has any concerns about what you are discussing. She does not answer. You ask her if she foresees any challenges with the treatment discussed, but again she does not answer. You ask the interpreter if there is additional context you are missing, and together with the patient's daughter, they explain that fasting is required during the holy month of Ramadan. Unfamiliar with the rules of Ramadan, you ask the interpreter for clarification, learning that it is generally acceptable not to fast for medical reasons. You discuss this with the patient, who still shows nonverbal signs of reluctance. Her daughter suggests she discuss it with her imam, which seems to give her relief. You explain the antibiotic dosing using concrete times which you write down and explicitly state the quantity and frequency of fluid intake. She is able to explain your instructions back to you, and her daughter, who usually helps her when visiting doctors, assures she will make sure your instructions are followed. Your patient has no further questions and thanks you for taking the time to care for her.

Teaching Points

1. Request the assistance of a certified healthcare interpreter whenever caring for patients with limited English proficiency who speak a language in which the medical provider has not been certified. Ad hoc interpreters and uncertified staff should not provide interpretation, as this challenges the patient's privacy, is prone to error, and is not permitted by federal law.
2. Low health literacy is widespread and difficult for physicians to identify. Use universal health literacy precautions, which include making eye contact, seeking the patient's understanding and using simple, jargon-free explanations.
3. Cultural humility involves exploring cultural differences with respect for the patient's autonomy and avoidance of judgment. Physicians exhibiting cultural humility use a communication repertoire that includes showing empathy and exploring the patient's perspective. They are situationally aware, paying attention to subtle cues of disagreement. They are able to adapt their communication style mid-encounter and have knowledge of core cultural issues that commonly result in conflict.

Discussion Questions

1. After explaining your assessment and treatment recommendations to your patient, she is silent for an unexpected period of time. What could this silence be signaling?
2. Your patient is unable to "teach back" the plan of care you discussed. What can you do?
3. How would you approach a patient who insists that their teenage child interpret the medical encounter?

References

1. National Academies of Sciences Engineering, and Medicine. Integrating Health Literacy, Cultural Competance, and Language Access Services: Workshop Summary. Washington, D.C.: The National Academies Press; 2016.
2. Ryan C. Language use in the United States: 2011. United States Census Bureau. 2013.
3. Diamond L, Grbic D, Genoff M, Gonzalez J, Sharaf R, Mikesell C, et al. Non-English-language proficiency of applicants to US residency programs. JAMA. 2014;312(22):2405–7.
4. Hampers LC, Cha S, Gutglass DJ, Binns HJ, Krug SE. Language barriers and resource utilization in a pediatric emergency department. Pediatrics. 1999;103(6 Pt 1):1253–6.
5. Ngai KM, Grudzen CR, Lee R, Tong VY, Richardson LD, Fernandez A. The association between limited English proficiency and unplanned emergency department revisit within 72 hours. Ann Emerg Med. 2016;68(2):213–21.
6. Institute of Medicine Committee on Health Literacy. Nielsen-Bohlman L, Panzer AM, Kindig DA, editors. Health literacy: a prescription to end confusion. Washington (DC): National Academies Press. 2004. https://pubmed.ncbi.nlm.nih.gov/25009856/.
7. Kutner M, Greenburg E, Jin Y, Paulsen C. The health literacy of America's adults: results from the 2003 national assessment of adult literacy. NCES 2006-483. National Center for Education Statistics. 2006.
8. McKee MM, Paasche-Orlow MK, Winters PC, Fiscella K, Zazove P, Sen A, et al. Assessing health literacy in deaf American sign language users. J Health Commun. 2015;20(Suppl 2):92–100.
9. Brice JH, Travers D, Cowden CS, Young MD, Sanhueza A, Dunston Y. Health literacy among Spanish-speaking patients in the emergency department. J Natl Med Assoc. 2008;100(11):1326–32.
10. Berkman ND, Dewalt DA, Pignone MP, Sheridan SL, Lohr KN, Lux L, et al. Literacy and health outcomes. Evid Rep/Technol Assess (Summ). 2004;(87):1–8.
11. Berkman ND, Sheridan SL, Donahue KE, Halpern DJ, Crotty K. Low health literacy and health outcomes: an updated systematic review. Ann Intern Med. 2011;155(2):97–107.
12. Tervalon M, Murray-Garcia J. Cultural humility versus cultural competence: a critical distinction in defining physician training outcomes in multicultural education. J Health Care Poor Underserved. 1998;9(2):117–25.
13. Arthur KC, Mangione-Smith R, Meischke H, Chuan Z, Strelitz B, Garcia MA, et al. Impact of English proficiency on care experiences in a pediatric emergency department. Acad Pediatr. 2015;15(2):218–24.
14. Carrasquillo O, Orav EJ, Brennan TA, Burstin HR. Impact of language barriers on patient satisfaction in an emergency department. J Gen Intern Med. 1999;14(2):82–7.
15. Mahmoud I, Hou XY, Chu K, Clark M, Eley R. Satisfaction with emergency department service among non-English-speaking background patients. Emerg Med Australas. 2014;26(3):256–61.
16. Samuels-Kalow ME, Stack AM, Porter SC. Parental language and dosing errors after discharge from the pediatric emergency department. Pediatr Emerg Care. 2013;29(9):982–7.
17. Schulson L, Novack V, Smulowitz PB, Dechen T, Landon BE. Emergency department care for patients with limited English proficiency: a retrospective cohort study. J Gen Intern Med. 2018;33(12):2113–9.
18. Waxman MA, Levitt MA. Are diagnostic testing and admission rates higher in non-English-speaking versus English-speaking patients in the emergency department? Ann Emerg Med. 2000;36(5):456–61.
19. Samuels-Kalow ME, Stack AM, Amico K, Porter SC. Parental language and return visits to the emergency department after discharge. Pediatr Emerg Care. 2017;33(6):402–4.
20. Gallagher RA, Porter S, Monuteaux MC, Stack AM. Unscheduled return visits to the emergency department: the impact of language. Pediatr Emerg Care. 2013;29(5):579–83.
21. Sarver J, Baker DW. Effect of language barriers on follow-up appointments after an emergency department visit. J Gen Intern Med. 2000;15(4):256–64.

22. Bagchi AD, Dale S, Verbitsky-Savitz N, Andrecheck S, Zavotsky K, Eisenstein R. Examining effectiveness of medical interpreters in emergency departments for Spanish-speaking patients with limited English proficiency: results of a randomized controlled trial. Ann Emerg Med. 2011;57(3):248–56 e1-4.
23. Gany F, Leng J, Shapiro E, Abramson D, Motola I, Shield DC, et al. Patient satisfaction with different interpreting methods: a randomized controlled trial. J Gen Intern Med. 2007;22(Suppl 2):312–8.
24. Garcia EA, Roy LC, Okada PJ, Perkins SD, Wiebe RA. A comparison of the influence of hospital-trained, ad hoc, and telephone interpreters on perceived satisfaction of limited English-proficient parents presenting to a pediatric emergency department. Pediatr Emerg Care. 2004;20(6):373–8.
25. Karliner LS, Perez-Stable EJ, Gregorich S. Easy access to professional interpreters in the hospital decreases readmission rates and healthcare expenditures for limited English proficient patients. J Gen Intern Med. 2016;31(2):S202–S3.
26. Bernstein J, Bernstein E, Dave A, Hardt E, James T, Linden J, et al. Trained medical interpreters in the emergency department: effects on services, subsequent charges, and follow-up. J Immigr Health. 2002;4(4):171–6.
27. Karliner LS, Jacobs EA, Chen AH, Mutha S. Do professional interpreters improve clinical care for patients with limited English proficiency? A systematic review of the literature. Health Serv Res. 2007;42(2):727–54.
28. Chen AH, Youdelman MK, Brooks J. The legal framework for language access in healthcare settings: title VI and beyond. J Gen Intern Med. 2007;22(Suppl 2):362–7.
29. Brenner JM, Baker EF, Iserson KV, Kluesner NH, Marshall KD, Vearrier L. Use of interpreter services in the emergency department. Ann Emerg Med. 2018;72:432.
30. Parker MM, Fernandez A, Moffet HH, Grant RW, Torreblanca A, Karter AJ. Association of patient-physician language concordance and glycemic control for limited-English proficiency Latinos with type 2 diabetes. JAMA Intern Med. 2017;177(3):380–7.
31. Hacker K, Choi YS, Trebino L, Hicks L, Friedman E, Blanchfield B, et al. Exploring the impact of language services on utilization and clinical outcomes for diabetics. PLoS One. 2012;7(6):e38507.
32. Jacobs EA, Sadowski LS, Rathouz PJ. The impact of an enhanced interpreter service intervention on hospital costs and patient satisfaction. J Gen Intern Med. 2007;22(Suppl 2):306–11.
33. Seijo R, Gomez H, Freidenberg J. Language as a communication barrier in medical care for Hispanic patients. Hisp J Behav Sci. 1991;13(4):363–76.
34. Ngo-Metzger Q, Sorkin DH, Phillips RS, Greenfield S, Massagli MP, Clarridge B, et al. Providing high-quality care for limited English proficient patients: the importance of language concordance and interpreter use. J Gen Intern Med. 2007;22(Suppl 2):324–30.
35. Yoon J, Grumbach K, Bindman AB. Access to Spanish-speaking physicians in California: supply, insurance, or both. J Am Board Fam Pract. 2004;17(3):165–72.
36. Zun LS, Sadoun T, Downey L. English-language competency of self-declared English-speaking Hispanic patients using written tests of health literacy. J Natl Med Assoc. 2006;98(6):912–7.
37. Diamond LC, Schenker Y, Curry L, Bradley EH, Fernandez A. Getting by: underuse of interpreters by resident physicians. J Gen Intern Med. 2009;24(2):256–62.
38. Baker DW, Parker RM, Williams MV, Coates WC, Pitkin K. Use and effectiveness of interpreters in an emergency department. JAMA. 1996;275(10):783–8.
39. Taira BR, Orue A. Language assistance for limited English proficiency patients in a public ED: determining the unmet need. BMC Health Serv Res. 2019;19(1):56.
40. Baker DW, Gazmararian JA, Williams MV, Scott T, Parker RM, Green D, et al. Functional health literacy and the risk of hospital admission among Medicare managed care enrollees. Am J Public Health. 2002;92(8):1278–83.
41. Ngoh LN. Health literacy: a barrier to pharmacist-patient communication and medication adherence. J Am Pharm Assoc: JAPhA. 2009;49(5):e132–46; quiz e47–9.
42. Griffey RT, Kennedy SK, McGownan L, Goodman M, Kaphingst KA. Is low health literacy associated with increased emergency department utilization and recidivism? Acad Emerg Med. 2014;21(10):1109–15.

43. Kelly PA, Haidet P. Physician overestimation of patient literacy: a potential source of health care disparities. Patient Educ Couns. 2007;66(1):119–22.
44. Bass PF 3rd, Wilson JF, Griffith CH, Barnett DR. Residents' ability to identify patients with poor literacy skills. Acad Med. 2002;77(10):1039–41.
45. Rogers ES, Wallace LS, Weiss BD. Misperceptions of medical understanding in low-literacy patients: implications for cancer prevention. Cancer Control: J Moffitt Cancer Center. 2006;13(3):225–9.
46. Parikh NS, Parker RM, Nurss JR, Baker DW, Williams MV. Shame and health literacy: the unspoken connection. Patient Educ Couns. 1996;27(1):33–9.
47. Easton P, Entwistle VA, Williams B. Health in the 'hidden population' of people with low literacy. A systematic review of the literature. BMC Publ Health. 2010;10:459.
48. Teal CR, Street RL. Critical elements of culturally competent communication in the medical encounter: a review and model. Soc Sci Med. 2009;68(3):533–43.
49. Betancourt JR. Cross-cultural medical education: conceptual approaches and frameworks for evaluation. Acad Med. 2003;78(6):560–9.
50. Shapiro J, Hollingshead J, Morrison EH. Primary care resident, faculty, and patient views of barriers to cultural competence, and the skills needed to overcome them. Med Educ. 2002;36(8):749–59.
51. Health literacy: report of the Council on Scientific Affairs. Ad Hoc Committee on Health Literacy for the Council on Scientific Affairs, American Medical Association. JAMA. 1999;281(6):552–7.
52. Taira BR. Improving communication with patients with limited English proficiency. JAMA Intern Med. 2018;178(5):605–6.
53. Samuels-Kalow ME, Stack AM, Porter SC. Effective discharge communication in the emergency department. Ann Emerg Med. 2012;60(2):152–9.
54. Ginde AA, Sullivan AF, Corel B, Caceres JA, Camargo CA Jr. Reevaluation of the effect of mandatory interpreter legislation on use of professional interpreters for ED patients with language barriers. Patient Educ Couns. 2010;81(2):204–6.
55. Chan YF, Alagappan K, Rella J, Bentley S, Soto-Greene M, Martin M. Interpreter services in emergency medicine. J Emerg Med. 2010;38(2):133–9.
56. Martin LR, Williams SL, Haskard KB, Dimatteo MR. The challenge of patient adherence. Ther Clin Risk Manag. 2005;1(3):189–99.
57. Schenker Y, Karter AJ, Schillinger D, Warton EM, Adler NE, Moffet HH, et al. The impact of limited English proficiency and physician language concordance on reports of clinical interactions among patients with diabetes: the DISTANCE study. Patient Educ Couns. 2010;81(2):222–8.
58. Morales R, Rodriguez L, Singh A, Stratta E, Mendoza L, Valerio MA, et al. National Survey of medical Spanish curriculum in U.S. medical schools. J Gen Intern Med. 2015;30(10):1434–9.
59. Valdini A, Early S, Augart C, Cleghorn GD, Miles HC. Spanish language immersion and reinforcement during residency: a model for rapid acquisition of competency. Teach Learn Med. 2009;21(3):261–6.
60. Prince D, Nelson M. Teaching Spanish to emergency medicine residents. Acad Emerg Med. 1995;2(1):32–6; discussion 36–7.
61. Mazor SS, Hampers LC, Chande VT, Krug SE. Teaching Spanish to pediatric emergency physicians: effects on patient satisfaction. Arch Pediatr Adolesc Med. 2002;156(7):693–5.
62. Diamond LC, Reuland DS. Describing physician language fluency: deconstructing medical Spanish. JAMA. 2009;301(4):426–8.
63. Munoz-Baell IM, Ruiz MT. Empowering the deaf. Let the deaf be deaf. J Epidemiol Community Health. 2000;54(1):40–4.
64. McKee MM, Barnett SL, Block RC, Pearson TA. Impact of communication on preventive services among deaf American sign language users. Am J Prev Med. 2011;41(1):75–9.
65. MacKinney TG, Walters D, Bird GL, Nattinger AB. Improvements in preventive care and communication for deaf patients. J Gen Intern Med. 1995;10(3):133–7.

66. Ebert DA, Heckerling PS. Communication with deaf patients. Knowledge, beliefs, and practices of physicians. JAMA. 1995;273(3):227–9.
67. Division of Civil Rights, Disability Rights Section. ADA business BRIEF: communicating with people who are deaf or hard of hearing in hospital settings. United States Department of Justice. 2003. https://www.ada.gov/hospcombr.htm.
68. DeWalt DA, Callahan LF, Hawk VH, Broucksou KA, Hink A, Rudd R, et al. Health literacy universal precautions toolkit. Rockville: Agency for Healthcare Research and Quality; 2010. p. 1–227.
69. DeWalt DA, Malone RM, Bryant ME, Kosnar MC, Corr KE, Rothman RL, et al. A heart failure self-management program for patients of all literacy levels: a randomized, controlled trial [ISRCTN11535170]. BMC Health Serv Res. 2006;6:30.
70. Eckman MH, Wise R, Leonard AC, Dixon E, Burrows C, Khan F, et al. Impact of health literacy on outcomes and effectiveness of an educational intervention in patients with chronic diseases. Patient Educ Couns. 2012;87(2):143–51.
71. Griffey RT, Shin N, Jones S, Aginam N, Gross M, Kinsella Y, et al. The impact of teach-back on comprehension of discharge instructions and satisfaction among emergency patients with limited health literacy: a randomized, controlled study. J Commun Healthc. 2015;8(1):10–21.
72. Slater BA, Huang Y, Dalawari P. The impact of teach-back method on retention of key domains of emergency department discharge instructions. J Emerg Med. 2017;53(5):e59–65.
73. Griffey RT, Melson AT, Lin MJ, Carpenter CR, Goodman MS, Kaphingst KA. Does numeracy correlate with measures of health literacy in the emergency department? Acad Emerg Med. 2014;21(2):147–53.
74. Wolf MS, Davis TC, Curtis LM, Bailey SC, Knox JP, Bergeron A, et al. A patient-centered prescription drug label to promote appropriate medication use and adherence. J Gen Intern Med. 2016;31(12):1482–9.
75. Kagawa-Singer M, Kassim-Lakha S. A strategy to reduce cross-cultural miscommunication and increase the likelihood of improving health outcomes. Acad Med. 2003;78(6):577–87.
76. Epstein RM, Franks P, Fiscella K, Shields CG, Meldrum SC, Kravitz RL, et al. Measuring patient-centered communication in patient–physician consultations: theoretical and practical issues. Soc Sci Med. 2005;61(7):1516–28.
77. Aita V, McIlvain H, Backer E, McVea K, Crabtree B. Patient-centered care and communication in primary care practice: what is involved? Patient Educ Couns. 2005;58(3):296–304.
78. Chapman EN, Kaatz A, Carnes M. Physicians and implicit bias: how doctors may unwittingly perpetuate health care disparities. J Gen Intern Med. 2013;28(11):1504–10.
79. Wegener D, Silva P, Petty R, Garcia-Marques T. The metacognition of bias regulation; In: P. Briñol & K. DeMarree, editors. Social metacognition. New York: Psychology Press; 2012, pp. 81–99.
80. Burgess D, van Ryn M, Dovidio J, Saha S. Reducing racial bias among health care providers: lessons from social-cognitive psychology. J Gen Intern Med. 2007;22(6):882–7.
81. Taira BR, Kim K, Mody N. Hospital and health system-level interventions to improve care for limited English proficiency patients: a systematic review. Jt Comm J Qual Patient Saf. 2019;45(6):446–58.
82. Grubbs V, Chen AH, Bindman AB, Vittinghoff E, Fernandez A. Effect of awareness of language law on language access in the health care setting. J Gen Intern Med. 2006;21(7):683–8.
83. Elderkin-Thompson V, Cohen Silver R, Waitzkin H. When nurses double as interpreters: a study of Spanish-speaking patients in a US primary care setting. Soc Sci Med. 2001;52(9):1343–58.
84. Joint Commission. Advancing effective communication, cultural competence, and patient- and family-centered care: A roadmap for hospitals. Joint Commission; 2010. https://www.jointcommission.org/-/media/tjc/documents/resources/patient-safety-topics/health-equity/aroadmapforhospitalsfinalversion727pdf.pdf?db=web&hash=AC3AC4BED1D973713C2CA6B2E5ACD01B.
85. Regenstein M, Trott J, West C, Huang J. In any language: improving the quality and availability of language services in hospitals. 2008.

86. Karliner LS, Napoles-Springer AM, Schillinger D, Bibbins-Domingo K, Perez-Stable EJ. Identification of limited English proficient patients in clinical care. J Gen Intern Med. 2008;23(10):1555–60.
87. Brach C, Keller D, Hernandez LM, Baur C, Parker R, Dreyer B, et al. Ten attributes of health literate health care organizations. NAM Perspectives. 2012;1–19.
88. Institute of Medicine Committee on Understanding and Eliminating Racial and Ethnic Disparities in Health Care. Unequal treatment: confronting racial and ethnic disparities in health care. Smedley BD, Stith AY, Nelson AR, editors. Washington, D.C.: National Academies Press; 2003.
89. Tang G, Lanza O, Rodriguez FM, Chang A. The Kaiser Permanente clinician cultural and linguistic assessment initiative: research and development in patient-provider language concordance. Am J Public Health. 2011;101(2):205–8.
90. Huang J, Jones KC, Regenstein M, Ramos C. Talking with patients: how hospitals use bilingual clinicians and staff to care for patients with language needs. Washington, D.C.: Department of Health Policy, School of Public Health and Health Services, The George Washington University; 2009.
91. The Joint Commission. The joint commission standards FAQs. https://www.jointcommission.org/standards_information/jcfaq.aspx. Accessed 20 Apr 2019.
92. Kuo DZ, O'Connor KG, Flores G, Minkovitz CS. Pediatricians' use of language services for families with limited English proficiency. Pediatrics. 2007;119(4):e920–7.
93. Wiener ES, Rivera MI. Bridging language barriers: how to work with an interpreter. Clin Pediatr Emerg Med. 2004;5(2):93–101.
94. Gordon HS, Street RL Jr, Sharf BF, Kelly PA, Souchek J. Racial differences in trust and lung cancer patients' perceptions of physician communication. J Clin Oncol. 2006;24(6):904–9.
95. Safran DG, Taira DA, Rogers WH, Kosinski M, Ware JE, Tarlov AR. Linking primary care performance to outcomes of care. J Fam Pract. 1998;47(3):213–20.
96. Shiu-Thornton S, Balabis J, Senturia K, Tamayo A, Oberle M. Disaster preparedness for limited English proficient communities: medical interpreters as cultural brokers and gatekeepers. Public Health Rep. 2007;122(4):466–71.
97. Taira BR, Preston-Suni, K. Caring for limited English proficiency patients. Foundations of Emergency Medicine. 2019. https://foundationsem.com/foundations-iii/. Accessed 22 Apr 2019.
98. The IDHEAL Social Determinants of Health Bedside Teaching Modules. Taira BR, Hsieh D, Ogunniyi A, editors. The Section of International and Domestic Health Equity and Leadership, UCLA Emergency Medicine; 2018. http://www.idheal-ucla.org/. Accessed 22 Apr 2019.

Neighborhood and the Built Environment

5

Daniel A. Dworkis and Erik S. Anderson

Key Points

- Access to healthcare resources, including acute care services, is not distributed equally among communities, and this structure shapes how communities interact with the healthcare system.
- Exposure to neighborhood risks such as violence, trauma, pollution, or alcohol sales increases both acute illness and exacerbations of chronic illnesses that can lead to emergency department visits.
- Emergency providers can glean important clinical clues by probing relevant historical details pertaining to a patient's built environment. The context provided in such a history can lead to improved acute care management and may reduce the need for emergency care.
- Observations and research on the built environment for patients served in the emergency department can lead to community-based interventions or policy recommendations that can improve population health.

Foundations

Background

The physical realities of where we live, work, and spend our time matter deeply for our health and need for emergency care. At least as far back as John Snow's investigation of the role of the Broad Street water pump during the cholera outbreak

D. A. Dworkis (✉)
Department of Emergency Medicine, Keck School of Medicine of USC, LAC+USC, Los Angeles, CA, USA

E. S. Anderson
Department of Emergency Medicine, Highland Hospital – Alameda Health System, Oakland, CA, USA
e-mail: esanderson@alamedahealthsystem.org

H. J. Alter et al. (eds.), *Social Emergency Medicine*,
https://doi.org/10.1007/978-3-030-65672-0_5

in 1850s London, investigators in medicine and public health have sought to understand how differences in the conditions of neighborhoods and local environments reflect and generate differences in our health. In more modern times, research on the role the physical environment plays in health has led to the identification by the Healthy People 2020 initiative of the neighborhood and built environment as one of the five key social determinants of health [1]. While most of the work on the role of the built environment within medicine has taken place in the realms of general medicine and pediatrics, emergency medicine providers are increasingly taking notice of the physical reality patients face outside the walls of the emergency department (ED), and of the role emergency physicians (EPs) can and should play in addressing this reality [2, 3].

This chapter focuses on understanding the built environment as a driver of emergency care, and how individual EPs and healthcare systems can work to explore and address structural issues affecting patients and their communities. We will consider several different potential emergencies related to the built environment with a particular focus on respiratory diseases like asthma and chronic obstructive pulmonary disease (COPD), where the links to the built environment have a strong evidence base and are conceptually straightforward.

The Centers for Disease Control and Prevention (CDC) defines the built environment as the "physical makeup of where we live, learn, work, and play—our homes, schools, businesses, streets and sidewalks, open spaces, and transportation options." [4] In this sense, the built environment is the nonhuman components of our surroundings: things like the quality of housing stock and roads; the availability of food, medicine, and healthcare; access to schools and outdoor spaces for exercise; and exposures to potential risks like violence, drugs, or toxic chemicals. Implicit in this idea is the concept that the built environment can vary by increments in both time and space and across multiple physical scales: the presence of mold in an entire apartment building might influence all its residents' probability of having an asthma attack, whereas insulation problems in a specific unit might additionally influence the probability of the inhabitants of that particular unit having breathing problems during the winter months.

Defining a "neighborhood" is somewhat more complicated and no universal definition exists. Generally speaking, neighborhoods are both spatial and cultural entities, combining the ideas of a particular geographic area and the repeated social interactions of people and communities in that geographic area [5]. For the purposes of research, neighborhoods are often defined imperfectly using proxies like zip codes or US census tracts or their divisions. While neighborhoods are often thought of as unified areas, some research has shown the need for emergency care can differ within a neighborhood. For example, in the Charlestown neighborhood within Boston, MA, there are the existence of geospatial differences at the census tract level in opioid-related ED visits [6]. For the purposes of this chapter, we focus on the geographic and sociodemographic rather than the interpersonal aspects of a neighborhood; we define a neighborhood as the local geographic area within which we will consider potential effects of the built environment on emergency health needs [5].

In evaluating the effects of neighborhoods and the built environment on emergency medicine, it is useful to consider the so-called first law of geography, described by Waldo Tobler: "Everything is related to everything else, but near things are more related than distant things." [7] In other words, the more we can understand where things are in space, the better understand how they may be related. Geospatial analysis, the use of geographic data and mapping tools to quantify and analyze the spatial relationships between things, is an increasingly important tool used in many of the studies cited in this chapter. A review of geospatial analysis is beyond this chapter's scope, but EPs interested in learning more about geospatial analysis and developing expertise might start by exploring QGIS, which is a free and open-source spatial software for geospatial analysis and mapping [8].

Evidence Basis

Neighborhoods differ substantially in their resources related to emergency care, and the built environment influences access to care at a variety of points. While the Emergency Medical Treatment and Labor Act of 1986 (EMTALA) guarantees all people in the US medical treatment during an emergency, access to this medical treatment is not distributed equitably. In some cases, simply getting to an ED involves overcoming significant structural barriers like distance or difficult terrain. In other cases, access to trained EPs may not be feasible. For example, recent work highlights inequities in access to emergency-trained providers in rural communities, and an even more striking disparity for American Indian/Alaskan Native communities [9–11]. In conditions like traumatic injury or stroke, where specialty resources like trauma surgery, anesthesia, neurology, and neurosurgery may be required, differences in access to high-level emergency care have also been observed in both rural and in more urban environments [12, 13].

Outside of an ED, differences in neighborhood-level access to other health care services such as pharmacies can influence the ability of certain communities to stay healthy and follow through with emergency discharge plans [14, 15]. The built environment also affects access to potentially lifesaving devices such as automated external defibrillators (AEDs) and opioid overdose reversal kits, which require rational, community-centric design to ensure access where and when they are most needed [16, 17]. One example of this modeled the effectiveness of placing AEDs inside 7-Eleven ® convenience stores, and found that placing AEDs inside these stores, which were designed to be open and accessible to passersby, could considerably improve public AED access [18]. Finally, the built environment where a patient lives can affect their ability to utilize certain medications, such as oxygen, as access to electricity and other essentials are not universal, especially in shelters and temporary housing [19].

Exposure to Risk

Neighborhoods also vary substantially in the levels of various types of risks their inhabitants face related to the built environment. Within the home, exposure to allergens like mold or cockroaches has been shown to increase asthma symptoms, and exposure to substandard housing conditions has been linked to increased asthma-related pediatric emergency department visits [20, 21]. Exposure to secondhand smoke has similarly been linked to increased asthma-related pediatric emergency department visits [22], and there is a strong socioeconomic gradient to childhood exposure to passive smoke, with children at or below the federal poverty level (FPL) exposed at more than four times the rate of children at or above 400% of the FPL [23]. Deficiencies in the structural characteristics of the built environment of the home are associated with heavy alcohol consumption [24].

Neighborhoods also determine an individual's risk of exposure to toxic chemicals, such as lead, that disproportionately affect poor communities and communities of color [25]. Contaminated drinking water, specifically lead contamination, is related to older housing and contaminated soil. In 2014, Flint, Michigan changed its water supply for a portion of the city that is comprised of a majority of Black people, and where 40% of residents live under the FPL. For several years afterward, residents complained of poor water quality, which was due to corroded pipelines that supplied homes in this community, arising from the change [26]. This lead-contaminated water crisis received national attention when lead levels in children under 6 years in this community were found to be substantially elevated—at levels that have been associated with cognitive, behavioral, and long-term cardiovascular complications including all-cause mortality [25, 27, 28].

A person's access to alcohol and healthy food is also determined largely by their neighborhood which has implications for chronic disease burden (e.g. obesity, metabolic syndrome, and cardiovascular disease) and can also impact individuals' safety [29]. In one study, Baylor Scott & White Health System in Texas tested a comprehensive program that provided healthy food options, as well as other services, for patients living in food deserts. Patients enrolled in the group that provided access to low-cost, healthy food options saw a decrease in ED visits over the subsequent year [30]. The presence of alcohol sales in a neighborhood has been linked to increased risk of cyclist or pedestrian injury [31]. The density of alcohol outlets in a neighborhood has also been shown to be related to excessive drinking behavior [32]. Several studies have noted a higher level of intimate partner violence, crimes, and adverse alcohol-related health outcomes associated with increased alcohol outlet density [33, 34]. In addition to access to food and nutrition, the air we breathe is tightly bound to where we live. Air pollutants and cold temperatures due to poor insulation have been linked to increased asthma symptoms, as has exposure to dust and smoke in the work environment [20].

Considering the role of the built environment in either access to care or exposure to risk can seem daunting, since the realities of a particular built environment might have existed long before an EP begans her practice and might be designed to last long after she retires. That said, organizations such as BuildHealthyPlaces.org are specifically dedicated to connecting public health experts, clinicians, community leaders, and businesses, to find creative solutions to develop communities in a way that promotes equity, justice, and improved health outcomes [35]. With training and multidisciplinary supportive systems, EPs can improve the built environment their patients inhabit and, in doing so, do better for their patients and the communities they serve.

Emergency Department and Beyond

Bedside

During the initial acute phase of an emergency evaluation, the EP should focus on caring for the patient, not on the patient's built environment outside the ED – with certain specific exceptions, such as the need to assess for toxins such as carbon monoxide, cyanide, or heavy metals exposures. An acute asthma exacerbation requires inhaled beta agonists, consideration of systemic steroids, and close monitoring, regardless of whether it was precipitated by a virus, difficulty reaching a pharmacy to obtain controller medication, or irritants like mold or second-hand smoke. That said, the details of the trigger need consideration when planning for a safe discharged and for how future exacerbations can be mitigated. A patient whose exacerbation was triggered by lack of access to a pharmacy could be helped by medication delivery services, or in the short term by simply discharge with an inhaler in hand. A patient whose exacerbation was related to housing issues like mold, however, might benefit from the help of an ED social worker, or a referral to a medical–legal partnership, which can help patients address dangerous housing conditions with legal action [36, 37].

Whatever the emergency, it is advisable that when taking a full history, EPs ask patients the details of the built environments to which they will return to when leaving the ED in order to create a rational discharge plan. For example, a patient with bilateral ankle fractures who temporarily requires a wheelchair before an interval operation cannot be discharged to a home that requires climbing three flights of stairs. In all cases, the discharge plan from the ED must make sense in the patient's environmental context. This requirement is set out in the Knowledge, Skills, and Abilities profiles from the American Board of Emergency Medicine, which states that EPs should be able to "[e]stablish and implement a comprehensive disposition plan that uses appropriate consultation resources; patient education regarding diagnosis; treatment plan; medications; and time- and location-specific disposition instructions" [38].

Hospital/Healthcare System

Hospitals can support neighborhoods and improve the built environment in a number of ways using interventions that are based in the ED, or in the larger hospital and health care system. Where specific resources already exist to address a particular need in the built environment, hospitals can use the ED as a referral source to such resources. For example, the Breathe Easy at Home Program, which operates primarily in Boston, allows EPs and other health professionals to identify patients with asthma whose housing conditions affect their asthma care; referral to the program triggers a home visit by the Boston Inspectional Services Department, which can work with property owners to address housing deficiencies [39]. In some cases, these types of referrals to specific resources can happen as part of a more general referral to an organization equipped to identify and address a patient's social needs, such as the national Health Leads program, the Health Advocates program in Alameda County, or the Kaiser Thrive Local Initiative [40–42]. For these types of referrals to succeed, however, hospital systems need to educate their EPs and ED staff about the communities their patients live and work in, and about what resources are available, both within and outside the hospital. Such training could take the form of meetings with community members, or engagement with community groups that provide services to improve neighborhood conditions.

Additionally, hospitals in some cases can directly fill gaps in the built environment their patients inhabit. In an area with minimal access to fresh produce or other healthy foods, hospitals can commit to host farmers markets and food banks, as does the Lindau Lab at the University of Chicago Medical Center. Feed1st, an initiative developed by Dr. Stacy Lindau, hosts a food pantry at the medical center, and at the same time connects patients with social services, job training, and other resource. The program has fed more than 20,000 individuals and 7000 households since opening in 2010 [43]. Other hospitals have started to grow food on their roofs in order to create health food pantries for patients [44]. If access to exercise facilities is an issue, hospitals could build on-campus walking trails or house exercise classes accessible to the community. Issues related to housing stock are often more complex due to greater capital requirements [45]; nevertheless, several hospital systems have started to address these issues directly by actively developing high-quality affordable housing and medical respite facilities [46].

Societal Level

Emergency physicians can marry their knowledge of emergency presentations to their links to the built environment to function as advocates for social change. Examples abound, but this change is likely to require multidisciplinary efforts involving not only doctors but also lawyers, urban planners, politicians, and most importantly, members of the involved community. One such example is the expanding evidence base surrounding blight mitigation in urban communities. EPs see the consequences of gun violence on a daily basis, particularly in urban communities, and an interesting strategy to reduce violence may lie in reshaping the built environment through blight mitigation. Several large cities, including Philadelphia, Detroit, and Los Angeles, have embarked on converting blighted and abandoned lots and parks into green spaces, with the express purpose of decreasing violence in vulnerable communities.

In a landmark paper, Branas et al. reported a randomized controlled trial in Philadelphia that examined the impact of converting blighted lots into urban green spaces [47]. The transformation they describe was specifically designed to be scalable, affordable, and meant to create community rather than disperse it through gentrification. The quantitative and ethnographic findings from the Branas study showed a significantly reduced incidence of gun violence and other police-related problems, as well as greater perceived safety among community members. Several of the beneficial effects from this intervention were most pronounced in some of

Philadelphia's most vulnerable neighborhoods. This study represents a clear and striking link between the built environment at a societal level and its interaction with the acute care system.

Recommendations for Emergency Medicine Practice

Basic

- EPs, as part of every discharge, should ensure that the patient goes into a built environment where they are capable of executing the discharge plan. If unsafe conditions are identified in a person's neighborhood or environment, EPs can connect patients with social services and specific resources or Help Desks in their community, and actively partner with social services in their ED.
- When an emergency presentation seems likely related to a particular factor in the patient's built environment, EPs can refer patients to extant resources in the hospital or ED to address these factors. These might include specific resources that require a provider referral, such as medical–legal partnerships or case management services.

Intermediate

- EPs can learn about the built environment their patients live and work in and can work with hospital administration to develop solutions for modifiable factors. For example, knowing that a pharmacy desert exists outside your hospital might lead toward developing a take-home prescription program from the ED [48].
- EPs interested in medical advocacy can work with social emergency medicine teams within ACEP, SAEM, and AAEM to focus efforts on advocating for neighborhood health issues like housing code reform or safe streets initiatives (redesigning streets for pedestrians and bicyclists).
- Hospitals and EDs can develop training programs to help their EPs learn more about the unique strengths and vulnerabilities of the communities and geographic areas their ED is likely to serve. This training should be part of orientation for permanent and intermittent staff [49].

Advanced

- Outside the ED, an individual EP might consider working with medical–legal partnerships to effect regulatory changes to the local built environment, such as through zoning or building codes, or potentially becoming involved in politics at local or larger scales [30, 39]
- At the hospital level, EPs can champion a project within their hospital that responds to potential limitations in the local built environment, such as

collaborating with a farmers' market to supply food for patients after ED discharge, or creating an on-campus walking trail.
- EPs interested in studying a particular aspect of the built environment and its role in emergency care can learn more about geospatial analysis and work with research teams to investigate specific connections to health.

Teaching Case

Clinical Case

Mr. W is a 60-year-old male who presents to your ED during an overnight shift via ALS ambulance with a chief complaint of "difficulty breathing." As you enter the room, Mr. W appears in moderate respiratory distress—he is tripoding and pulling at the EKG leads and oxygen saturation monitors your nursing team is trying to place on him. Report from the paramedics indicate he has a history of chronic obstructive pulmonary disease (COPD), hypertension, and diabetes, and he was picked up from his apartment, where the patient had called 911 for his breathing troubles. His initial vitals include a temperature of 37.0 °C, heart rate 105 bpm, blood pressure 180/90 mmHg, respiratory rate 30/minute, and oxygen saturation 88% on room air. An initial finger-stick blood glucose was 180.

Mr. W's initial exam, in addition to his obvious respiratory distress, is notable for decreased breath movement throughout all lung fields and significant use of accessory muscles for inspiration. There is no lower extremity edema, and your point of care ultrasound shows lung sliding bilaterally with no evident B-lines. Your leading diagnosis is a COPD exacerbation, and you start Mr. W on noninvasive positive pressure ventilation and administer inhaled beta agonists and anticholinergics, as well as systemic steroids.

As he starts to improve, you note in your EMR that he was recently discharged from the inpatient medical service after a similar ED visit, and has had five ED visits for similar exacerbations in the last 6 months. When he's able to talk, you ask him what happened since that last visit, and he tells you that he lives immediately adjacent to a highway which is a thoroughfare for long-haul trucks. The city has identified his neighborhood's unsafe levels of air pollution, but he has been unable to find other low-income housing options in his community. As you admit Mr. W to a telemetry bed with the inpatient medical team, you contact your hospital's medical–legal help desk and ask them to see the patient while in the hospital. The help desk is able to identify housing stipends that are available to him through the city, and he is paired with a case manager to help him to find alternative housing.

Teaching Points

1. Early recognition of risks in Mr. W's built environment can lead to early engagement of social work and the mobilization of resources to improve discharge planning for the admitting team.
2. Asking patients about specific barriers can help uncover issues in their built environment that can be modified by leveraging ED and hospital-based resources.

Discussion Questions

1. What resources would you engage in your emergency department for this patient?
2. What tools would help you care for this patient if his respiratory status was such that he could potentially be managed as an outpatient?
3. Are there steps that the hospital could take to improve care for this patient and decrease future unscheduled visits for his COPD exacerbations?
4. How could EPs translate this common clinical scenario into specific policy recommendations in their community?

References

1. Healthy people 2020: social determinants of health. Available at https://www.healthypeople.gov/2020/topics-objectives/topic/social-determinants-of-health. Accessed 04/2019.
2. Anderson ES, Lippert S, Newberry J, Bernstein E, Alter HJ, Wang NE. Addressing social determinants of health from the emergency department through social emergency medicine. West J Emerg Med. 2016;17(4):487–9.
3. Dworkis DA, Peak DA, Ahn J, Joseph TA, Bernstein E, Nadel ES. Reaching out of the box: effective emergency care requires looking outside the emergency department. West J Emerg Med. 2016;17(4):484–6.
4. The built environment assessment tool manual. Centers for Disease Control and Prevention. Available at https://www.cdc.gov/nccdphp/dnpao/state-local-programs/built-environment-assessment/. Accessed 04/2019.
5. Diez Roux AV. Investigating neighborhood and area effects on health. Am J Public Health. 2001;91(11):1783–9.
6. Dworkis DA, Taylor LA, Peak DA, Bearnot B. Geospatial analysis of emergency department visits for targeting community-based responses to the opioid epidemic. PLoS One. 2017;12(3):e0175115.
7. Miller HJ. Tobler's first law and spatial analysis. Ann Assoc Amer Geographers. 2004;94(2):284–9.
8. QGIS: A Free and Open Source Geographic Information System. https://qgis.org/en/site/. Accessed 10/10/2019.
9. Hall MK, Burns K, Carius M, Erickson M, Hall J, Venkatesh A. State of the National Emergency Department Workforce: who provides care where? Ann Emerg Med. 2018;72(3):302–7.
10. Bernard K, Hasegawa K, Sullivan A, Camargo C. A profile of Indian health service emergency departments. Ann Emerg Med. 2017;69(6):705–710.e4.
11. Carr BG, Bowman AJ, Wolff CS, Mullen MT, Holena DN, Branas CC, et al. Disparities in access to trauma care in the United States: a population-based analysis. Injury. 2017;48(2):332–8.
12. Anderson ES, Greenwood-Ericksen M, Wang NE, Dworkis DA. Closing the gap: improving access to trauma care in New Mexico (2007-2017), Am J Emerg Med. 2019. 37:2028.
13. Khan JA, Casper M, Asimos AW, Clarkson L, Enright D, Fehrs LJ, et al. Geographic and sociodemographic disparities in drive times to Joint Commission-certified primary stroke centers in North Carolina, South Carolina, and Georgia. Prev Chronic Dis. 2011;8(4):A79.
14. Amstislavski P, Matthews A, Sheffield S, Maroko AR, Weedon J. Medication deserts: survey of neighborhood disparities in availability of prescription medications. Int J Health Geogr. 2012 Nov 9;11:48.
15. Hensley C, Heaton PC, Kahn RS, Luder HR, Frede SM, Beck AF. Poverty, transportation access, and medication nonadherence. Pediatrics. 2018;141(4):e20173402.
16. Levy MJ, Seaman KG, Millin MG, Bissel RA, Jenkins JL. A poor association between out-of-hospital cardiac arrest location and public automated external defibrillator placement. Prehosp Disaster Med. 2013;28(4):342–7.
17. Dworkis DA, Weiner SG, Liao VT, Rabickow D, Goldberg SA. Geospatial clustering of opioid-related emergency medical services runs for public deployment of naloxone. West J Emerg Med. 2018;19(4):641–8.

18. Huang CY, Wen TH. Optimal installation locations for automated external defibrillators in Taipei 7-Eleven stores: using GIS and a genetic algorithm with a new stirring operator. Comput Math Methods Med. 2014;2014:241435.
19. Dworkis DA, Brown SS, Gaeta JM, Brown JW, Gonzalez MH. Barriers to outpatient respiratory therapy among adult residents of emergency shelters. Ann Intern Med. 2012;157(9):679–80.
20. Gautier C, Charpin D. Environmental triggers and avoidance in the management of asthma. J Asthma Allergy. 2017;10:47–56.
21. Beck AF, Huang B, Chundur R, Kahn RS. Housing code violation density associated with emergency department and hospital use by children with asthma. Health Aff (Millwood). 2014;33(11):1993–2002.
22. Merianos AL, Dixon CA, Mahabee-Gittens EM. Secondhand smoke exposure, illness severity, and resource utilization in pediatric emergency department patients with respiratory illnesses. J Asthma. 2017;54(8):798–806.
23. Secondhand Smoke Exposure Among Nonsmoking Youth, 2013-1016. https://www.cdc.gov/nchs/products/databriefs/db348.htm. Accessed 9 Aug 2020.
24. Bernstein KT, Galea S, Ahern J, Tracy M, Vlahov D. The built environment and alcohol consumption in urban neighborhoods. Drug Alcohol Depend. 2007;91(2–3):244–52.
25. Lanphera B, Rauch S, Auinger P, Allen R, Hornung R. Low-level lead exposure and mortality in US adults: a population-based cohort study. Lancet Publ Health. 2018;3(4):E177–84.
26. Kennedy M. Lead-laced water in Flint: a step-by-step look at the makings of a crisis. National Public Radio. April 20, 2016. https://www.npr.org/sections/thetwo-way/2016/04/20/465545378/lead-laced-water-in-flint-a-step-by-step-look-at-the-makings-of-a-crisis. Accessed 7/30/2020.
27. Kennedy C, Yard E, Dignam T, et al. Blood Lead levels among children aged <6 years — Flint, Michigan, 2013–2016. MMWR Morb Mortal Wkly Rep. 2016;65
28. Clay K, Portnykh M, Severnini E. The legacy lead deposition in soils and its impact on cognitive function in preschool-aged children in the United States. Econ Hum Biol. 2019;33:181–92.
29. Kelli H, Kim J, Samman Tahhan A, Liu C, Ko YA, Hammadah M, et al. Living in food deserts and adverse cardiovascular outcomes in patients with cardiovascular disease. J Am Heart Assoc. 2019;8:e010694.
30. Wesson D, Kitzman H, Halloran K, Tecson K. Innovative population health model associated with reduced emergency department use and inpatient hospitalizations. Health Aff. 2018;37(4):543.
31. DiMaggio C, Mooney S, Frangos S, Wall S. Spatial analysis of the association of alcohol outlets and alcohol-related pedestrian/bicyclist injuries in New York City. Inj Epidemiol. 2016;3(1):11.
32. Ahern J, Margerison-Zilko C, Hubbard A, Galea S. Alcohol outlets and binge drinking in urban neighborhoods: the implications of nonlinearity for intervention and policy. Am J Public Health. 2013;103(4):e81–7.
33. Campbell CA, Hahn RA, Elder R, Brewer R, Chattopadhyay S, Fielding J, et al. The effectiveness of limiting alcohol outlet density as a means of reducing excessive alcohol consumption and alcohol related harms. Am J Prev Med. 2009;37(6):556–69.
34. Livingston M. A longitudinal analysis of alcohol outlet density and domestic violence. Addiction. 2011;106(5):919–25.
35. Build Healthy Places. Buildhealthyplaces.org. Accessed 5/23/19.
36. Zuckerman B, Sandel M, Smith L, Lawton E. Why pediatricians need lawyers to keep children healthy. Pediatrics. 2004;114(1):224–8.
37. National Center for Medical-Legal Partnership. Available at: https://medical-legalpartnership.org/. Accessed 04/2019.
38. Knowledge Skills and Abilities, American Board of Emergency Medicine. Available at: https://www.abem.org/public/publications/knowlege-skills-and-abilities-(ksas). Accessed 10/2015.
39. Breathe Easy at Home. Available at http://www.bphc.org/whatwedo/healthy-homes-environment/asthma/Pages/Breathe-Easy-at-Home.aspx. Accessed 04/2019.
40. Health Leads: About us. Available at https://healthleadsusa.org/about-us/. Accessed 04/2019.

41. Losonczy LI, Hsieh D, Wang M, et al. The Highland Health Advocates: a preliminary evaluation of a novel programme addressing the social needs of emergency department patients. Emerg Med J. 2017;34(9):599–605.
42. Johnson SR. Kaiser to launch social care network. Modern Healthcare. Available at https://www.modernhealthcare.com/care-delivery/kaiser-launch-social-care-network Accessed 05/2019.
43. Alexis Florence. Effects of food insecurity felt throughout Hyde Park community. The Chicago Maroon. https://www.chicagomaroon.com/article/2019/10/30/effects-food-insecurity-felt-throughout-hyde-park/. Accessed 7/30/2020.
44. BMC Preventive Food Pantry. Available at https://www.bmc.org/nourishing-our-community/preventive-food-pantry Accessed 05/1/2019.
45. Perdue WC, Stone LA, Gostin LO. The built environment and its relationship to the public's health: the legal framework. Am J Public Health. 2003;93(9):1390–4.
46. Bon Secour joins effort to build, sustain affordable housing projects. https://www.baltimoresun.com/health/bs-hs-bon-secours-housing-20190220-story.html. Accessed 9 Aug 2020.
47. Branas CC, South E, Kondo MC, Hohl BC, Bourgois P, Wiebe DJ, et al. Citywide cluster randomized trial to restore blighted vacant land and its effects on violence, crime, and fear. Proc Natl Acad Sci U S A. 2018;115(12):2946–51.
48. Hall AB, Novotny A, Bhisitkul D, et al. Association of emergency department albuterol dispensing with pediatric asthma revisits and readmissions. J Asthma. 2017;54:498–503.
49. Hsieh DT, Coates WC. Poverty simulation: an experiential learning tool for teaching social determinants of health. AEM Educ Train. 2018;2(1):51–4.

Queering the ED: Lesbian, Gay, Bisexual, and Queer Health in Emergency Medicine

6

Rebecca Karb and Shannon McNamara

Key Points

- A history of discrimination from the medical community and beyond often precedes a patient's emergency department (ED) visit. EPs and staff should be mindful of their own potential implicit or explicit biases.
- A simple way for EPs to establish trust is to use gender neutral and inclusive language.
- Sexual identity is not a proxy for sexual practices. If relevant to the chief complaint, EPs need to ask about sexual practice.
- Healthcare systems should incorporate gender identity and sexual orientation into institutional frameworks (such as electronic medical records, mission statements).

Foundations

Background

Our task of "queering" the ED is to critically examine the ways in which gender and sexuality enter into the practice of emergency medicine and how we can challenge our current beliefs and practices with the goal of providing better care to our LGBQ patients. In this chapter, we use the acronym "LGBQ" as an umbrella term meant to encompass patients who identify as lesbian, gay, bisexual, or queer. We

R. Karb (✉)
Department of Emergency Medicine, Rhode Island Hospital/Brown University, Providence, RI, USA

S. McNamara
Department of Emergency Medicine, NYU School of Medicine, New York, NY, USA
e-mail: shannon.mcnamara@nyulangone.org

H. J. Alter et al. (eds.), *Social Emergency Medicine*,
https://doi.org/10.1007/978-3-030-65672-0_6

recognize that there has been, and will continue to be, evolving terminology as we strive for representation and inclusivity. In clinical practice, an individual patient's preferred terminology should always be elicited, respected, and used preferentially during the patient encounter [1]. While it may be convenient to collapse subgroups together based on the nature of the structural violence and systemic oppression they face, it can create a problematic perception that the group is homogenous. On the contrary, the subpopulations represented by "LGBQ" are diverse and face their own unique struggles with health and within the healthcare system. Sexual identities also intersect with other differences—such as race, ethnicity, socioeconomic status, religion, and geography—creating complex and textured lived experiences.

Sexual orientation and gender identity have purposefully been separated in this textbook. While there is certainly intersectionality and shared cultural experiences, sexual orientation is distinct from gender identity. Sexual orientation refers to sexual or romantic feelings that a person might have for people of the same gender, a different gender, or more than one gender. Gender identity refers to one's concept of self as male, female, a combination of both, or neither; and can be the same or different from the assigned sex at birth. A transgender or gender nonconforming individual may be gay, straight, bisexual, or asexual. It is important to not sexualize gender by making assumptions about sexual orientation based on a person's gender identity or expression. Nonetheless, much of the health and healthcare disparities literature consolidates data on sexual orientation and gender identity ("LGBTQ", where T represents transgender). Whenever literature is cited in this chapter, we will report the language used in the original source to most accurately reflect the research. Although there are obstacles to accurate measurement, the current estimate is approximately 3.5% of adults in the US identify as LGBQ—just over eight million people. Furthermore, 8.2% of adults report same-sex behavior and 11% report same-sex attraction [2].

While a comprehensive historical review is beyond the scope of this chapter, a basic knowledge of the history of oppression, discrimination, and violence faced by the LGBQ community within the healthcare system is essential to understanding current LGBQ health disparities. Figure 6.1 presents an (incomplete) timeline of important events related to LGBQ health over the past 75 years. First, it is valuable to recognize that there is evidence of same-sex love and attraction in almost every documented culture and recorded as far back as Ancient Greece and Egypt [3]. However, the word "homosexual" was not coined until 1869 by Hungarian writer and journalist Karl Kertbeny. The American Psychiatric Association's (APA) first *Diagnostic and Statistical Manual of Mental Disorders* (DSM-1), published in 1952, designated homosexuality as a "sociopathic personality disturbance" [3]. Twenty years later in 1973, after persistent organizing efforts and educational campaigns led by the LGBQ community, homosexuality was removed as a pathology in DSM-II. It was not until 2000 that the APA took an official stance against reparative therapies [4]. Despite this apparent advance, conversion therapy and other "reparative" treatments continue to be recommended or provided by healthcare professionals.

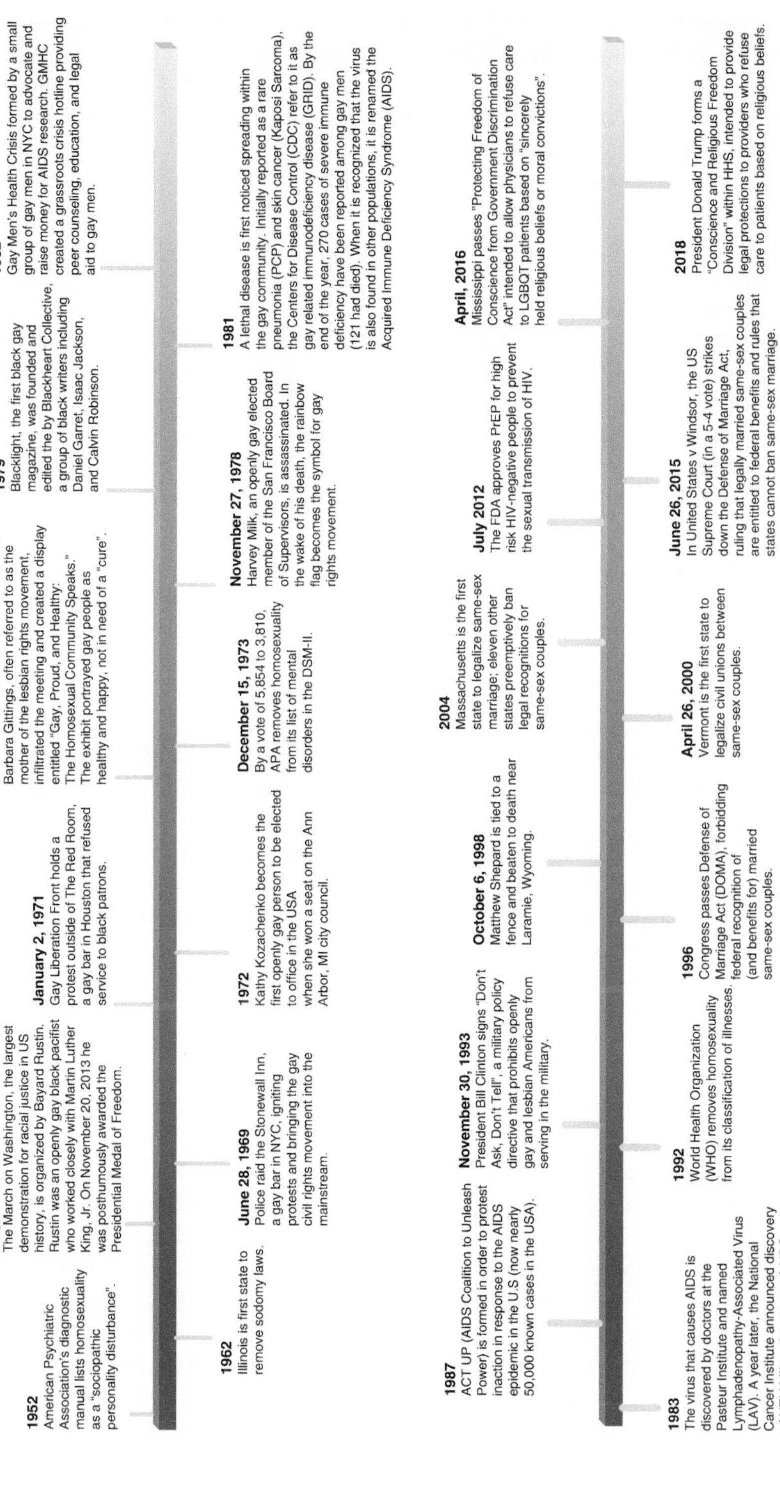

1952
American Psychiatric Association's diagnostic manual lists homosexuality as a "sociopathic personality disturbance".
1962
Illinois is first state to remove sodomy laws.
August 28, 1963
The March on Washington, the largest demonstration for racial justice in US history, is organized by Bayard Rustin. Rustin was an openly gay black pacifist who worked closely with Martin Luther King, Jr. On November 20, 2013 he was posthumously awarded the Presidential Medal of Freedom.
June 28, 1969
Police raid the Stonewall Inn, a gay bar in NYC, igniting protests and bringing the gay civil rights movement into the mainstream.
January 2, 1971
Gay Liberation Front holds a protest outside of The Red Room, a gay bar in Houston that refused service to black patrons.
1972
Kathy Kozachenko becomes the first openly gay person to be elected to office in the USA when she won a seat on the Ann Arbor, MI city council.
May 2, 1972
At the APA's conference in Dallas, Texas, Barbara Gittings, often referred to as the mother of the lesbian rights movement, infiltrated the meeting and created a display entitled "Gay, Proud, and Healthy: The Homosexual Community Speaks." The exhibit portrayed gay people as healthy and happy, not in need of a "cure".
December 15, 1973
By a vote of 5,854 to 3,810, APA removes homosexuality from its list of mental disorders in the DSM-II.
November 27, 1978
Harvey Milk, an openly gay elected member of the San Francisco Board of Supervisors, is assassinated. In the wake of his death, the rainbow flag becomes the symbol for gay rights movement.
1979
Blacklight, the first black gay magazine, was founded and edited the by Blackheart Collective, a group of black writers including Daniel Garret, Isaac Jackson, and Calvin Robinson.
1981
A lethal disease is first noticed spreading within the gay community. Initially reported as a rare pneumonia (PCP) and skin cancer (Kaposi Sarcoma), the Centers for Disease Control (CDC) refer to it as gay related immunodeficiency disease (GRID). By the end of the year, 270 cases of severe immune deficiency have been reported among gay men (121 had died). When it is recognized that the virus is also found in other populations, it is renamed the Acquired Immune Deficiency Syndrome (AIDS).
1982
Gay Men's Health Crisis formed by a small group of gay men in NYC to advocate and raise money for AIDS research. GMHC created a grassroots crisis hotline providing peer counseling, education, and legal aid to gay men.
1983
The virus that causes AIDS is discovered by doctors at the Pasteur Institute and named Lymphadenopathy-Associated Virus (LAV). A year later, the National Cancer Institute announced discovery of HTLV-III as the cause of AIDS. It is concluded (in 1986) that these are the same virus, renamed HIV.
1987
ACT UP (AIDS Coalition to Unleash Power) is formed in order to protest inaction in response to the AIDS epidemic in the U.S (now nearly 50,000 known cases in the USA).
1992
World Health Organization (WHO) removes homosexuality from its classification of illnesses.
November 30, 1993
President Bill Clinton signs "Don't Ask, Don't Tell", a military policy directive that prohibits openly gay and lesbian Americans from serving in the military.
1996
Congress passes Defense of Marriage Act (DOMA), forbidding federal recognition of (and benefits for) married same-sex couples.
October 6, 1998
Matthew Shepard is tied to a fence and beaten to death near Laramie, Wyoming.
April 26, 2000
Vermont is the first state to legalize civil unions between same-sex couples.
2004
Massachusetts is the first state to legalize same-sex marriage; eleven other states preemptively ban legal recognitions for same-sex couples.
July 2012
The FDA approves PrEP for high risk HIV-negative people to prevent the sexual transmission of HIV.
June 26, 2015
In United States v Windsor, the US Supreme Court (in a 5-4 vote) strikes down the Defense of Marriage Act, ruling that legally married same-sex couples are entitled to federal benefits and rules that states cannot ban same-sex marriage.
April, 2016
Mississippi passes "Protecting Freedom of Conscience from Government Discrimination Act" intended to allow physicians to refuse care to LGBQT patients based on "sincerely held religious beliefs or moral convictions".
2018
President Donald Trump forms a "Conscience and Religious Freedom Division" within HHS, intended to provide legal protections to providers who refuse care to patients based on religious beliefs.

Fig. 6.1 An Incomplete Timeline of Modern US LGBQ History

Throughout the 1980s and 1990s, the AIDS crisis exacerbated the longstanding homophobia in healthcare [1]. While many demographics were affected by AIDS, the community of men who have sex with men (MSM) was the hardest hit. The Centers for Disease Control and Prevention (CDC) began tracking HIV/AIDS cases and deaths in 1981; by the end of 2000 over 750,000 people in the US had been diagnosed with AIDS, of whom almost 500,000 had died [5]. There was widespread discrimination within the healthcare system and by healthcare providers towards patients with HIV [1, 5]. The governmental response to the crisis in terms of research investment was viewed by many as inadequate, leading to a delay in appropriate treatment, prevention, and education [1, 5].

Throughout time, LGBQ people have demonstrated tremendous strength and resilience in the face of structural, physical, and sexual violence. The LGBQ community has led significant progress in breaking down structural barriers and educating the public and healthcare community. However, disparities in health and healthcare persist.

Evidence Basis

The "minority stress model" provides a useful framework for understanding LGBQ health disparities. Beginning in youth, LGBQ people live with the daily stress of structural stigma, discrimination, and marginalization. Chronic exposure to these daily stressors accumulates over the life course, and the resulting "wear and tear" on the body ultimately manifests in poorer mental and physical health outcomes [6–10]. A robust literature on experiences of discrimination and health has documented the effects on multiple physiological levels, from cellular functioning and gene expression (e.g., DNA methylation and histone modifications) to neuroendocrine dysregulation (e.g., cortisol patterns) to a range of health behaviors and outcomes (e.g., cardiovascular disease, diabetes, depression, smoking, substance use/misuse, and medication nonadherence) [11–18].

Early experiences of shame, rejection, and isolation can begin at home. LGBQ youth who report high levels of family rejection are 8 times more likely to have attempted suicide, 6 times more likely to be depressed, and 3 times more likely to use illegal drugs compared to those LGBQ youth who reported no or low levels of family rejection [19]. Family rejection also increases the risk of homelessness. Approximately 40% of homeless youth identify as LGBTQ and it is estimated that as many as 80,000 LGBTQ youth experience homelessness each year [20, 21]. By contrast, family acceptance/support has been shown to be associated with positive self-esteem and good general health [22]. Research to date has supported the conclusion that the mental health disparities among LGBQ youth are not inherent to sexual identity but rather result from societal stigma and familial rejection.

LGBQ youth may also face significant challenges at school. Experiences of homophobic bullying are associated with lower educational achievement, depression, suicidality, social isolation, and substance use [23]. More recently, attention has focused on the potential protective factors that may mitigate the effects of exposure to homophobic bullying. Positive influences from the family or home may work as a buffer against negative impacts of both homophobic bullying and aggression that a youth may experience in school [24].

Experiences of shame and rejection both at home and school can lead to social isolation and harmful coping behaviors. LGBT youth are more likely than straight youth to report misuse of prescription opioids and sedatives [17], and are less likely than straight youth to engage in physical activity or team sports [25]. One-third of LGB youth engage in hazardous weight control behaviors, such as fasting more than 24 hours, using diet pills, or vomiting or using laxatives [26]. LGBT youth are twice as likely to have suicidal ideation and four times more likely to make a serious suicide attempt compared to heterosexual youth [27]. This last number is almost certainly an underestimate, as the sexual orientation of youth who complete suicide is often unknown.

The local sociopolitical environment also may be related to the health of LGBQ individuals. In a recent national study, health disparities among LGBQ people were greatest in communities with low levels of approval of same-sex marriage [28]. A study of same-sex couples prior to the Supreme Court ruling affirming same-sex marriage found that couples living in states with legally sanctioned marriage reported higher levels of self-rated health compared to those living in states with constitutional amendments against same-sex marriage [29]. It has also been shown that LGBQ adults who were raised in highly stigmatizing communities (as measured by LGBQ representation in local government, employment and nondiscrimination policies, and public opinion) exhibited blunted cortisol responses to a laboratory stressor [30].

LGBTQ individuals have a high lifetime risk of being a victim of a violent crime; 38% of gay men and 13% of lesbian women report hate crimes against their person or property [31]. In a recent national survey, 58% of LGBT respondents reported being subjected to jokes or slurs, 26% reported being threatened or physically attacked, and 21% reported being treated unfairly by an employer [32]. Over 80% of LBGT youth report experiencing verbal harassment at school, while 38% report having been physically assaulted [33]. Despite the long history of physical and sexual violence against the LGBQ community, it was not until 2009 that sexual orientation and gender identity were first included under federal "hate crimes" with the passage of the Matthew Shepard Act.

At the other end of the age spectrum, older LGBQ adults face unique challenges. Most came of age at a time of far less societal acceptance and with fewer available resources and role models. Of older LGBT individuals, 63–65% report experiencing physical violence related to their sexual identity at some point during their lifetime [34]. They experience high rates of internalized stigma, often leading to poorer mental health outcomes [35]. LGBT older adults are also more likely to be economically disadvantaged—a result of early and cumulative structural discrimination across the lifespan—exposing them to the higher overall mortality known to be associated with low socioeconomic status [36]. Older gay men living with HIV are more likely to live alone, have poor social support, and are at increased risk of depression [34].

Other subgroups of the LGBQ population also face unique health and healthcare needs. Men who have sex with men have an increased risk of HIV/AIDS, STIs, and anal cancers [37]. Lesbian and bisexual women have twice the risk of obesity compared to straight women as well as an increased risk of gynecological and breast cancers [14, 38, 39]. In addition to lower mammography rates, lesbian women on average have higher rates of some risk factors for breast cancer, including greater alcohol use and lower likelihood of childbearing [39]. Lesbian and bisexual women are also more likely to smoke and use illicit drugs compared to straight women [37].

Health disparities are compounded by unequal exposure to other well-documented adverse social determinants of health, such as low socioeconomic status. According to an analysis of the 2006–2010 National Survey of Family Growth, more than one-quarter (28%) of lesbian and bisexual women experience poverty, compared with 21% of straight women. Just over 1 in 5 gay and bisexual men (23%) experience poverty, compared to 15% of straight men [40].

LGBQ people of color are exposed to intersecting dynamics of discrimination that place them at greater risk of poor health outcomes. In one study, Latino men reported the highest number of negative family reactions related to their sexual orientation in adolescence [19]. Non-White lesbians report the poorest self-rated health compared to White lesbians, non-White straight women, and men [8]. The LGBQ community itself is not immune to ingroup racial discrimination and inequality. For example, lesbians of color were systematically marginalized and silenced within the feminist movement throughout the 60s and 70s, and continue to struggle for representation in the gay rights movement.

Emergency Department and Beyond

The healthcare system—historically and presently—is a significant contributing source of health disparities among the LGBQ community. Many LGBQ individuals report difficulty finding a healthcare environment in which they feel safe and respected; there remains a dearth of providers who are both welcoming and knowledgeable about the unique healthcare needs of LGBQ patients [3]. In a recent survey, 27% of medical students had observed judgmental attitudes and behaviors toward LGBQ patients from physicians [41]. Even in the absence of overt discrimination, physicians can deliver suboptimal care due to unconscious bias or simply a lack of knowledge and comfort with LGBQ-specific health issues. Negative interactions with(in) the healthcare system often lead LGBQ patients to delay care or even avoid care altogether [42].

Bedside

An emergency department (ED) encounter can be stressful under the best of circumstances, but this can be exacerbated by negative encounters with nurses, physicians, and other staff. Even before stepping foot in the ED, many patients are already burdened with the cumulative weight of their prior negative healthcare experiences. Small microaggressions at the bedside can accumulate over time and can influence patient trust in providers. For example, this excerpt from a book on the experiences of queer and trans patients highlights how seemingly "harmless" assumptions made by providers can be isolating for patients.

> Finally, a middle-aged nurse with lime-green glasses comes over to offer me a heated blanket and, apparently, some comfort. "This must be really hard on you," she says, laying the blanket over my legs. "But at least your mom's here."
>
> For a fleeting moment, I actually feel embarrassed for her. Until I don't.
>
> "Uh, no. That's definitely not my mom."
>
> "What…?"
>
> "This is my wife. And she's five years older than I am."
>
> As the realization hits, her face falls. She scans her brain for a comeback. "It's.. it's just that you look so *young*," she says. "Which is *good*! You're lucky!"
>
> But my irritation has nothing to do with vanity and everything to do with her assumption: this is an ugly case of heteronormativity. Refusing to consider that we might be queer, this nurse reached into her brain for the closest heterosexual explanation for the intimacy between us, picking—for whatever reason—'mother and daughter'. We are clearly close in age and don't look alike, yet she'd stuffed us into a box that obviously didn't fit. Even in the era of same-sex marriage, rainbow families, and out-and-proud celebrities, it's still the case that everyone is presumed straight (innocent) until proven gay (guilty) [43].

When LGBQ patients present to the ED, the long history of institutional discrimination by the medical system precedes their arrival and may cloud the patient-clinician interaction. LGBQ patients may have had a negative experience with filling out registration forms or interacting with hospital staff before they reach the treating clinician [44–46]. Stories about patients who have experienced outright discrimination or whose families were denied hospital visitation because of their LGBQ relationship are well known in popular culture [47]. It is the responsibility of the treating clinician to establish trust in the patient encounter with the acknowledgement that creating that trust may be more difficult than with a typical patient encounter [48].

All emergency clinicians can use inclusive and gender-neutral language to establish trust with LGBQ patients. As it is often difficult to tell which patients are LGBQ, this language is useful when speaking with all patients. When first meeting a patient, refer to the patient by their full first and last name. Avoid using gendered titles like Mr. or Mrs. If a significant discrepancy between the documented name of the patient and the presenting gender of the patient exists, it is reasonable to confirm the last name and date of birth and ask, "What would you like to be called?" This allows patients a chance to take ownership of their title, as documented sex and gender identity/presentation may not correlate.

- When asking about patient's partner(s), avoid using gendered language. Instead, use terms such as "partner(s)" or "significant other(s)" until the patient clarifies the gender of their partner(s) and the nature of the relationship.
- When asking about visitors with a patient, avoid assuming the nature of the relationship. Ask open ended questions, such as "Who is here with you today?"

Avoid "Is this your mother/sister/wife?" Allow the patient or visitor to identify themselves and their relationship to the patient [48].
- Of the 650,000 same-sex couples in the US, 19% have children under the age of 18 [49]. Do not ask, unless medically relevant, which parent is the "real" or "biological" parent; treat both parents as equal caregivers.

These communication strategies can avoid many of the unintentional pitfalls clinicians experience when they accidentally misgender a patient or make the wrong assumptions about a patient's relationships. Assuming a bisexual woman has a male partner, or asking a gay man if his husband is his brother can further undermine the patient's trust in the clinical team and the healthcare system. Even with the best intentions, missteps may still occur. After making a mistake, the best approach is to acknowledge it directly, offer the patient a genuine apology for any harm caused, and move on with the clinical encounter. An example: "I apologize for using the wrong pronoun/name/terms. I did not mean to disrespect you" [50].

Sexual *identity* and sexual *practices* do not always align. While labels such as "gay" or "lesbian" can be heuristically useful, it is important to keep in mind that sexual identity may be distinct from sexual practices, and both may be dynamically fluid over the life course. Sexual identity should not be used as a proxy for behavioral risk factors (or lack thereof). It is important for healthcare providers to ask about both sexual orientation *and* sexual practices—if a patient's chief complaint warrants inquiry—in order to provide the most appropriate care. It is equally important to balance this care and necessary information gathering with respecting privacy. Especially in the ED, where due to limited resources, histories may be conducted in hallways or other less private areas, emergency practioners should be mindful about where, how, and why they are asking about sexual behavior and sexual history. While it may be tempting to ask questions out of curiosity to learn more about LGBQ people, LGBQ patients may want to keep their personal and medical histories private, just like everyone else. We must reflect on whether the questions being asked are to obtain necessary information to deliver care, or to satisfy a curiosity, and avoid asking unnecessary questions [50].

After moving the patient to a private space and determining that a focused sexual history is relevant to the ED visit, the questions below can help guide history taking for patients of any sexual orientation. As the sexual orientation of a patient is generally unknown prior to asking about their sexual practices, it is ideal to use gender neutral language initially and to specifically inquire about the gender of the patient's partner or partners. A sexual history should focus on the actual sexual behaviors of the patient, not only their stated sexual orientation. As many patients may have sexual partners that identify as transgender or gender nonbinary, the gender identity of a patient or their partner may not correlate with sexual anatomy. Discussions about risk for sexually transmitted infection and pregnancy need to be tailored to individual patient's sexual behaviors.

LGBQ Sexual History Questions:

Are you currently sexually active?
Have you ever been sexually active?
Tell me about the gender of your partner or partners.
When having sex, do you have vaginal, anal, and/or oral sex?
If relevant: do you use condoms or other barriers when having vaginal, anal, or oral sex? How often?
Do you and your partner(s) use any other protection against STIs? If no, Could you tell me the reason why not? If yes, what kind of protection do you use, how often? [51].
***If relevant and if the patient is sexually active with a partner or partners capable of producing pregnancy:*
Do you think you might like to have (more) children at some point?
If the patient is considering future parenthood: When do you think that might be?
How important is it to you to prevent pregnancy (until then)? [51]

Risk reduction strategies for STI and pregnancy prevention can be patient centered by focusing on the patient's goals and the patient's specific practices and partner(s). Regardless of gender or sexual identity, the clinically relevant information in a sexual history is what anatomy each person has, what sexual behaviors they engage in, and the level of individual agency present.

In addition to the sexual history, the family history may be a challenging part of the encounter for LGBQ patients. Often when conducting a family history, a clinician will ask about a patient's mother and father's health. Children from same-sex families are often conceived using sperm donors, egg donors, or gestational carriers. It may be more accurate to ask: "I'd like to learn more about your genetic risk factors for disease. Please tell me what you know about your genetic history."

Emergency clinicians are often champions of patient equality – treating any patient, any disease process, anytime, and proudly treating all patients as equals. In the context of health disparities and social emergency medicine, the framework for the individual encounter between clinicians and LGBQ patients must be one of **health equity**, not simply equality. Health equity is defined as "the principle underlying a commitment to reduce—and, ultimately, eliminate—disparities in health and in its determinants, including social determinants. Pursuing health equity means striving for the highest possible standard of health for all people and giving special attention to the needs of those at greatest risk of poor health, based on social conditions" [52]. This may mean it requires more time and investment with LGBQ patients from the treating clinician to obtain the same level of trust as with non-LGBQ patients. By being open and flexible to patients' needs, clinicians can work to disrupt the effects of structural stigma and discrimination in a tangible way through the clinical encounter.

Through a commitment to justice and with proper research and education, emergency physicians can work to transform the healthcare encounter from a source of shame and rejection to a source of affirmation and empowerment. By becoming competent in the care of LGBQ patients, we can transform the doctor–patient relationship from a risk factor to a protective factor.

Hospital/Healthcare System

Cultural change within a profession begins with education. A survey of medical school deans at 176 allopathic and osteopathic medical schools found that the median reported time dedicated to teaching LGBT-related content in the medical curriculum was five hours. One-third of schools reported zero hours during clinical years, and 43.9% of institutions rated their curricular LGBT content as only “fair” [53]. In 2014, a survey of EM program directors characterized the prevalence of content and needs related to LGBT education, barriers to curricula, and program demographics associated with inclusion of LGBT educational material. Only 26% reported that a dedicated LGBT lecture had ever been presented, while 33% reported incorporating LGBT topics into other components of the didactic curriculum. The average amount of time spent on LGBT health was 45 minutes per year. Programs with LGBT faculty and residents expressed more support of inclusion of LGBT-focused material into training curricula compared to programs without LGBT faculty [54].

The Human Rights Campaign (HRC), a LGBTQ civil rights organization, has developed a “Healthcare Equality Index” (HEI) to score hospitals and other healthcare facilities on their compliance with best practices in LGBTQ health practices. In 2018, 626 facilities participated [55]. The HEI rates hospitals on nondiscrimination policies and staff training, patient services and support, employee benefits and policies, patient and community engagement, and responsible citizenship [56]. Similarly, The Association for American Medical Colleges (AAMC) has published best practices for developing a healthy institutional climate for LGBT faculty, students, residents, and administrators. In addition, they discuss the role of medical education in addressing health disparities and offer specific curricula for teaching core competencies related to LGBT health in medical schools [56]. The Gay and Lesbian Medical Association released a document outlining best practices for creating a climate of inclusion for LGBTQ health professionals and students [57]. These include LGBQ inclusion in mission statements, new employee orientations, and CME training requirements.

Recent research has established the utility and acceptability of routine collection and display of sexual orientation and gender identity in medical records. In qualitative interviews of patients and ED providers, LGBQ patients were much less likely to refuse to provide sexual orientation than ED providers expected [58]. Discordant views between providers and patients regarding collection of sexual orientation highlights the discomfort that many providers have in asking about sexual practices, in contrast to the willingness of the LGBQ community to be seen and normalized within the healthcare system. Gathering data on sexual orientation in clinical settings and in EHRs helps us better understand LGBQ health, including disparities in insurance coverage, access to care, diagnosis, and treatment [59]. Moreover, making sexual orientation and gender identity readily visible to providers in the EHR can mitigate misidentification and serve as a reminder of its importance in the healthcare encounter.

Societal Level

As illustrated above, the etiology of LBGQ health disparities can be traced to broader social, political, and economic conditions. As cultural changes have led to greater acceptance of LGBQ individuals and families, we are optimistic that reductions in health disparities will follow. Yet, disparities persist.

Over the past decade, there has been a long overdue recognition of the unmet healthcare needs of the LGBQ community accompanied by a renewed focus on research and action. In Healthy People 2020—the nation's roadmap for improving health over the next decade—the US committed for the first time to eliminating LGBT health disparities. Healthcare providers should "appropriately inquire about and be supportive of a patient's sexual orientation to enhance the patient–provider interaction and regular use of care" [60]. In 2016, the National Institutes of Health (NIH) designated sexual and gender minorities (SGM) as a health disparities population alongside racial/ethnic minorities, socioeconomically disadvantaged populations, and underserved rural populations for the purpose of research and grant funding [61, 62]: "In doing so, the NIH recognizes that more research in SGM health is critical to better understanding both the well-being of and the potentially undiscovered health disparities experienced by this population" [61].

A number of relatively recent legal changes have shaped access to and quality of health services for LGBQ Americans. In June 2013, the Supreme Court's ruling in United States v. Windsor overturned a portion of Defense of Marriage Act (DOMA) and required the federal government to recognize legal same-sex marriages for the first time. This decision has had ripple effects on LGBQ health, as marriage is tied to a range of federal benefits including tax deductions and access to health insurance [63]. The Affordable Care Act (ACA), passed in 2010, extended coverage to millions of uninsured persons through the expansion of Medicaid and the creation of new federally subsidized health insurance marketplaces in all states. It also included new federal regulations barring discrimination in insurance provision based on sexual orientation and gender identity. In addition, restrictions on coverage based on preexisting conditions (for example, HIV or mental illness) that historically disproportionately affected the LGBQ community, were eliminated. In 2010, the Department of Health and Human Services issued a policy stating that hospitals needed to allow patients to designate visitors regardless of sexual orientation, gender identity, or any other nonclinical factors. The Centers for Medicare and Medicaid Services also issued guidance noting that same-sex couples have the same rights as all patients to use an advanced directive to name a representative to make decisions on their behalf [64].

These changes have begun to narrow the longstanding disparities in insurance coverage for LGBQ individuals [65]. However, there have been recent setbacks as well. Beginning in 2019 the federal government attempted to expand the availability of specialized insurance plans that provide exemptions from key protections for sexual orientation and gender identity. "Conscience and religious" exemptions for healthcare providers have also been liberalized, which may limit access to care and treatment for LGBT people, particularly in resource limited settings. A comprehensive approach to improving LGBQ health must include advocacy at the local, state, and national level to ensure equity in access to quality care.

Recommendations for Emergency Medicine Practice

Effective care for LGBQ patients in the ED requires an understanding not only of specific health risks but also of the larger sociopolitical context in which health disparities emerge. Moreover, providers need to be self-reflective and open to exploring personal biases (both explicit and implicit) in order to develop the skills needed for welcoming and respectful healthcare delivery. The following are concrete, tangible steps that EM providers can take to improve their clinical care of the LGBQ patient.

Basic

- Do not use sexual identity as a proxy for history taking. Ask the patient about sexual practices in ways which are nonjudgmental and affirming.
- Respect and reflect the terminology used by your patient. For example, if a male patient refers to the person accompanying him as his "husband," do not refer to him as the patient's "partner" or "friend." Try to stay current on evolving terminology. If you are not sure what language to use, ask the patient.
- When taking a history, avoid gender-specific language. For example, instead of asking "Do you have a wife or girlfriend?" you might ask "Are you in a relationship?"
- If you make a mistake, recognize it, apologize, and move on.

Intermediate

- Emergency providers can be allies to LGBQ patients by helping to create safe spaces by speaking up when they hear discriminatory language or witness discriminatory behavior.
- Bring up issues of biases in care when teaching residents, medical students, and staff. Encourage providers to reflect on personal biases that may be impacting the care they provide, and do so for your own care.
- Integrate LGBQ issues into resident conferences and simulations. This should be done on an ongoing basis to reflect the most current knowledge of health and healthcare needs as well as changes in terminology.
- Assess your own implicit bias. Take a free, evidence-based test at Project Implicit (https://implicit.harvard.edu/implicit/) [66]. Complete the free, self-guided case scenarios at the National LGBT Health Education Center titled "Learning to Address Implicit Bias Towards LGBTQ Patients" [67]. https://www.lgbthealtheducation.org/publication/learning-to-address-implicit-bias-towards-lgbtq-patients-case-scenarios/

Advanced

- Work with your hospital or healthcare organization to incorporate sexual orientation and gender identity into institutional frameworks. For example, advocate for LGBQ inclusion in mission statements, new employee orientations, and CME training requirements.
- Advocate for the implementation of EMR-based systems for identifying sexual orientation and gender identity. While there have been concerns about electronic identification leading to increased stigma or discrimination, studies show these datapoints are a catalyst for new provider trainings and improved cultural competence [59].
- Participate in community-level efforts to raise awareness and advocate for LGBQ rights.

Teaching Case

Clinical Case

Luis Garcia is a one and a half-year-old male presenting to the ED with a chief complaint of fever and cough. The nurse first assesses the patient and his family in triage. Vital signs show: temperature 38 C, HR 132, RR 32 and oxygen saturation of 94% on room air. The nurse's triage note documents scattered wheezing and mild retractions and reports that the history was given by the child's mother. The triage history notes that symptoms started yesterday and worsened today. He has never wheezed before.

You enter the exam room and find Luis sitting on a woman's lap. Another woman of similar age is also in the room. He is drinking a bottle when you enter the room and you notice mild subcostal retractions.

After introducing yourself, you inquire about the identity of Luis's caregivers by asking "Who is here today with Luis?" The woman holding Luis responds "We're Luis's parents." You shake his parents' hands and learn more about the history of this illness. You note that at 18 months, Luis is a little old for bronchiolitis, but also young to have a first episode of asthma or bronchospasm [68]. You are curious about Luis's birth history and familial genetic history of bronchospastic disease. You also know that as a family with same-sex parents, this family has likely experienced many intrusive questions about Luis's birth history and inquiries about which parent is his "real mom." You want to honor the equal role of both caregivers while obtaining important information about this child's health. You know that like any other child, Luis may have been conceived using either one of his parent's genetic material, donor egg and/or sperm, or been adopted.

You decide to explain your rationale and ask permission before going forward. "I notice that Luis is wheezing today. I'm concerned that this could be an early sign of asthma, but I'm not sure. I'd like to learn more about Luis's birth history and ask some more questions about his genetic family history. Would that be all right?" Both parents nod. You proceed to ask about his gestational age at birth and any complications. You ask about his genetic heritage and if there was any known history of asthma or atopic dermatitis. His parents note that he was conceived using a known sperm donor and neither the donor nor the mother who carried him using her eggs have a history of asthma or atopic dermatitis.

You proceed to examine Luis. You decide to treat with supportive care for bronchiolitis with nasal suction, PO fluids, oxygen, fever management and close reassessment. You discuss your assessment and plan with his parents and answer their questions. After the visit, you document the history in the electronic health record, including that you spoke to both parents and inquired about Luis's genetic family history. You notice that the EHR template is not set up to make writing in this information easy, and you make a note to talk to your ED director to see if there's a better template available that's more inclusive for LGBQ families.

Teaching Points

1. Use open-ended questions to inquire about visitors with patients.
2. When asking history questions that may be perceived as invasive, first explain your rationale and how they relate to the goals of care for the patient. Ask permission to build trust.
3. LGBQ health encompasses much more than sexual health. As a clinician, it's essential to work to build trust and acknowledge the effects of minority stress and the systemic barriers these patients may face as part of their LGBQ experience.

Discussion Questions

1. Have you ever made a mistake when assuming the relationship of a patient and a visitor? What happened? How did you recover from it to continue the clinical encounter?
2. How do you approach obtaining a potentially sensitive patient history about a topic that may carry stigma, like sexual health, mental illness, or infertility?
3. What parts of this case and chapter overall affirmed your practice? What challenged your practice? Why?
4. How do you think an LGBQ-centered approach of not assuming family relationships, asking open-ended history questions and explaining the rationale for certain history questions would be received by non-LGBQ patients?

References

1. Eckstrand KL, Sciolla AF. History of health disparities among people who are or may be LGBT, gender nonconforming, and/or born with DSD. In: Hollenbach AD, Eckstrand KL, Dreger A, editors. Implementing curricular and institutional climate changes to improve health care for individuals who are LGBT, gender nonconforming, or born with DSD: a resource for medical educators. Association for American Medical Colleges, Washington DC; 2014.
2. Gates GJ. How many people are lesbian, gay, bisexual and transgender? The Williams Institute. https://williamsinstitute.law.ucla.edu/wp-content/uploads/Gates-How-Many-People-LGBT-Apr-2011.pdf. Accessed 3/1/19.
3. Morris BJ. History of lesbian, gay, bisexual and transgender social movements. American Psychological Association. 2017. https://www.apa.org/pi/lgbt/resources/history. Accessed 4/1/2019.

4. Makadon HJ, Goldhammer H, Davis JA. Providing optimal health care for LGBT people: changing the clinical environment and educating professionals. In: Makadon HJ, Mayer KH, Potter J, Goldhammer H, editors. Fenway guide to lesbian, gay, bisexual, and transgender health. 2nd ed: American College of Physicians, Philadelphia, PA; 2015.
5. CDC. HIV and AIDS, United States 1981-2000. MMWR Weekly. 2001;50(21):430–4. https://www.cdc.gov/mmwr/preview/mmwrhtml/mm5021a2.htm.
6. Krieger N. Embodying inequality: a review of concepts, measures, and methods for studying health consequences of discrimination. Int J Health Serv. 1999;29(2):295–352.
7. Meyer IH. Minority stress and mental health in gay men. J Health Soc Behav. 1995;36(1):38–56.
8. Mays VM, Cochran SD. Mental health correlates of perceived discrimination among lesbian, gay, and bisexual adults in the United States. Am J Public Health. 2001;91:1869–76.
9. Meyer IH. Prejudice, social stress, and mental health in lesbian, gay, and bisexual populations: conceptual issues and research evidence. Psychol Bull. 2003;129:674–97.
10. Frost DM, Lehavot K, Meyer IH. Minority stress and physical health among sexual minority individuals. J Behav Med. 2015;38:1–8.
11. Hsieh N, Ruther M. Sexual minority health and health risk factors: intersection effects of gender, race, and sexual identity. Am J Prev Med. 2016;50:746–55.
12. Zaneta MT, Kuzawa CW. Biological memories of past environments: epigenetic pathways to health disparities. Epigenetics. 2011;6(7):798.
13. Pascoe EA, Richman S. Perceived discrimination and health: a meta-analytic review. Psychol Bull. 2009;135(4):531–54.
14. Conron KJ, Mimiaga MJ, Landers SJ. A population-based study of sexual orientation identity and gender differences in adult health. Am J Public Health. 2010;100(10):1953–60.
15. Lee H, Turney K. Investigating the relationship between perceived discrimination, social status, and mental health. Soc Ment Health. 2012;2(1):1–20.
16. Hatzenbuehler ML, McLaughlin KA. Structural stigma and hypothalamic–pituitary–adrenocortical axis reactivity in lesbian, gay, and bisexual young adults. Ann Behav Med. 2014;47:39–47.
17. Kecojevic A, Wong CF, Schrager SM, Silva K, Bloom JJ, Iverson E, et al. Initiation into prescription drug misuse: differences between lesbian, gay, bisexual, transgender (LGBT) and heterosexual high-risk young adults in Los Angeles and New York. Addict Behav. Nov 2012;37(11):1289–93.
18. Everett BG, Rosario M, McLaughlin KA, Austin SB. Sexual orientation and gender differences in markers of inflammation and immune functioning. Ann Behav Med. 2014;47:57–70.
19. Ryan C, Huebner D, Diaz RM, Sanchez J. Family rejection as a predictor of negative health outcomes in white and Latino lesbian, gay, and bisexual young adults. Pediatrics. 2009;123(1):346.
20. Durso LE, Gates GJ. Serving our youth: findings from a national survey of service providers working with lesbian, gay, bisexual, and transgender youth who are homeless or at risk of becoming homeless. The Williams Institute. 2012. https://williamsinstitute.law.ucla.edu/research/safe-schools-and-youth/serving-our-youth-july-2012/. Accessed 4/1/19.
21. National Alliance to End Homelessness. 2012. LGBTQ youth national policy statement. http://www.endhomelessness.org/page/-/files/4552_file_LGBTQ_Youth_National_Policy_Statement_April_2012_Final.pdf. Accessed 4/1/19.
22. Ryan C, Russell ST, Huebner D, Diaz R, Sanchez J. Family acceptance in adolescence and the health of LGBT young adults. J Child Adolesc Psychiatry Nurs. 2010;23(4):205–13.
23. Tucker JS, Ewing BA, Espelage DL, Green HD Jr, de la Haye K, Pollard MS. Longitudinal associations of homophobic name-calling victimization with psychological distress and alcohol use during adolescence. J Adolesc Health. 2016;59(1):110–5.
24. Bowes L, Maughan B, Caspi A, Moffitt TE, Arseneault L. Families promote emotional and behavioural resilience to bullying: evidence of an environmental effect. J Child Psychol Psychiatry. 2010;51(7):809–17.
25. Calzo JP, Roberts AL, Corliss HL, Blood EA, Kroshus E, Austin SB. Physical activity disparities in heterosexual and sexual minority youth ages 12-22 years old: roles of childhood gender nonconformity and athletic self-esteem. Ann Behav Med. 2014;47(1):17–27.

26. Hadland SE, Austin SB, Goodenow CS, Calzo JP. Weight misperception and unhealthy weight control behaviors among sexual minorities in the general adolescent population. J Adolesc Health. 2014;54(3):296–303.
27. Marshal MP, Dietz LJ, Friedman MS, Stall R, Smith HA, McGinley J, et al. Suicidality and depression disparities between sexual minority and heterosexual youth: a meta-analytic review. J Adolesc Health. 2011;49(2):115–23.
28. Hatzenbuehler ML, Flores AR, Gates GJ. Social attitudes regarding same-sex marriage and LGBT health disparities: results from a national probability sample. J Soc Issues. 2017;73(3):508–28.
29. Kail BL, Acosta KL, Wright ER. State-level marriage equality and the health of same-sex couples. Am J Public Health. 2015;105(6):1101–6.
30. Hatzenbuehler ML, McLaughlin KA. Structural stigma and hypothalamic-pituitary-adrenocortical axis reactivity in lesbian, gay, and bisexual young adults. Ann Behav Med. 2014;47(1):39–47.
31. Herek GM. Hate crimes and stigma-related experiences among sexual minority adults in the United States: prevalence estimates from a national probability sample. J Interpers Violence. 2008;24(1):54–74.
32. Pew Research Center. A survey of LGBT Americans: attitudes, experiences and values in changing times. https://www.pewsocialtrends.org/2013/06/13/a-survey-of-lgbt-americans/3/. Accessed 4/1/19.
33. Kosciw JG, Greytak EA, Bartkiewicz MJ, et al. The 2011 National School Climate Survey: the experiences of lesbian, gay, bisexual, and transgender youth in our nation's schools. New York: Gay, Lesbian & Straight Education Network; 2012. https://www.glsen.org/sites/default/files/2011%20National%20School%20Climate%20Survey%20Full%20Report.pdf
34. Fredriksen-Goldsen KI. Resilience and disparities among lesbian, gay, bisexual, and transgender older adults. Publ Policy Aging Rep. 2011;21(3):3–7.
35. Emlet C. Social, economic, and health disparities among LGBT older adults. Gen: J Am Soc Aging. 2016;40(2):24–32.
36. Fredriksen-Goldsen KI, Hoy-Ellis CP, Muraco A, Goldsen J, Kim H-J. The health and well-being of LGBT older adults: disparities, risks, and resilience across the life course. In: Orel NA, Fruhauf CA, editors. The lives of LGBT older adults: understanding challenges and resilience. Washington, DC: American Psychological Association; 2015. p. 25–53. https://doi.org/10.1037/14436-002.
37. Operario D, Gamarel KE, Grin BM. Sexual minority health disparities in adult men and women in the United States: National Health and Nutrition Examination Survey, 2001–2010. Am J Public Health. 2015;105(10):e27–34.
38. Dean L, Meye IH, Robinson K, Sell RL, Sember R, Silenzio VM, et al. Lesbian, gay, bisexual and transgender health: findings and concerns. J Gay Lesbian Med Assoc. 2000;4(3):102–51.
39. Cochran SD, Mays VM, Bowen D, Gage S, Bybee D, Roberts SJ, et al. Cancer-related risk indicators and preventive screening behaviors among lesbians and bisexual women. Am J Public Health. 2001;91(4):591–7.
40. Badgett MV, Durso LE, Schneebaum A. New patterns of poverty in the lesbian, gay, and bisexual community. The Williams Institute. 2013. https://williamsinstitute.law.ucla.edu/research/census-lgbt-demographics-studies/lgbt-poverty-update-june-2013/. Accessed 4/1/19.
41. Nuyen BA, Scholz R, Hernandez RA, Graff N. LGBT health issues immersion day: measuring the impact of an LGBT health education intervention. J Invest Med. 63(1):106.
42. Múle NJ, Ross LE, Deeprose B, Jackson BE, Daley A, Travers A, et al. Promoting LGBT health and wellbeing through inclusive policy development. Int J Equity Health. 2009;8:18.
43. Crawshaw C. Sick of it: one patient's adventures in heteronormativity. In: Sharman Z, editor. The remedy: queer and trans voices on health and health care. Vancouver, BC: Arsenal Pulp Press; 2016. p. 157–64.
44. Callahan EJ, Henderson CA, Ton H, MacDonald S. Utilizing the electronic health record as a tool for reducing LGBT health disparities: an institutional approach.In: Eckstrand KL,

Ehrenfeld JM, editors. Lesbian, gay, bisexual and transgender healthcare. Springer; 2016. p. 81–91.
45. National LGBT Health Education Center. LGBT health readiness assessments in health centers: key findings. https://www.lgbthealtheducation.org/publication/lgbt-health-readiness-assessments-health-centers-key-findings/. Accessed 1 May 2019.
46. National LGBT Health Education Center. Ready, set, go! Guidelines and tips for collecting patient data on sexual orientation and gender identity (SO/GI). 2018. https://www.lgbthealtheducation.org/publication/ready-set-go-guidelines-tips-collecting-patient-data-sexual-orientation-gender-identity/. Accessed 1 May 2019.
47. Human Rights Campaign. Health care quality index. 2018. https://www.hrc.org/hei. Accessed 1 May 2019.
48. Streed C. Medical history. In: Eckstrand KL, Ehrenfeld JM, editors. Lesbian, gay, bisexual and transgender healthcare: A Clinical Guide to Preventive, Primary, and Specialist Care, Springer International, Switzerland; 2016. p. 65–80.
49. Gates GJ. LGBT parenting in the United States: UCLA: The Williams Institute; 2013. https://escholarship.org/uc/item/9xs6g8xx. Accessed 1 May 2019.
50. National LGBT Health Education Center. Providing inclusive services and care for LGBT people. https://www.lgbthealtheducation.org/publication/learning-guide/. Accessed 1 May 2019.
51. Callegari LS, Aiken AR, Dehlendorf C, Cason P, Borrero S. Addressing potential pitfalls of reproductive life planning with patient-centered counseling. Am J Obstet Gynecol. 2017;216(2):129–34. https://doi.org/10.1016/j.ajog.2016.10.004.
52. Braveman P. What are health disparities and health equity? We need to be clear. Public Health Rep. 2014;129(2):5–8. https://doi.org/10.1177/00333549141291S203.
53. Obedin-Maliver J, Goldsmith ES, Stewart L. Lesbian, gay, bisexual, and transgender–related content in undergraduate medical education. JAMA. 2011;306(9):971–8.
54. Moll J, Krieger P, Moreno-Walton L, Lee B, Slaven E, James T, et al. The prevalence of lesbian, gay, bisexual, and transgender health education and training in emergency medicine residency programs: what do we know? Acad Emerg Med. 2014;21(5):608–13.
55. Human Rights Campaign. HEI 2019 rating system and methodology. 2015. https://www.hrc.org/hei/hei-scoring-criteria. Accessed 1 May 2019.
56. Hollenbach AD, editor. Implementing curricular and institutional climate changes to improve health care for individuals who are LGBT, gender nonconforming, or born with DSD: a resource for medical educators. 1st ed. Washington, DC: Association for American Medical Colleges; 2014.
57. Snowdon S. Recommendations for enhancing the climate for LGBT students and employees in health professional schools: a GLMA white paper. Washington, DC: GLMA; 2013.
58. Haider AH, Schneider EB, Kodadek LM, Adler RR, Ranjit A, Torain M, et al. Emergency department query for patient-centered approaches to sexual orientation and gender identity: the EQUALITY study. JAMA Intern Med. 2017;177(6):819–28. https://doi.org/10.1001/jamainternmed.2017.0906.
59. Cahill S, Makadon H. Sexual orientation and gender identity data collection in clinical settings and in electronic health records: a key to ending LGBT health disparities. LGBT Health. 2014;1(1):34–41. https://doi.org/10.1089/lgbt.2013.0001.
60. Lesbian, Gay, Bisexual and Transgender Health. Healthy people 2020. https://www.healthypeople.gov/2020/topics-objectives/topic/lesbian-gay-bisexual-and-transgender-health. Accessed 3/15/19.
61. Pérez-Stable EJ. Director's message: sexual and gender minorities formally designated as a health disparity population for research purposes. Bethesda: National Institutes of Health; 2016. www.nimhd.nih.gov/about/directors-corner/message.html.
62. Office of Disease Prevention and Health Promotion. Healthy people 2020: lesbian, gay, bisexual, and transgender health. https://www.healthypeople.gov/2020/topics-objectives/topic/lesbian-gay-bisexual-and-transgender-health. Accessed 1 May 2019.
63. Kate J, Ranji U, Beamesderfer A, Salganicoff A, Dawson L. Health and access to care and coverage for lesbian, gay, bisexual, and transgender individuals in the U.S., Kaiser Foundation,

May 3 2018. https://www.kff.org/disparities-policy/issue-brief/health-and-access-to-care-and-coverage-for-lesbian-gay-bisexual-and-transgender-individuals-in-the-u-s/.
64. US Department of Health & Human Services. Access to healthcare: non-discrimination. 2017. https://www.hhs.gov/programs/topic-sites/lgbt/accesstohealthcare/nondiscrimination/index.html. Accessed 12 Feb 2020.
65. Gates, GJ. In U.S., LGBT more likely than non-LGBT to be uninsured.https://news.gallup.com/poll/175445/lgbt-likely-non-lgbt-uninsured.aspx. Accessed 4/15/19.
66. Project Implicit. https://implicit.harvard.edu/implicit/. Accessed 1 May 2019.
67. National LGBT Health Education Center. Learning to address implicit bias towards LGBTQ patients: case scenarios. 2018. https://www.lgbthealtheducation.org/publication/learning-to-address-implicit-bias-towards-lgbtq-patients-case-scenarios/. Accessed 1 May 2019.
68. Dunn M, Zorc J, Kreindler J, Tyler L, Devon P, et al. CHOP inpatient clinical pathway for treatment of the child with bronchiolitis. 2018. https://www.chop.edu/clinical-pathway/bronchiolitis-inpatient-treatment-clinical-pathway. Accessed 1 May 2019.

Competent ED Care of Gender-Diverse Patients

7

Makini Chisolm-Straker and Adrian D. Daul

Key Points

- There is not yet a formal, systematic recognition of the transgender and gender nonconforming (TGGNC) population living in the US. Like the US census, most healthcare research and electronic health records only collect binary sex or gender data so TGGNC people in the US are systematically made invisible.
- A simple, gender-affirming practice for emergency medicine (EM) clinicians is to universally ask patients their name (or how they would like to be addressed or called) and pronouns, and then use the correct names and pronouns. Gender affirmation supports patients' mental health and also protects their social safety.
- EM clinicians should be familiar with the social, medical, and/or surgical gender-affirming practices of TGGNC people and be able to competently address complications.
- TGGNC-relevant education and training of all EM staff, including nonclinical staff, will improve emergency healthcare of TGGNC patients.

M. Chisolm-Straker (✉)
Department of Emergency Medicine, Icahn School of Medicine at Mount Sinai, Mount Sinai Queens, New York City, NY, USA
e-mail: makini.chisolm-straker@mountsinai.org

A. D. Daul
Department of Emergency Medicine, Cooley Dickinson Hospital, Northampton, MA, USA

H. J. Alter et al. (eds.), *Social Emergency Medicine*,
https://doi.org/10.1007/978-3-030-65672-0_7

Foundations

Background

Transgender and gender nonconforming (TGGNC)[1] individuals have a gender identity that is different from that which was assumed or assigned, commonly based upon genitalia, at their birth. Transgender and gender nonconforming are separate gender identities, though they can overlap for the individuals who hold these identities. For example, some transgender individuals also identify as gender nonconforming. Gender identity is different from sexual orientation, though the two are often conflated or grouped together in the acronym "LGBT." TGGNC people, like cisgender people, can identify with any sexual orientation, including heterosexual, bisexual, pansexual, lesbian, or gay (Table 7.1).

Evidence Basis

TGGNC people have disproportionately high rates of negative and traumatizing life experiences that adversely impact their physical and psychological health, which often compounds their need for healthcare access. However, TGGNC individuals face significant barriers to receiving healthcare, beyond the inability to afford care [1]. Multiple studies across various medical specialties, including emergency medicine, have demonstrated that TGGNC patients experience negative interactions with healthcare practitioners and institutions [2–8]. Although most of the general ED population (92%) are satisfied with their care, TGGNC patients disproportionately report dissatisfaction with ED care [2–4, 9] and report misgendering, being mocked by ED clinicians and staff, being outed by staff, being asked inappropriate questions, and even being assaulted by care practitioners [2–4]. Some TGGNC individuals report not seeking care to avoid discriminatory interactions in yet another societal institution [2, 8].

Training about physical health, behavioral health, and social issues that affect TGGNC people is critical for clinicians yet has been traditionally left out of standard medical education curricula. The estimated number of TGGNC in the US is similar to the number of people living with HIV in the US: 1.2 million [10]. Basic information about HIV and its relevant medications and emergency complications is common knowledge among EM clinicians—it is a standard part of EM education and training. The same cannot be said of TGGNC-relevant healthcare. At the time of this writing, there is no systematic, formalized training about the care of TGGNC patients in health professional schools and residency training programs [11–14]. TGGNC patients find that their clinicians are not versed in the medical care that is relevant to gender-affirming practices, surgeries, and medications, so patients avoid or delay care, and/or have to teach their clinicians [1, 2].

---

[1] Not all "TGGNC" people have a "trans" identity. Some people whose life experience meets the definition for transgender identify exclusively with a particular gender. For example, a person who was female-assigned at birth might identify as a man (and *not* a transman). In this chapter, "TGGNC" includes people who identify as any gender that was not assigned to them at birth.

Table 7.1 Definition of terms

Term	Definition
Cisgender	Gender identity aligns with sex genitalia; gender identity matches gender assumed at birth
Intersex (or "disorders" or "differences of sex development"[a])	People whose gonadal, genital or other sex characteristic development varies from stereotypical female or male sex development; usually related to a congential difference in development
Transman (female-to-male (FTM))	Assigned female gender at birth, but gender identity is male
Gender-affirming	Behaviors or interventions that affirm a transgender person's gender identity (e.g. hormone use, choice of clothing)
Gender dysphoria	Distress that arises from incongruence between one's gender identity and one's assumed sex at birth (including physical traits and gender role)
Gender expression	Behavior, clothing, and/or personal traits that communicate gender (though expression and identity may not be the same)
Gender identity	Personal or subjective sense of self as belonging to a particular gender
Gender nonconforming/ gender nonbinary/ genderqueer	Terms for people who do not subscribe to binary "male" or "female" gender distinctions and may identify with both, neither, or a combination of male and female genders
Transwoman (male-to-female (MTF))	Assigned male gender at birth, but gender identity is female
Misgender	To incorrectly gender someone (in speech or in writing) by using the wrong name and/or pronouns.
Outing	Rather than "coming out," in which an individual purposefully tells about their gender identity (and/or sexual orientation), "outing" is when someone else tells or behaves in a way that discloses an individual's gender identity (and/or sexual orientation) without that person's express consent.
Queer	Umbrella term for people who do not identify as heterosexual and/or cisgender
Sex (Natal sex)	Genetic and physical traits associated with maleness or femaleness
Sex marker	The sex binary is used on formal identification and insurance cards, and the "F" or "M" identifies the holder as "female" or "male," respectively. "Sex marker" and "gender marker" are often used interchangeably, although sex and gender have different meanings.
Transgender	Actual gender identity does not align with gender assumed based upon genitalia at birth
Transitioning	Shifting one's gender expression to be more or less masculine/feminine
Transfeminine	Birth-assigned male individuals that identify as girls, women, or gender nonconforming
Transmasculine	Birth-assigned female individuals that identify as boys, men, or gender nonconforming
Transphobia	Prejudice against and dislike of transgender and genderqueer people
Transsexual	Historically used to refer to gender-affirming expressions of identity different than that which was assigned at birth
Two-Spirit	This term was coined in 1990, at the 3rd Annual Inter-tribal Native American, First Nations, Gay and Lesbian American Conference, and is used by some Indigenous people to communicate that they have a masculine and feminine spirit (this conceptualization includes diverse gender identities and same-gender attraction)[b].

Jalali and Sauer [47]

Printed with permission of Dr. Jalali; adapted by Drs. Adrian Daul and Makini Chisolm-Straker.

[a]In 2006, the medical community stopped using the term "intersex" and started using "Disorders of Sexual Development." But this terminology is not uniformly, or even commonly, accepted among the people to whom it is meant to be applied, as it pathologizes many people with healthy bodies. Interested readers can start to learn more here: https://ihra.org.au/allies/

[b]"Two-Spirit Community." Re:Searching for LGBTQ2S+ Health. 2019. Available at: https://lgbtqhealth.ca/community/two-spirit.php. Accessed 28 April 2019.

The following sections aim to familiarize EM clinicians with the basics of TGGNC-specific emergency medicine needs using evidence-based information when available, and otherwise, best-practice guidelines. Because inclusive gender data has not been systematically collected, potential health outcomes are largely postulated based upon expert opinion and reasoned extrapolations from cisgender data. However, clinician and patient experiences, and available evidence, indicate that gender-affirming interventions (medical, surgical, and social) improve the mental health of TGGNC people, and that overall, when performed safely, gender-affirming interventions impart more benefit than risk [15–18].

Emergency Department and Beyond

Bedside

First and foremost, emergency clinicians can foster an environment of respect and safety that ensures TGGNC patients are treated with the same dignity as any ED patient. Unfortunately, TGGNC patients consistently report negative experiences in emergency departments including overt discrimination and mockery, as well as shaming or disgust from their treatment team [2–4]. To make emergency departments safe for TGGNC people, clinicians will need to combat their own explicit and implicit biases,[2] and constructively intervene when colleagues behave in ways that perpetuate stigma. Samuels et al. identified three specific areas for improving the ED experiences of TGGNC patients: communication, privacy, and competency [3].

Communication

Communication is of the utmost import to high-quality healthcare provision and yet, when caring for TGGNC patients, clinicians may find communication challenging on many fronts. One challenge is that the emergency clinician may not know when they are caring for a TGGNC patient. Often this information is not contained in or successfully communicated via the electronic health record (EHR). Some

[2] "Bias is the application of an attitude or preconceived notion (stereotype) to form a preference toward or against something or someone, which can manifest through behavior. Bias is explicit when the holder of the bias is conscious or aware that he or she has this preference or partiality. 'Implicit' or 'unconscious bias' is an unconscious attitude or partiality that 'is not readily apparent to the individual and can differ markedly from a person's explicit and expressed beliefs' (Sabin et al., 2009)" [Chisolm-Straker & Straker, 2015]. Implicit bias can affect clinical decisions and provider-patient interactions.

Sabin, J.A., Nosek, B.A., Greenwald, A.G. and Rivara, F.P. (2009), "Physicians' implicit and explicit attitudes about race by MD race, ethnicity, and gender," Journal of Health Care for the Poor and Underserved, Vol. 20 No. 3, pp. 896–913, doi: https://doi.org/10.1353/hpu.0.0185.

Chisolm-Straker M, Straker H. (2017) "Implicit bias in US medicine: complex findings and incomplete conclusions," International Journal of Human Rights in Healthcare, Vol. 10 Iss 1 pp. 43–55.

transgender patients, whose current name, sex marker, and gender expression socially conform to their gender identity, may be treated without a practitioner being aware that the patient had a different sex assumed at birth. For example, imagine Laverne Cox is an ED patient. Ms. Cox is an actress and transwoman well known for her role on *Orange is the New Black*. Given her feminine name and appearance, she is likely to be perceived as a cisgender female by a clinician. Some TGGNC patients readily disclose their identity to clinicians, others do not. Some do not share this information if they do not discern clinical relevance or if they fear it will provoke maltreatment [1, 2]. Others who have not legally transitioned may not feel empowered to ask clinicians to use their chosen name and gender identity. Astute clinicians may be clued in about a patient's TGGNC identity when they notice a "mismatch" between the patient's name, sex marker, and the patient's gender expression, or when they recognize gender-affirming medications/surgeries/practices during a patient encounter.

If not clearly stated in the EHR, a simple and best practice for clinicians is to ask what name and pronouns a patient uses. Using the correct name and pronouns for a TGGNC patient is a basic and *profoundly* gender-affirming practice. Asking allows the clinician to avoid misgendering, which is a distressing experience for TGGNC people. Many words traditionally used to greet patients are inherently gendered including "sir," "ma'am," "Mr.," and "Ms." For a transwoman to be misgendered and addressed as "Mr." is distressing and likely to negatively impact the entire encounter. Instead, clinicians can initially address patients by last name. "Hi, is your last name 'Hanley-Okua'?" And then follow up with: "What name would you like me to call you by? What pronouns do you use?" Or alternatively, "Hi I am Dr. Jansson and I use he and him pronouns; tell me what name you go by and what pronouns you use." At times, it can be helpful to specify pronoun options: she/hers, he/his, they/theirs. Occasionally, a patient may be upset by these questions or may need some additional explanation: "These are questions that all patients are now being asked to help clinicians, like me, avoid assumptions that can have a negative effect on patients' health and healthcare."

Even when a patient's gender identity is not directly relevant to the clinical presentation, a patient's experience and willingness to seek needed care in the future will be informed by how they were addressed and treated by the medical team. When a practitioner uses an incorrectly gendered term for a patient, it is best to simply acknowledge and correct the error. Unless the patient requests otherwise, the entire clinical care team should *always* use the patient's correct name and pronouns, regardless of whether the patient is present for the discussion. This information should be relayed to all team members, including technicians, phlebotomists, consultants, and transporters. This is particularly important when the electronic health record fails to communicate the correct name and pronouns to use.

Often the very ways clinicians have been taught to elicit histories rely on assumptions that are a barrier to communication with TGGNC people. For example, a question when assessing a patient's sexual health risks such as "Are you using condoms?" makes assumptions about body parts, sexual partners, and/or sexual practices.

Practitioners can instead use more open-ended questions (e.g. "How do protect yourself from sexually transmitted infections?"). Clinicians also need to practice respectful strategies for asking about gender-affirming therapies. As part of best practice with all ED patients, the clinical relevance of any sensitive exam should be explained to the patient and their express verbal consent should be ensured before proceeding. As is always the case, patients with capacity have the right to decline any aspect of care, including parts of the exam. Readers can review the case at the end of the chapter for more guidance on how to communicate about sensitive aspects of the history and exam.

Privacy

All patients deserve privacy for their history and examination and yet, in the reality of many emergency departments, true privacy can be hard to secure. When asking about names, pronouns, and medical/surgical history, it is useful to explicitly acknowledge the lack of privacy and use a lower volume. Sensitive physical exams (including exposure of the chest for EKG) should be performed in private. Routine practices such as fully exposing polytrauma patients to examine for clandestine injury can be distressing for TGGNC patients. Ensuring privacy during the sensitive parts of the history and exam will prevent the inadvertent "outing" of a TGGNC patient to other ED patients or staff members who are not participating in the patient's care. Practitioners should bear in mind that being "outed" in public spaces poses a safety threat for TGGNC patients.

Sensitive history and exams should only be performed if they are clinically relevant. Curiosity is an inappropriate reason to ask a patient about their gender-affirming practices or conduct exams that are not clinically relevant. For example, if a patient presents with a laceration to the arm, their gender-affirming practices are not relevant to the care they should receive. Inquiring into unnecessary history can be traumatizing for the patient and risks outing the patient to those who may overhear. Moreover, it is unacceptable to use a TGGNC patient for teaching purposes without their permission, especially when the "teaching" is simply about the presence of a TGGNC person in the ED. In one study of TGGNC patients that accessed ED care, a participant shared, "I have also had doctors/nurses call over other people on duty to come look at me for no reason. It made me feel like an animal in a zoo" [2].

Competency

Most practitioners have had little training on TGGNC health and health needs [19], and TGGNC people are often put in the position of having to educate their clinicians in order to receive appropriate care [1–3]. Basic knowledge of the gender-affirming social, medical, and surgical interventions TGGNC people use is critical and necessary for EM clinicians. Practitioners must understand why TGGNC people adopt these interventions and practices: to reduce gender dysphoria and/or have their gender expression convey their gender identity. Among a population where suicide is epidemic, these practices and interventions may be life-saving and should be framed as such.

Social (Nonmedical, Nonsurgical) Gender-Affirming Practices

Transition is the social process by which a TGGNC person shifts their gender expression to align with their internal sense of gender identity. TGGNC people may express their gender in the clothing, hairstyle, cosmetics, and accessories they choose, as well as through mannerisms. Other social practices TGGNC people may adopt to modify their gender expression include binding, packing, and tucking. As part of gender transition, some TGGNC people choose to change their legal name/sex. Legally changing one's name and sex marker are two separate processes and can be an onerous challenge. Although requirements can vary by state, changing one's legal name requires a court order and often entails paperwork, fees, and/or placing notices in the newspaper to announce the name change. Many people may not be able to navigate or afford this process. Requirements to change one's sex marker also vary by state and often require a physician's note verifying that a TGGNC person has "completed" a gender transition. This formally and systematically values medical and surgical gender affirmation over social gender-affirming practices. Medical and surgical interventions are not financially and/or medically feasible for all TGGNC people, nor are they uniformly desired [20]. If and once name/gender change has been achieved, a person must then comb through all aspects of their life to update their name and gender marker, including bank accounts, credit cards, loans, titles, utilities, professional degrees, licensures, and insurances.

The costs of transitioning are enormous. Insurance does not always cover gender-affirming medical or surgical interventions. In addition to time and fiscal inputs, the emotional costs include a process that requires innumerable episodes of coming out as TGGNC and also has the potential to exact crushing social loss. Many TGGNC people who transition experience loss of job, family, and/or friends as well as threats to safety. Thus, the decision to transition is quite complex and unique for each individual. For some TGGNC people, social gender-affirming practices are the only way they desire to or have the ability to modify their gender expression.

Binding

Definition

Binding describes the practice of tightly compressing the breasts against the chest wall to create a masculine contour of the chest. For transmasculine people, binding mitigates the gender dysphoria associated with having breasts. Binding can be accomplished using a variety of materials including doubled up sports bras, elastic wrap, and commercially made binders.

Possible Complications

Binding can result in chest pain, dyspnea, broken ribs, and skin avulsion and breakdown if people bind too tightly, for too long, or with unsafe material (e.g., plastic wrap, duct tape) [21].

Appropriate Clinical Interventions

Symptomatic binding should be addressed using harm reduction principles.[3] For the person who binds, this practice has mental health and safety purposes as it mitigates gender dysphoria and allows one to blend in when accessing masculine spaces (e.g., men's restrooms). Thus, simply directing a patient to stop binding is not a safe or feasible option, and doing so communicates a lack of caring about the patient's well-being and safety. Instead, clinicians should recommend safer binding practices such as using properly fitted (not too small) sports bras or commercial binders; limiting daily use to less than eight consecutive hours; recommending against nighttime wear; and suggesting occasional "off days," in which the breast tissue is unbound.

Packing

Packing is a practice some transmasculine people adopt to create a masculine contour to the groin. Packing devices, worn to simulate male genitalia, may be commercially- or home-made. Sometimes these devices may also be designed to allow a transman to urinate while standing ("stand-to-pee" device), which may be important to safely accessing men's restrooms. In the ED, packing devices are most likely to be incidentally discovered during the exposure of a multi-trauma or critically ill patient. In this case, the packing device should be discretely stored with the rest of the patient's personal items. The use of packing devices is not commonly associated with serious medical complications.

Tucking

Definition

Tucking is a practice some transfeminine people adopt to create a more feminine appearance to the groin. This is accomplished by pushing the testicles into the inguinal canals and/or wrapping the penis between the legs. Tape or tight-fitting underwear is used to hold the genitalia in place [22].

Possible Complications

The potential risks of tucking include reduced fertility (tucking renders the testicles unable to move closer to/farther from the body to regulate temperature), fungal infections (due to naturally moist conditions in this region), skin irritation (related to repeated use/removal of tape), urinary tract infections (related to urination avoidance because urination requires one to un-tuck), and chronic pain in the genitalia.

[3] Harm reduction refers to policies, programs, and practices that aim to reduce harms associated with a behavior or action, in people that are unable or unwilling to stop. The defining features are the focus on the prevention or mitigation of harm, rather than on prevention of the behavior itself. Harm reduction focuses on and prioritizes the person, not the behavior.

Appropriate Clinical Interventions
Dance belts (designed to support and conceal the shape of male genitals) or gaffs (designed solely to conceal male genitalia) may be used to accomplish feminization of the groin contour – although perhaps not as convincingly as tucking – and also avoid some of the risks associated with tucking. Patients who have symptoms related to tucking practices can also be counseled to try alternating between dance belt/gaff and tucking.

Gender-Affirming Medications

Gender-affirming medications, including hormones, are another method used to modify gender expression. There are few instances in which an EM clinician should stop or change a patient's gender-affirming medications. Additionally, for those TGGNC patients who are boarding in the emergency department (including psychiatric emergency patients), gender-affirming medications should be continued unless there is a specific contraindication. People who identify as gender nonconforming and desire medication-assisted transition may also be on low doses of virulizing or feminizing medications described below. Many TGGNC people do not have adequate health insurance and/or do not have access to qualified licensed medical practitioners. Still, their mental health is improved with the capacity to alleviate gender dysphoria and express their gender identity. Use of medications from unlicensed providers allows TGGNC people without other means to have more agency in gender expression, but puts them at increased risk of adverse side effects if they are exposed to inappropriate formulations or dosing. Hence it is prudent to inquire about who prescribes gender-affirming medications to a TGGNC patient. A brief summary about medications follows; for in-depth information on specific medical and surgical interventions, practitioners can access "Guidelines for the Primary and Gender-Affirming Care of Transgender and Gender Nonbinary People" on the Center of Excellence for Transgender Health website [23]. Whether gender-affirming hormones expose TGGNC people to increased risk for mortality- and morbidity-related cardiovascular disease remains to be determined with certainty; the best studies to date are small and have considerable methodological limitations, consequently limiting the implications of their findings [18].

Transmasculine Gender-Affirming Medications

Bioidentical testosterone is used as a single virulizing agent. This is most often delivered as an injection (intramuscularly or subcutaneously), although topical preparations also exist. Testosterone therapy causes voice deepening, clitoral enlargement, male pattern of hair growth/loss, increased muscle mass, and cessation of menses. Polycythemia and derangements of lipid metabolism are common adverse effects.

Transfeminine Gender-Affirming Medications

The primary estrogen prescribed for feminizing therapy is 17-beta estradiol, which is the bioidentical hormone. This can be delivered via oral, topical, or injectable routes. Estrogen therapy causes breast enlargement, feminine adipose distribution, decreased erections, and testicular atrophy with potential for irreversible infertility. The most common serious adverse side effect that EM providers need to be aware of is the risk of venous thromboembolism (VTE), and still, VTE is uncommon in this population. The risk of VTE is higher with other preparations such as ethinyl estradiol, which can be found in certain oral hormonal therapies prescribed to cisgender girls and women. The risk of VTE is higher with injectable estrogen compared to oral and transdermal preparations; risk is also higher if patients are getting estrogen "off the street" or overdosing (it can be tempting to take extra hormone in hopes of speeding up one's physical transition).

In addition to estrogen, transwomen who still have testicles may also use an anti-androgen agent such as high dose spironolactone (e.g. 200–400 mg daily) to lower testosterone to desired levels. As spironolactone is a potassium-sparing diuretic, hyperkalemia is a medication effect that EM clinicians may encounter.

Gender-Affirming Surgeries

Gender-affirming surgeries are less prevalent among TGGNC people than is medication use. These surgeries are often expensive and can require a long recovery period. Some TGGNC people simply have no desire for permanent physical alteration of their body. As with any surgery, these are often more prone to complications than medical interventions. While there are a variety of "minor" procedures, the major surgeries are often grouped into "top" or breast surgeries and "bottom" or genital surgeries.

Transmasculine Gender-Affirming Surgeries

Among transmasculine people, top surgery, or double mastectomy with chest reconstruction, is the most common surgery. Bottom surgeries include hysterectomy/oophorectomy, metoidioplasty (elongation of the clitoris with or without urethral lengthening), phalloplasty (creation of a phallus), and scrotoplasty (creation of a scrotum).

Transfeminine Gender-Affirming Surgeries

Among some transfeminine people, top surgery, or breast augmentation, may be desired if there is unsatisfactory breast growth after a 1–2 years of gender-affirming hormones. Bottom surgeries include penectomy (removal of penis), orchiectomy (removal of testicles), and vaginoplasty (creation of a neovagina). Some transfeminine people may have other procedures such as facial "feminization," "feminizing" vocal cord surgery, and/or body contouring. The concept of "feminizing" implies that certain ways of looking and sounding are for women, despite the obvious fact that women have a variety of body shapes and voice pitches. Still, transfeminine people with more "feminine" physiques and voices may more easily have their gender identity respected than transfeminine people that have more "masculine" bodies or voices.

- *Facial "Feminization"*

  Surgeries may include forehead and brow reshaping, jaw and chin contouring, rhinoplasty, hairline advancement, and tracheal shaving (reduction in the visibility of the "Adam's apple").

- *Voice "Feminization"*
  In this type of surgery, the vocal cords' length is shortened to produce a higher vocal pitch. Patients may have vocal coaching with a speech and language pathologist before such surgery, to "optimize" results.
- *Body Contouring*
  Body contouring is most commonly used by transwomen who seek a commonly recognized "feminine" shape. While this can be done with clothing modifications (e.g., corsets, shapewear), it can also be done surgically. For example, surgeons may shift abdominal fat to the gluteal and hip region to create the appearance of wider hips and a narrower waist.
  One unsafe body contouring practice used among some transfeminine people is free silicone injections. Nonmedical free silicone injections into the thighs, buttocks, and/or hips are a means of immediate relief of body dysphoria symptoms for transwomen. Free silicone injections may be particularly appealing for transwomen who are unable to access gender-affirming hormones and/or surgeries for medical or financial reasons. Free silicone can cause a serious local soft tissue inflammatory reaction and, if injected intravascularly, can result in embolic disease. The common practice of using nonmedical grade silicone for these injections, as well as having injections performed by nonmedical, inexperienced practitioners, increases the risk for devastating outcomes. Silicone pulmonary embolism, which clinically presents similarly to fat emboli syndrome, can lead to dyspnea, chest pain, hemoptysis, alveolar hemorrhage, acute respiratory distress syndrome, and devastating neurologic sequelae. Care is supportive and patients experiencing sequelae often require admission to a medical intensive care unit.

Insurance coverage for gender-affirming care – medications and surgeries – is company dependent. Some private insurers cover it and others do not. For people on Medicaid, coverage varies from state to state, with a handful of states explicitly excluding coverage for gender affirmation-related care [24]. Some TGGNC people, lacking coverage for needed care, may use unlicensed or unsafe medications and/or procedures, which can have negative health impacts. That said, gender-affirming care is increasingly considered a "medical necessity" and coverage expanded significantly with the passage of the Affordable Care Act [24].

Hospital/Healthcare System

Nondiscrimination Policy

Lambda Legal and the Human Rights Campaign coauthored a landmark publication to guide best practice in hospital policies called *Creating equal access to quality health care for transgender patients: Transgender-affirming hospital policies* [25]. The first best practice they recommend is to adopt and broadly announce a hospital-wide nondiscrimination policy that explicitly includes gender identity and gender expression.

Privacy

Gender identity information should be collected in a private setting. In multiple studies, TGGNC patients have described having their gender identity revealed to other patients as registration personnel asked them about their gender identity with other patients nearby. Not only is this a breach of privacy, but also it poses safety issues including threatening patients' ability to safely access gendered spaces like restrooms. TGGNC people may fear violent attacks if their identity is revealed [26, 27]. In one study, a participant shared, "I revealed my status, which no one knows usually until I tell them....when I tried to use the woman's restroom before I left, they threatened to call security on me. It was humiliating. I would die before I went back there again" [2]. Despite misconception, transgender and genderqueer people using the restroom are substantially more likely to be attacked by cisgender people than cisgender people are to be by TGGNC people [1, 26].

Electronic Health Record & Charting

Systematic, inclusive gender documentation in health records is an important step toward evidence-based care of this patient population. Inclusive gender information should be universally collected from patients and meaningfully displayed in the health record. This can get complicated when a patient has a name and/or gender identity that conflicts with their legal documents. Achieving this goal requires a capable electronic health record and staff properly trained to input the data. Even deciding who within the healthcare team should collect or have access to this information requires careful consideration. Literature indicates that, among literate populations, patients prefer when they can enter this data themselves [28].

For the clinician documenting an encounter with a patient, it is best practice to use the patient's pronouns throughout the document. Sex assumed at birth should be referenced only when it is medically relevant, or when the patient requests that it is documented so that other care team members are aware.

Rooming

Patients should be roomed according to their gender identity. This best practice policy is also supported by the Lambda report [25]. Complaints from other patients do not constitute grounds for an exception to best-practice room assignment, akin to nondiscrimination policies that protect against racial discrimination.

Billing

An ED visit can lead to serious fiscal consequences for an uninsured patient or for a patient whose insurance company refuses to pay for care it deems inappropriate based on the legally recognized gender of the patient. For example, insurance companies have refused to cover gynecologic care for legally recognized men who were assigned a female gender at birth. Awareness of this reality can help practitioners partner with patients to find the best and safest diagnostic and treatment options.

Although the visibility of gender diverse people is on the rise, the average emergency clinician cannot be expected to be an expert in TGGNC health. Beyond ensuring a respectful environment of care, a basic familiarity with the gender-affirming social, medical, and surgical practices of TGGNC people can improve the safety, efficacy, and perception of emergency care in this patient population. Knowledge about best-practice and evidence-based care of TGGNC-patients is a core competency for all EM clinicians.

Societal Level

Lack of Systematic Recognition

There is not yet a formal, systematic recognition of the TGGNC population living in the US. Data collected about sex and/or gender are almost always binary, including only "female" and "male." The federal census only counts female and male *sexes*; in fact, the census is not concerned with *gender* at all. What do TGGNC people do? The options are to choose something that does not completely reflect their identity or experience. Not being counted perpetuates the untruth that such people do not exist. Like the US census, most healthcare research and electronic health records only collect binary sex or gender data. So, TGGNC people in the US are devalued by systematically being made invisible. This presents an ongoing barrier to an evidence-based understanding of the long-term health impact of gender-affirming interventions and how these interact with other facets of a person's health.

Systemic Discrimination, Trauma, and Poverty Negatively Impact Health

While TGGNC identities are not routinely collected in most national databases, there are an estimated 1–1.3 million TGGNC people in the US [29–31]. TGGNC people often experience discrimination in school settings, the criminal justice system, the workplace, and society at large, all of which can impact their health negatively [1, 32, 33]. TGGNC people are bullied in school, verbally harassed while using the restroom, and endure daily micro-aggressions.[4] In the 2015 US Transgender Survey, 52% of those perceived as transgender in primary school were not allowed to wear gender-affirming clothing; 12% experienced harassment when using a public restroom [1]. Violence is a common experience in this community. In a 2011 national survey of TGGNC people, 61% reported physical assault, and 64% sexual assault, at least once [32]. Law enforcement does not reliably offer supportive recourse. In fact, it is not uncommon for TGGNC people to experience trauma from law enforcement agents: 58% of 2011 national survey participants reported verbal harassment, misgendering, physical assault, or sexual assault by police officers [32]. A staggering 40% of TGGNC people report at least one suicide attempt compared to the national average of 4% [32, 34]. TGGNC people are more likely than cisgender people to use illicit substances or alcohol, and many describe their use as a way to cope with the disproportionately high rates of discrimination and trauma they experience [32].

---

[4] Microaggression: "a comment or action that subtly and often unconsciously or unintentionally expresses a prejudiced attitude toward a member of a marginalized group" (Merriam-Webster, 2019).

Despite achieving higher rates of advanced education, TGGNC people are disproportionately un- or underemployed in comparison to the general population: while 5% of the general population are unemployed, 15% of TGGNC people are unemployed [1]. Denied access to mainstream and legal employment opportunities, TGGNC people are more likely to live in extreme poverty compared to the general population. In a 2011 national survey, TGGNC study participants were four times more likely to have an annual household income of less than $10,000 in comparison to the general US population [30]. To survive, TGGNC individuals are more likely to turn to the "underground economy" (e.g., commercial sex work or selling drugs) to provide for themselves; hence they are more likely to be incarcerated [32].

TGGNC people of color disproportionately experience even higher rates of trauma and hardship than their White TGGNC peers. For example, in the 2015 US Transgender Survey, about 18% of participants lived in poverty, but nearly 42% of TGGNC people of color lived in poverty [1]. For some racial groups, the disparities are even more stark: While 1.4% of the survey's respondents were living with HIV (0.3% are living with HIV in the general US population), the rate among Black TGGNC people was 6.7%; for Black transwomen, the rate was 19% [1]. The rate of murder of TGGNC people—especially transwomen of color—is disproportionately higher than any other population [35]. Such disparities are not directly due to race or ethnicity, but result from the intersectionality of marginalized experiences. As evidenced above, systems of oppression, like racism[5] and transphobia, compound harm to individuals with membership in multiple oppressed groups.

In 2017 the National Institutes of Health (NIH) recognized gender "minorities," including transgender populations, as a health disparity population for the purposes of NIH research [36, 37]. But TGGNC people's status in the US is precarious and their rights change based upon the federal administration [38, 39]. Consequently, federal funds and practical capacity to study health issues that specifically affect this population are currently inadequate, limiting evidence-based advancements in TGGNC-healthcare.

The Old Is New Again

The presence of TGGNC people is not novel. People of all genders have been recognized for centuries [40–43]. For example, in some Indigenous nations on the North American continent, a third[6] gender exists to recognize those who do not identify solely as men or women. Such people were valued members of their communities, before European colonization and genocide. In the twentieth century, European Christian influences and impositions yielded a significant loss of community standing for (those who would today be called) Two-Spirit people [44].

[5] Racism can be observed "as a pattern of deeply entrenched and culturally sanctioned beliefs, practices, and policies which, regardless of intent, serve to provide or defend the advantages of Whites and disadvantages to groups assigned to other racial or ethnic categories" (van Ryn et al., 2011).

van Ryn, M. and Saha, S. (2011), "Exploring unconscious bias in disparities research and medical education", JAMA, Vol. 306 No. 9, pp. 995–6.

[6] Some Indigenous nations recognize more than three genders. Wilbur M, Keene A (hosts). All My Relations & Indigenous Feminism. All My Relations. Episode 1. 26 February 2019. https://www.buzzsprout.com/262196/973365-ep-1-all-my-relations-Indigenous-feminism

Marriages between Two-Spirit people and their spouses were no longer legally recognized and Two-Spirit people were forced into the gender binary. Many of those who did not conform lived in secret or killed themselves [44]. With the "gay rights"[7] and "Red Power"[8] movements that started in the 1960s, a reclaiming of culture and respect for gender diversity is reemerging.

It is largely communities of color that have documented TGGNC peoples throughout time, and TGGNC people of color are disproportionately represented among those with negative experiences and health outcomes. But, they are not representatively included in the limited health outcomes' research about TGGNC people [45]. This may be because many of these research endeavors rely on TGGNC individuals already connected to social service and healthcare organizations. Connection to services can out individuals or call attention to "otherness," which may be even more unsafe for TGGNC people of color in the US. Nonetheless, that TGGNC people have been documented throughout time lends credence to the conceptualizations of (1) gender as existing on a spectrum (not as a binary) and (2) diversity in gender identity and expression as a natural, expected phenomenon.

Recommendations for Emergency Medicine Practice

Basic

- The existence of gender diversity around the world has been documented for centuries and yet, until recently, the TGGNC population in the US has been largely invisible, including within the healthcare system, owing partly to the lack of systematic collection of data. *EM practitioners should be able to discuss how this serves to reinforce ongoing disparity and stigma, both in healthcare and at the societal level.*
- Best clinical practice is that EM clinicians universally ask patients their name (or how they would like to be addressed and called) and pronouns, and then *use the correct names and pronouns.*
- *EM clinicians need to be clinically competent in caring for TGGNC patients.* TGGNC people may use social, medical, and/or surgical gender-affirming interventions that sometimes result in health complications. EM residency curricula should include this content.
- *EM practitioners should respect the importance of gender-affirming practices and partner with patients to reduce harm, when it occurs.* Gender affirmation supports patients' good mental health, and may also protect their social safety, which impacts physical health.

[7] While the 1960s "gay rights" movement in the USA largely focused on white, cisgender gay men, its ethos theoretically included other queer people.

[8] A Native American social movement demanding self-determination of Indigenous people in the USA.

Intermediate

- *Develop written hospital policies that explicitly support TGGNC patients.* Formal hospital support normalizes clinicians' gender-inclusive and gender-affirming efforts, and facilitates TGGNC-patient safety and self-efficacy. Policy-makers can refer to the Lambda report for specific recommendations [25].
- *EHRs should capture and communicate the gender diversity of patients,* which is an important component of improving the evidence base for clinicians caring for TGGNC patients.
- *All EM staff, including nonclinical staff, should receive TGGNC-relevant education and training* to improve TGGNC-patient experiences, and ultimately make the ED a safer place for this marginalized population.

Advanced

- *EM clinician-researchers should include all genders in any research endeavors.* For example, when asking participants to share their demographics, the options might be "man," "woman," "transman," "transwoman," "nonbinary," and "other."
- *EM clinician-educators teaching medical professionals should purposefully point out false gender binaries whenever citing gender-noninclusive research.* By calling attention to the exclusion, education leaders make explicit their recognition of TGGNC people, the gaps in knowledge, and a willingness to accept gender nonbinary data.
- *EM clinicians should support state and federal legislation that improves TGGNC-patients' access to gender-affirming healthcare.* In this way, EM clinicians can improve patients' access to primary and preventative care, and may help decrease negative health impacts from the use of unlicensed practices and medications.

Teaching Case

Clinical Case

"James S."

A 34-year-old transman presents to the emergency department with a chief complaint of abdominal pain. Vital signs in the electronic medical record are a temperature of 37.9 °C, blood pressure of 127/72 mm Hg, heart rate of 87 beats per minute, respiratory rate of 18 breaths per minute, and a pulse oximeter of 97% on room air.

Practitioner *Hi, I'm Dr. Martinique and I'll be taking care of you today. What name should I call you by?*

JS *James is fine.*

JS reports one day of right lower quadrant pain with associated nausea and vomiting. He denies fever, chills, or a change in urinary/bowel habits.

Practitioner *James, I need to ask you some questions about your sexual health and gender in order to figure out what might be causing this pain.*

	I see from the electronic record that you identify as a man and were assigned "female" at birth. Is this correct?
JS	*Yep.*
Practitioner	*I see that you are on testosterone. Who prescribes the testosterone and do you take it differently than prescribed?*
JS	*I normally go to a gender clinic and they give me the injection every two weeks – I missed my last couple injections because my insurance fell through.*
Practitioner	*Have you had any gender-affirming surgeries, like top or bottom surgery?*
JS	*Just top surgery.*
Practitioner	*Can you describe who you are having sex with and what parts of your body you use during sex?*
JS	*I only have one partner. I've been with him for the past two months. We only have oral sex.*
Practitioner	*Okay, just to make sure I understand: You're sexually active with one male partner. You have sexual contact in which he puts his mouth on your genitals and vice versa.*
JS	*Yes.*
Practitioner	*How do you protect yourself from STIs?*
JS	*We don't use anything. He told me he's clean.*
Practitioner	*Have you noticed any discharge, bleeding, or pain in the lower genital or anal region?*
JS.	*No.*

Given that the patient has his natal female anatomy, the initial differential diagnosis in this case is very broad and includes ovarian cyst, ovarian torsion, pelvic inflammatory disease, tubo-ovarian abscess, and ectopic pregnancy, in addition to appendicitis.

Practitioner	*James, there are many things that could be causing your symptoms today and it's important that we look for the most serious causes. To understand what is going on, I need to examine you carefully including a lower genital exam to see if your symptoms could be coming from there. Is that okay with you?*
JS	*Yes.*
Practitioner	*Sometimes people use different words when they talk about their genitals. I am happy to use whatever words will make you most comfortable. What words do you use for your genitals?*
JS	*You can just say "vagina."*
Practitioner	*Ok. Anytime I perform a genital exam, I have to have a chaperone. What gender of chaperone would you prefer? Is there a support person you would like to be here during the exam?*

The exam is notable for an obese abdomen with tenderness in right lower quadrant and right inguinal region without guarding. On pelvic exam, there is no cervical motion tenderness or adnexal tenderness. Lab work shows leukocytosis with left shift, normal urinalysis, and negative urine HCG. CT scan shows uncomplicated appendicitis. The clinician orders antibiotics and consults general surgery. Ultimately, the patient goes to the operating room for appendectomy.

Practitioner	*Hello Dr. DeJesus, this is Dr. Martinique in the ED. I'm calling to let you know about a patient I have here with a CT-confirmed appy. Do you have a few moments to talk?*
Consultant	*Oh hi, sure, go ahead.*
Practitioner	*Great; the medical record number is A12345. Mr. JS is a 34 year old transman who came in with a day of nausea, vomiting, and right lower quad pain. Pretty classic story.*
Consultant	*I'm sorry, "transman" means what?*
Practitioner	*He was assigned a female sex at birth, but currently identifies as a man. You'll notice that the sex marker is female but he goes by "James" and uses "he/him" pronouns.*
Consultant	*Oh. Uh, are you sure this isn't her ovaries or something GYN then?*
Practitioner	***His** ovaries were assessed, and are healthy. The appendicitis is confirmed on CT; he already got antibiotics.*
Consultant	*Oh. Is there anything special we need to do for this? Do we need a GYN or medicine consult? Any chance this is an ectopic pregnancy?*
Practitioner	*The urine pregnancy test was negative. Seems to be a straightforward, uncomplicated appendicitis.*
Consultant	*I see. Ok. So…basically he has appendicitis and needs surgery; when there is a male bed available, he gets it.*
Practitioner	*Yep! He's in slot 13 in the ED. I told him to expect your team to come down to talk with him some more. I'll be here till seven. See you soon.*
Consultant	*Ok. I'll send the admitting PA. Thanks.*
Practitioner	*Take care.*

Teaching Points

1. Avoiding assumptions: How to ask about sex organs/genitalia, sexual partners, and sexual practices.

   The dialogue between the clinician and the patient demonstrates open-ended questioning to assess genital symptoms and sexual health. TGGNC patients may be uncomfortable talking about their natal anatomy using the standard medical terms (e.g., a transman may feel uncomfortable if a clinician asks about his vagina). Clinicians can use broad, unisex terms like "chest" or "lower area"/"lower genital area" to ask about symptoms in those regions. Alternatively, a practitioner can ask the patient what words they use for anatomical parts and then use those. If directly relevant to the chief complaint, the clinician should collect a surgical history to determine what anatomy the patient has.

   Often the way clinicians have been taught to elicit sexual health information is laced with assumptions about partners and behaviors. For example, asking a patient if they have sex with "men, women, or both" reinforces a binary concept of gender. The following questions recommended by the Fenway Institute can guide practitioners on how to ask sexual history questions without making assumptions [46]:

   - Are you having sex?
   - Who are you having sex with?

- What types of sex are you having? What parts of your body do you use for sex?
- How do you protect yourself from STIs?

2. How to approach a sensitive exam.

   Physical exams can be anxiety inducing for TGGNC patients. This is particularly true for chest (including electrocardiograms) and genital exams. Given that many TGGNC patients report having received unnecessary exams by practitioners, it is essential to explain the purpose of any invasive exam and ensure the patient expressly consents. Visual inspection and bimanual exam are all that is necessary for many gynecologic concerns. Practitioners should only use speculum exams with patient consent and when absolutely necessary to advance an emergency diagnosis or therapeutic intervention. These exams should take place in a private room (not in hallways). The clinician should allow a support person to stay in the exam room. If the patient is not accompanied by a support person, the clinician should ask the patient what gender chaperone they would prefer. Practitioners should use the patient's preferred terms for body parts and consider giving an anxiolytic for those with severe anxiety.

3. How to document the encounter.

   Use the patient's pronouns throughout the narrative sections of your chart.

HPI

34-year-old transgender man (on testosterone therapy, no history of gonad/genital surgery) presenting with right lower quadrant pain with associated nausea and vomiting, but no diarrhea, fevers, chills. Denies urinary symptoms, vaginal/anal discharge, pain, or bleeding. He has one male sexual partner and engages in unprotected oral sex.

Assessment and Plan

34-year-old transman with right lower quadrant pain and tenderness. My leading differential diagnostic consideration is appendicitis. Also consider GYN pathology like ruptured ovarian cyst/torsion, tubo-ovarian abscess, ectopic pregnancy, pelvic inflammatory disease but less likely given unremarkable pelvic exam. Pain control initiated. Lab work and cross-sectional imaging ordered.

Discussion Questions

1. Compare and contrast the sexual health questions used by the practitioner in this case with how you were trained to ask these questions. What hidden assumptions can you find in the ways you were trained to elicit this information?
2. Discuss how you might address a situation in which the patient's nurse seemed uncomfortable about the patient's transgender identity and consistently used female pronouns when talking about this patient.

Acknowledgments The authors would like to thank B Bradburd, MA, for his expert review of this chapter.

References

1. James SE, Herman JL, Rankin S, Keisling M, Mottet L, Anafi M. The report of the 2015 U.S. transgender survey. Washington, D.C.: National Center for Transgender Equality; 2016.
2. Chisolm-Straker M, Jardine L, Bennouna C, Morency-Brassard N, Coy L, Egemba MO, et al. Transgender and gender nonconforming in emergency departments: a qualitative report of patient experiences. Transgender Health. 2017;2(1):8–16. https://doi.org/10.1089/trgh.2016.0026.
3. Samuels EA, Tape C, Garber N, Bowman S, Choo EK. "Sometimes you feel like the freak show": a qualitative assessment of emergency care experiences among transgender and gender-nonconforming patients. Ann Emerg Med. 2018;71(2):170–182.e1. https://doi.org/10.1016/j.annemergmed.2017.05.002. Epub 2017. Jul 14. PMID: 28712604.
4. Bauer GR, Scheim AI, Deutsch MB, Massarella C. Reported emergency department avoidance, use, and experiences of transgender persons in Ontario, Canada: results from a respondent-driven sampling survey. Ann Emerg Med. 2014;63:713–20.
5. Clark TC, Lucassen MFG, Bullen P, Fleming TM, Robinson EM, et al. The health and well-being of transgender high school students: results from the New Zealand adolescent health survey (Youth'12). J Adolesc Health. 2014;55:93–9.
6. Whittle S, Turner L, Combs R, Rhodes S. Transgender EuroStudy: legal survey and focus on the transgender experience of health care. ILGA-Europe. 2008. Available at: www.ilga-europe.org/resources/ilgaeuropereports-and-other-materials/transgender-eurostudy-legalsurvey-and-focus. Accessed 24 Jan 2017.
7. Pitts MK, Couch M, Mulcare H, Croy S, Mitchell A. Transgender people in Australia and New Zealand: health, well -being and access to health services. Feminism Pyschol. 2009;19:475–95.
8. Brown JF, Fu J. Emergency department avoidance by transgender persons: another broken thread in the 'safety net' of emergency medicine care. Ann Emerg Med. 2014;63:721–2.
9. Marco CA, Davis A, Chang S, Mann D, Olson JE. ED patient satisfaction: factors associated with satisfaction with care. Am J Emerg Med. 2015;33:1708–9.
10. CDC. Prevalence of diagnosed and undiagnosed HIV infection — United States, 2008–2012. MMWR. 2015;64:657–62.
11. Liang JJ, Gardner IH, Walker JA, Safer JD. Observed deficiencies in medical student knowledge of transgender and intersex health. Endocr Pract. 2017;23:897–906.
12. Rondahl G. Students' inadequate knowledge about lesbian, gay, bisexual and transgender persons. Int J Nurs Educ Scholarsh. 2009;6:Article 11.
13. White W, Brenman S, Paradis E, Goldsmith ES, Lunn MR, Obedin-Maliver J, et al. Lesbian, gay, bisexual, and transgender patient care: medical students' preparedness and comfort. Teach Learn Med. 2015;27:254–63.
14. Obedin-Maliver J, Goldsmith ES, Stewart L, White W, Tran E, Brenman S, et al. Lesbian, gay, bisexual, and transgender-related content in undergraduate medical education. JAMA. 2011;306:971–7.
15. Wilson EC, Chen YH, Arayasirikul S, Wenzel C, Raymond HF. Connecting the dots: examining transgender women's utilization of transition-related medical care and associations with mental health, substance use, and HIV. J Urban Health. 2015;92(1):182–92.
16. Meier C, Fitzgerald KM, Pardo ST, Babcock J. The effects of hormonal gender affirmation treatment on mental health in female-to-male transsexuals. J Gay Lesbian Ment Health. 2011;15:281–99.
17. Reisner SL, Radix A, Deutsch MB. Integrated and gender-affirming transgender clinical care and research. J Acquir Immune Defic Syndr. 2016;72(Suppl 3):S235–42.
18. Streed CG, Harfouch O, Marvel F, Blumenthal RS, Martin SS, Mukherjee M. Cardiovascular disease among transgender adults receiving hormone therapy: a narrative review. Ann Intern Med. 2017;167(4):256. https://doi.org/10.7326/M17-0577.
19. Chisolm-Straker M, Willging C, Daul ADD, McNamara S, Sante SC, Shattuck DG, et al. Transgender and gender-nonconforming patients in the emergency department: what physicians know, think, and do. Ann Emerg Med. 71(2):183–8.
20. Murphy Tim. "Breaking down the binary." Brown Alumni Magazine. July/August 2018;118(6).

21. Jarrett BA, Corbet AL, Gardner IH, Weinand JD, Peitzmeier SM. Chest binding and care seeking among transmasculine adults: a cross-sectional study. Transg Health. 2018;3(1):170–8. https://doi.org/10.1089/trgh.2018.0017.
22. Erickson-Schroth L. Trans bodies, trans selves: a resource for the transgender community. Oxford, New York: Oxford University Press; 2014.
23. UCSF Transgender Care. "Guidelines for the primary and gender-affirming care of transgender and gender nonbinary people." 2016. Available at: https://transcare.ucsf.edu/guidelines. Accessed 29 Nov 2019.
24. Lambda Legal. "FAQ: equal access to health care." N.d. Available at: https://www.lambdalegal.org/know-your-rights/article/trans-related-care-faq. Accessed 29 Nov 2019.
25. Creating equal access to quality health care for transgender patients: Transgender-affirming Hospital Policies. Lambda Legal, Human Rights Campaign, Hogan Lovells, New York City Bar. Revised May 2016. Available at: https://www.lambdalegal.org/publications/fs_transgender-affirming-hospital-policies, Accessed 1 Apr2019.
26. Barnett B. Anti-trans 'bathroom bills' are based on lies. Here's The Research To Show It. *Huffington Post.* 11 Sept 2018. Retrieved on 25 Mar 2019, from https://www.huffingtonpost.com/entry/opinion-transgender-bathroom-crime_us_5b96c5b0e4b0511db3e52825.
27. What experts say. National Center for Transgender Equality. https://transequality.org/what-experts-say. Retrieved 25 Mar 2019.
28. German D, Kodadek L, Shields R, Peterson S, Synder C, Schneider E, et al. Implementing sexual orientation and gender identity data collection in emergency departments: patient and staff perspectives. LGBT Health. 2016;3(6):416–23. https://doi.org/10.1089/lgbt.2016.0069.
29. Conron KJ, Scott G, Stowell GS, Landers SJ. Transgender health in Massachusetts: results from a household probability sample of adults. Am J Public Health. 2012;102:118–22.
30. Meerwijk EL, Sevelius JM. Transgender population size in the United States: a meta-regression of population-based probability samples. Am J Public Health. 2017;107:216.
31. Crissman HP, Berger MB, Graham LF, Dalton VK. Transgender demographics: a household probability sample of US adults, 2014. Am J Public Health. 2016;107:213–5.
32. Grant JM, Mottet LA, Tanis J. Injustice at every turn, a report of the National Transgender Discrimination Survey. The National Center for Transgender Equality and the National Gay and Lesbian Task Force. 2011. Available at: www.thetaskforce.org/static_html/downloads/reports/reports/ntds_full.pdf. Accessed 24 Jan 2017.
33. Lombardi EL, Wilchins RA, Priesing D, Malouf D. Gender violence: transgender experiences with violence and discrimination. J Homosex. 2001;42:89–101.
34. Lipari R, Piscopo K, Kroutil LA, Miller GK. Suicidal thoughts and behavior among adults: results from the 2014 National Survey on Drug Use and Health. 2015. Available at: https://www.samhsa.gov/data/sites/default/files/NSDUH-FRR2-2014/NSDUH-DR-FRR2-2014.htm. Accessed 8 Jan 2019.
35. Unerased: counting transgender lives. Available at: https://mic.com/unerased?fbclid=IwAR3hkfUIJFOF0NGa6FG6ZqcGFt7_RJtry66ZXtr-bwvpYTNlurtDUr9HzLM. Accessed 24 Feb 2019.
36. National Institutes of Health, Sexual and Gender Minority Research Office. Available at: https://dpcpsi.nih.gov/sgmro. Accessed 24 Feb 2019.
37. Eliseo J. Pérez-Stable. Director's Message, 6 October 2016. "Sexual and gender minorities formally designated as a health disparity population for research purposes." National Institute on Minority Health and Health Disparities, U.S. Department of Health & Human Services. Available at: https://www.nimhd.nih.gov/about/directors-corner/messages/message_10-06-16.html. Accessed 10 Apr 2019.
38. Military Service by Transgender Individuals, Presidential Memorandum for the Secretary of Defense and the Secretary of Homeland Security. National Security and Defense. August 25, 2017. Available at: https://www.whitehouse.gov/presidential-actions/presidential-memorandum-secretary-defense-secretary-homeland-security/. Accessed 24 Feb 2019.
39. Peters JW, Becker J, Hirschfeld Davis J. "Trump rescinds rules on bathrooms for transgender students." The New York Times. 22 February 2017. Available at: https://www.nytimes.com/2017/02/22/us/politics/devos-sessions-transgender-students-rights.html. Accessed 24 Feb 2019.

40. Diavolo L. "Gender variance around the world over time." Teen Vogue. 21 June 2017. Available at: https://www.teenvogue.com/story/gender-variance-around-the-world. Accessed 14 Apr 2019.
41. Independent Lens. A map of gender-diverse cultures. 11 August 2015. Available at: http://wwwpbsorg/independentlens/content/two-spirits_map-html/ Accessed 14 Apr 2019.
42. Avery Martens. "Transgender people have always existed." ACLU Ohio. 10 June 2016. Available at: https://www.acluohio.org/archives/blog-posts/transgender-people-have-always-existed. Accessed 14 Apr 2019.
43. UNDP, USAID. Being LGBT in Asia: Thailand country report. Bangkeok. 2014.
44. Williams WL. "The 'two-spirit' people of Indigenous North Americans." 11 Oct 2010. The Guardian.
45. Reisner SL, Gamarel KE, Dunham E, Hopwood R, Hwahng S. Female-to-male transmasculine adult health: a mixed-methods community-based needs assessment. J Am Psychiatr Nurses Assoc. 2013;19:293–303.
46. Gelman M, van Wagenen A, Potter J. Principles for taking an LGBTQ-inclusive health history and conducting a culturally competent physical exam. In: Fenway guide to lesbian, gay, bisexual, and transgender health. 2nd ed. Philadelphia: American College of Physicians; 2015.
47. Jalali S, Sauer LM. Improving care for lesbian, gay, bisexual, and transgender patients in the emergency department. Ann Emerg Med. 2015. https://doi.org/10.1016/j.annemergmed.2015.02.004.

Part III

Health Care Delivery

Access to Care: Access Is a Prerequisite to Quality

8

Karin Verlaine Rhodes and Margaret E. Samuels-Kalow

Key Points

- Access to community-based care plays a key role in individuals' choice of whether and when to seek emergency department (ED) care: evidence regarding how access to non-ED care affects use of the ED remains mixed.
- Insurance coverage does not equate to access to care. There are many barriers in addition to insurance coverage that influence access to healthcare.
- Patients' access to post-ED follow-up care should contribute to emergency provider decision-making during the ED visit and at the time of disposition.

Foundations

Background

Access as a Multidimensional Concept

Penchansky and Thomas (1981) developed a theoretical model of access based on five dimensions: availability, accessibility, accommodation, affordability, and acceptability [1]. *Availability* refers to the adequacy of the supply, by volume and type, of physicians and facilities to meet demand. *Accessibility* is the relationship between the location of the healthcare providers/facilities and the location of patients, taking into account patient transportation resources, travel time, distance, and cost. *Accommodation* is the relationship between the organization of the supply, including appointment systems, hours of operation, walk-in facilities, and telephone

K. V. Rhodes (✉)
Department of Emergency Medicine, Donald and Barbara Zucker School of Medicine at Hofstra Northwell, Manhasset, NY, USA

M. E. Samuels-Kalow
Department of Emergency Medicine, Massachusetts General Hospital, Boston, MA, USA
e-mail: Msamuels-kalow@partners.org

H. J. Alter et al. (eds.), *Social Emergency Medicine*,
https://doi.org/10.1007/978-3-030-65672-0_8

services, and its appropriateness to patient demand. *Affordability* refers to the relationship of health care prices and providers' insurance requirements to patients' ability to pay and their existing health insurance. *Acceptability* is the relationship between attitudes and characteristics of patients and providers. More recently, Saurman modified the Penchansky framework and added *Awareness*, which involves effective communication about the existence and availability of a service, as well as knowledge of when and how to use it [2]. For ED patients all six of these dimensions influence both ED use and access to follow-up care after an ED visit.

Health insurance is not included in the theoretical model above, yet with few exceptions (e.g. emergency departments and free clinics), insurance coverage is a prerequisite for access. In the US in 2018, 27.5 million people (8.5%) had no health insurance at some point during the year, a slight increase from 2017 (7.9%) [3]. Private health insurance is a sector comprised of multiple companies (e.g. Kaiser; Blue Cross; Anthem). It is largely employer-based and covered 67.3% of insured individuals in 2018. Public (government funded) insurance covered 34.4% of Americans in 2018. The primary forms of public insurance in the US are Medicare and Medicaid. Medicaid is a joint federal and state-funded program for poor individuals. Medicaid eligibility varies from state to state: most states have chosen to expand Medicaid under the Affordable Care Act for individuals under 133% of the federal poverty level. Together with the Children's Health Insurance Program (CHIP), Medicaid provides health coverage to over 72.5 million Americans, including children, pregnant women, parents, and individuals with disabilities including some seniors over age 65 [4], making it the single largest source of health coverage in the US. Medicare covers individuals age 65 and above and some individuals under age 65 with specific chronic conditions (e.g., ESRD on dialysis) [4].

While a lack of insurance hinders access to health care [5] and negatively impacts health outcomes, access to care also varies based on insurance type. For example, research has shown that individuals covered by Medicaid have less timely access to care than those with Medicare or private insurance [5] and that individuals with public insurance may have worse health outcomes in some instances [6]. Patients with public insurance, such as Medicaid, face specific challenges around health care access [7]. Measures of availability or potential access assess whether sufficient resources are available for Medicaid enrollees to obtain care. This includes not only the overall supply of practitioners, but also the proportion of practitioners who participate in Medicaid, and the proportion of these practitioners who are accepting new Medicaid patients.

In this chapter we first provide historical data on the advocacy role that emergency physicians have played in informing health policy and laws that protect access to emergency care. Then we focus on the relationship between emergency care and access to the rest of the health and social services system, with particular emphasis on how emergency physicians have developed and refined methodology for measuring access to outpatient care. Finally, we discuss the ways in which emergency medicine can continue to evolve and embrace new models of care delivery and care coordination that may reduce fragmentation and improve access and quality for all patients. We define access to care broadly as the individual and community resources that impact decisions about when, whether, and how to seek care, along with the choice of care setting and the ability to connect to longitudinal sources of care following ED care delivery.

Emergency Medical Treatment and Active Labor Act (EMTALA)

EMTALA is the legacy of the 1946 Hospital Survey and Construction Act, which required hospitals accepting federal funds for modernization to provide charity care for those who were unable to pay "without consideration as to race, color, creed, national origin" [8, 9]. Despite this mandate, refusal to provide care to uninsured patients was rampant in the 1970s and 1980s. Emergency physicians brought to the public's attention that hospitals were refusing care and/or "dumping" critically ill and unstable uninsured patients into the public hospital system [10]. The fact that the majority of ED to ED transfers were of minority, low-income, or uninsured patients raised concerns of systematic discrimination. EMTALA was enacted in 1986 to address these unsafe and discriminatory practices. It requires EDs in hospitals that participate in Medicare to provide a medical screening exam (MSE) and emergency treatment for all individuals regardless of ability to pay. Hospitals can transfer patients to another ED only if it provides a higher level of care: a service that the patient requires for emergency treatment, but that is not offered at the current hospital [8]. EMTALA now extends to on-call specialists, requiring hospitals with needed specialized services to accept patient transfers. It is important to note that EMTALA does not extend to follow-up care once an emergency situation is stabilized, nor does it require hospitals to provide needed care for non-emergency conditions [11]. Importantly, EMTALA has not been sufficient to improve meaningful access to care or reduce disparities in outcomes for vulnerable populations. More work is needed to evaluate how EMTALA – and emergency medicine – can do more to address inequities in access that are associated with disparities in health outcomes [12].

Despite the nationwide mandate of EMTALA, it is important to consider that geographic access to emergency care is not equally distributed and has been shown to disproportionately impact people of color and those who are poor. For example, although shorter transport times are associated with reduced mortality [13], geographic access to trauma care is widely variable across states [14] and cities [15]. Trauma centers and hospitals in rural areas and areas with higher proportions of individuals who are black, uninsured, or poor are more likely to close [16, 17]. ED closures are associated with time delays (increased driving time) and increased mortality [18, 19].

Access to Outpatient Care and ED Utilization

Inability to access timely primary and specialty care is frequently described as a factor that influences ED use. Access barriers can include those around service provision (limited provider capacity or availability); insurance status (provider unwillingness to accept public insurance such as Medicaid, lack of insurance or challenges affording premiums or co-payments); communication (inability to reach a primary care provider in a timely fashion or language and other barriers to scheduling appointments such as lack of a telephone); transportation (lack of private/public transportation, ambulance service only able to take patients to the ED); and timing (lack of extended and weekend hours for outpatient health care services). Research has shown that people of color have unequal access to medical care compared to those who are white, and this is reflected in disparate health outcomes [20]. In addition, people of color may feel patient–provider discordance is a barrier to access [21]. Understanding the challenges patients face in accessing non-ED care is important to improve the patient experience, and to design interventions that are effective in improving access to care across the health system.

Evidence Basis

Primary Care

Many studies have attempted to determine if increasing access to primary care providers (PCPs) leads to reductions in ED use. In some studies, increased access to primary care is associated with decreased ED utilization. One study found that patients in areas with lower spatial access to primary care had higher rates of preventable ED use [22] and another found that expanded primary care access was associated with reduced ED visits [23]. Other studies have found inconclusive evidence that increasing after-hours care or access to PCPs reduces ED visits [24, 25]. For example, in California there was no consistent association between access to Federally Qualified Health Centers (FQHCs) and ED use by non-elderly adults who were uninsured or had Medicaid [26].

There have been multiple evaluations of Oregon's Medicaid expansion for childless adults via the Oregon Health Plan (OHP). One aspect of the expansion included enrollment by lottery into an "OHP Standard Plan" that imposed cost-containment strategies (primary care and ED co-payments, small monthly premiums) which could be viewed as access barriers. One evaluation found that for individuals in the Standard Plan, ED use decreased, but visit service intensity increased, so that overall expenditures were unchanged [27]. A second study also found reductions in ED use but interpreted these reductions as potential deferrals of necessary care [28]. Three studies of the effect of the overall Oregon Medicaid expansion on ED use were mixed: two found no significant change in ED visits over time [29, 30] while another found increased ED use after 18 months of enrollment across all ED visits (except those resulting in admission) [16, 17]. Several years after Medicaid expansion there was a decrease in ED use by the Medicaid population that corresponded with the start of Oregon's Coordinated Care Organizations, which actively reach out to new Medicaid beneficiaries to enroll them in primary care [31].

To date, the literature on whether expanding primary care access impacts ED use fails to support the idea that primary care expansion alone is the answer to reducing ED use and health care costs. One analysis of the National Health Service (NHS) in England found that access to expanded primary care hours (including evenings and weekends) resulted in fewer patient self-referred ED visits, but that overall ED visits were not significantly decreased. In addition, the cost of these additional primary care appointments equaled $4.8 million, whereas the 11,000 avoided ED visits were projected to cost $1.1 million [23, 32]. These results may not be generalizable to areas outside of England and the NHS, but the analysis highlights the complexity of primary care expansion as a means to reduce ED use and associated costs. One recent study found that adherence to primary care was associated with reduced ED use [33], and another found that linking uninsured patients to primary care in the ED may result in subsequent ED visit reductions [34]. A qualitative study with adults of low socioeconomic status found that many actually prefer to use the ED because they perceive it as care that is more accessible and of higher quality [35].

Pediatric Care

There have also been mixed reports on the association between access to pediatric care and ED use. In one qualitative study, primary care physicians described significant barriers in obtaining specialty care for children with public insurance. They identified use of the ED as a strategy for mitigating these barriers, as specialists were more willing to see children if they were referred from the ED [7]. Geographic provider density has been shown to be inversely related to pediatric ED use [36]. Similarly, self-reported access barriers (e.g., trouble finding a doctor or making an appointment; transportation access) are associated with multiple ED visits for children [37]. Although use of outpatient pediatric care has been associated with lower rates of ED use [38], one study showed that having an in-network pediatrician was associated with increased risk of high-frequency, low-acuity use of the pediatric ED [39]. The impact of access to pediatric providers may have differential effects based on insurance type: one study found that increasing access to pediatricians for Medicaid-insured children, compared to children with private insurance, was associated with fewer ED visits [40]. Overall, the literature to date supports the idea that increasing access to pediatric care may result in reduced ED use, although as studies in adult populations have highlighted, the impact on health care expenditures and health outcomes is an open question.

Limitations

Overall, these studies are limited by the lack of a standard definition of access. They are also limited by the variety and quality of their research methods (e.g. pre-post designs, single center studies) and the effect of temporal changes apart from primary care or coverage expansion [32], making it difficult to draw definitive conclusions. Many studies examine "low-acuity" or potentially preventable ED utilization, yet both of these concepts are difficult to define [41].

Measuring Access to Care

The most common methods of measuring access to care have been national household and physician surveys such as the National Health Interview Survey, the Medical Expenditure Panel Survey, and the National Ambulatory Care Survey. These provide valuable national estimates and state-level comparisons that can be tracked over time but have some important limitations that can lead to an underreporting of access problems, particularly for disadvantaged groups. National surveys can mask large variation in access to care across states due to state-level variations in insurance eligibility and physician payment rates. Within states there may be significant variation due to urban/rural population and provider workforce distributions. Among other major limitations of provider surveys are low response rates, and the potential for social desirability bias. For example a provider who is asked whether they accept patients with a given characteristic (e.g., transgender, minority, substance use) or insurance status might give the answer that they perceive to be more socially acceptable or might decide not to complete the survey at all. In addition, some surveyed providers may not actually know whether their clinic is accepting new Medicaid patients [42].

Emergency physicians pioneered and refined the use of simulated patient methodology (also known as mystery shopper or audit studies) for measuring underinsured patients' access to outpatient care. In the typical simulated patient study, trained supervised research assistants pose as patients contacting physician practices to request appointments. Variations of the script are used to measure whether availability of appointments differs by insurance status or other specific patient characteristics. While more complicated to execute and more expensive than physician surveys, a major advantage of simulated patient studies is that they obtain information in "real time" from office staff who book appointments in a manner that minimizes risk of social desirability and recall biases, while protecting the confidentiality of human subjects. The purpose of simulated patient methodology is not to study individuals or individual clinics but the health system itself [43]. By observing what people do, not what they say they do, researchers are able to measure real-world behavior. Further, by pairing calls and altering just one trait, the design allows researchers to rigorously control for other patient factors and examine the influence of a single variable of interest (insurance status, for instance) on access to care.

Using simulated patient methodology, the Medicaid Access Study Group found that Medicaid patients with minor health problems have few options for outpatient care and, as a result, are frequently directed to seek treatment in EDs [44]. In 2005, Asplin et al. found that in nine US cities only 64% of privately insured, 24% of uninsured, and 34% of publicly insured adults were able to get timely primary care appointments after an ED visit for a potentially life-threatening condition [45]. Access challenges were particularly profound for patients who needed mental health follow-up [46]. Bisgaier et al. found children identified as insured by Medicaid or the CHIP were offered fewer specialty care appointments (33% vs. 89%) and experienced longer wait times for these appointments (42 days vs. 20 days) compared to children with private insurance [47]. Additional simulated patient studies have verified and quantified other disparities and capacity problems in the health care delivery system that had previously been reported only anecdotally [45–47]. Notably, the US Office of the Inspector General recently endorsed these studies as the most valid and "direct means for states to monitor access in Medicaid Managed Care organizations" [48].

Simulated patient methodology is best at measuring access, defined as appointment availability, as opposed to quality. Additionally, the methodology does not identify reasons for the observed behaviors. Qualitative methods, such as in-depth interviews or focus groups, can be used to better understand the reasoning behind behaviors. Simulated patient studies are therefore best used in conjunction with other measures of access, such as information from provider and household surveys, qualitative feedback from patients, and data on actual service use.

Improving Follow-Up and Transitions in Care After an ED Visit

Some patients face significant barriers to accessing appropriate timely primary and specialty care follow-up. Although the majority of studies examining barriers to effective transitions of care from the ED have focused on being underinsured as a barrier to access, there are numerous other barriers that can make appropriate follow-up care difficult. There is increasing attention being paid to the importance of patients' perspectives on care transitions, and patient reported outcome measures have been developed to evaluate care transitions from the ED to home [49]. Work by Sabbatini et al. supports use of a patient-reported measure of transitional care, the Care Transitions Measure – 3 (CTM-3) in the ED setting, finding it to be associated with outcomes after an ED visit, including ED return visits and medication adherence [50].

Recently, with the support of the Emergency Care Coordinating Council (ECCC), the National Quality Forum (NQF) convened a multidisciplinary panel of experts to review the evidence and develop standards and guidelines for transitions of care into and out of the ED [51]. The Panel identified a set of priority measures and concepts to improve care transitions, including: (1) development of new infrastructure and linkages to support ED transitions that are patient-centered (e.g., investments in ED-based care managers, navigators, and social workers); (2) enhancements to health information technology (HIT) and interoperability between HIT systems that can support high-quality ED transitions in care and shared decision-making between providers and patients during ED transitions in care; (3) new payment models to facilitate quality improvement in ED transitions (e.g., global budgets that incentivize coordinated care); (4) a research agenda that can identify patients at highest risk for facing problems related to access and care transitions, and design and evaluate effective interventions [52]. These recommendations reinforce the need for ED providers to work within their local health systems and communities on care coordination as patients transition into and out of the ED, in an effort to reduce access barriers and the poor health outcomes associated with fragmented care [32].

Emergency Department and Beyond

Bedside

Whether patients can easily access non-ED care impacts whether they choose to seek ED care and affects patient and provider decision-making during the visit. Emergency care physicians should be aware of and attentive to local access challenges around obtaining post-ED follow-up care, especially for patients with Medicaid insurance and those who are uninsured. In general, information about insurance and primary care provision is obtained at the time of registration, although it may be incorrect. ED providers should confirm this information (including with the patient) in order to inform discharge planning. In addition, it is critically important to advocate for and coordinate with the social work and financial service resources that can assist low income patients with insurance applications, and provide realistic referrals for patients without insurance (e.g., FQHCs or community health centers) to ensure access to needed follow-up care. Some EDs have trialed "visit passes" that allow ED patients to obtain a specialist visit within the same healthcare system without insurance. Others have created post-ED clinics to provide follow-up care directly [50]. Case management and care coordinators can be used to address multiple types of access barriers (e.g., obtaining transportation, establishing insurance coverage), and interpreter services are also critical for patients with limited English proficiency. ED providers should understand the local availability of those resources and how health-related social needs can inhibit a patient's access to both health care and social services.

Hospital/Healthcare System

Some public and private insurers have attempted to reduce low-acuity ED use by increasing the cost of ED care to patients, in anticipation that this would encourage use of other sources of care. Introduction of copayments has been associated with small decreases in ED use, without changes in overall expenditures [27]. A study examining randomization to different health insurance plans found that cost sharing did not decrease potentially inappropriate hospitalization [53]. Increasing ED visit costs may reduce some low acuity ED use, but may also reduce high acuity ED use and increases barriers to ED utilization for *needed* care for the most vulnerable patients. In addition, inconsistency in the literature around ED use and access to outpatient care suggests that simply increasing the availability of clinics is unlikely to significantly reduce ED utilization [23, 32].

Potential improvements could include having a same day co-located primary care setting to address low acuity conditions and improving the value of acute unscheduled care in the ED by linking patients to more comprehensive health and social services during their visit [54]. ED visits can be used to screen for and address barriers to access, which are closely tied to an individual's ability to adhere to a treatment plan and to their overall health. Currently, functions such as care coordination are not covered by traditional payment models, and, because they are time consuming, may be discouraged by productivity-based metrics [55]. Regardless of insurance status, communication between ED care and post-ED care providers is fragmented. However, this is particularly true for patients who are underinsured, e.g., uninsured or have Medicaid insurance [56]. Studies have identified that underinsured patients can face formidable barriers to accessing needed follow-up care, including being directed to public hospital EDs without transfer paperwork or records [57]. Hospitals, healthcare systems, health insurers, and state health agencies need to ensure the availability of needed post-ED follow-up care both for privately and publicly insured patients and for those without insurance. Subspecialty clinics, such as asthma or sickle cell centers, may be able to efficiently address the needs of patients with exacerbations of common chronic specialty conditions [52].

Similarly, post-ED care specialty clinics for wound care, trauma, or orthopedic injuries are a potential model to increase access and improve care transitions after acute injuries. However, new models are needed to improve coordination between EDs and on-going community-based social services and primary care.

The increased use of health information technology could be used to assist with coordinating follow-up care for patients [55]. A growing number of emergency physicians are involved the use of telemedicine to increase access to acute and emergency care at lower cost [58]. Emergency and outpatient physicians should work together to improve bidirectional notification and communication and make it easy for both ED and outpatient providers to work with patients and families regarding hospitalization decisions. This involves the development of easy-to-use interoperable HIT and systems for information exchange, as well as new models of care that could include hospital ED discharge centers that have appropriate screening, social work, and care management resources that can support the ED provider, assist with follow-up after the visit, and improve overall hospital quality and efficiency [59].

In addition, there are also other emergency care innovations with the potential to improve access to the most appropriate care at the right time and place. Specifically, there are new emerging roles for Emergency Medical Technicians (EMTs) and community paramedics. The Center for Medicare and Medicaid Innovation (CMMI) has just proposed a payment reform demonstration project called Emergency Medical Triage, Treatment and Transport (ET3) [60]. To date, most ambulance services are only reimbursed if they transport patients directly to an ED, even if the patient preference or condition suggests that other dispositions may be medically appropriate and more patient-centered. The ET3 demonstration model will allow emergency medical systems (EMS) personnel to assess and transport patients to the most appropriate setting, for example primary care settings, urgent care settings, community health centers, or dialysis centers. Alternatively, patients can be assessed and treated in their own homes, with the use of telemedicine and appropriate medical control [60]. These innovations in emergency care have the potential to increase access, improve quality, and reduce the cost of acute care, but will require careful assessment for safety and clinical outcomes.

Societal Level

Ultimately, access to health care is a societal and political issue, as evidenced by current and ongoing debates about how to increase access to care via healthcare reform. Emergency providers have traditionally been advocates for increased access. On federal, state, and local levels, emergency physicians have been active in collecting and publishing data measuring and monitoring access to care and documenting the impact of lack of access on patient health. We need to also be active in translating that data for policy makers, including providing stories of the human impact of current policies. In addition, reducing the proportion of uninsured patients and financially incentivizing outpatient physicians to accept Medicaid and provide care for the uninsured would improve access to care following an ED visit. Advocacy to avoid closing EDs in already vulnerable neighborhoods and for ongoing federal support for local community health centers and FQHCs is important for reducing geographic barriers.

In addition, as emergency physicians, we need to advocate for the environmental and place-based public health interventions and innovations that affect access and thus health outcomes. For example, a study of new Medicaid enrollees found low walkability to neighborhood resources was associated with decreased odds of having a usual source of care [61]. Improved neighborhood transportation links and accessibility of outpatient care may improve access to non-ED sources of care. In summary, emergency medicine must take a leadership role in shaping healthcare system delivery reform while assuring affordability, quality, and patient safety. In doing so, we must focus on improving access to appropriate and timely healthcare, which will help to remedy current inequities and disparities in health outcomes.

Recommendations for Emergency Medicine Practice

Basic

- Understand local barriers to accessing primary and specialty care and consider that race, ethnicity, and immigration status may negatively affect this access.
- Identify patients' insurance coverage and presence or absence of a primary care provider as a part of discharge planning.
- Know about free or low-cost accessible programs for post-ED primary and subspecialty care within the local health system.
- Advocate for and involve ED social work and discharge coordinators to facilitate appropriate ED follow-up care.
- Inform patients about interpreter service resources for scheduling outpatient appointments.

Intermediate

- Use case management, care coordination, and other institutional resources to screen for and address patients' health-related social needs, such as transportation and insurance, which are critical to ensuring access to outpatient care.
- Advocate for reducing barriers to ED access and post-ED care, for example, geographically equitable ED access and improved mental health follow-up resources, in the local community.

Advanced

- Measure or be aware of local measures of access for both primary and specialty care.
- Develop integrated systems of care to help patients access timely, high-quality care across the care continuum.
- Create initiatives that facilitate the ED as a connector to social resources, care coordination, and community-based organizations.

Teaching Case

Clinical Case

E.C. is an 8–month-old female who presented to the ED on a winter evening. Her mother speaks Spanish and the history was obtained through a phone interpreter. E.C. was a full-term, previously healthy child with one day of fever, with a maximum temperature of 38.5°C. Her mother noted slightly decreased oral intake for solids, but she was still drinking liquids with normal urine output. Her mother noted a fever in the afternoon and took her to the local health center, which has an "urgent care" walk-in area for pediatric patients during the day. The center told her mother that she could no longer be seen there. She was told to present to the ED.

E.C.'s mother was unable to explain to the ED staff why the clinic would not see her child. Initially, the clinical team thought it was because the child was thought to be too ill for a clinic visit, and she was triaged immediately to an ED room. In the ED, E.C.'s vital signs were notable for fever but otherwise within normal limits for age. Her exam demonstrated no focal source of fever, and was otherwise reassuring.

Ultimately, the clinical team called and discovered that her insurance was now part of a new plan that the clinic would no longer accept. Her fever was thought to be most likely due to a viral syndrome. She was treated with acetaminophen with improvement in her fever, drank well in the ED, and the team planned to discharge her with PCP follow-up. Because she had only one day of fever, the clinical team felt she did not need urine screening for a urinary tract infection that day, but that she would need it if the fever persisted. However, because of concerns that she would not be able to access timely PCP follow-up, the decision was made to place a urine bag for urinalysis testing in the ED. The urine bag was not filled until 3 h later. One hour following the urine collection, the urine result was negative for infection. The mother expressed considerable frustration at the prolonged ED stay. Ultimately, the patient was referred to the hospital's pediatric resident clinic to obtain a new PCP.

Teaching Points

1. Changing insurance plans and networks of care can make it challenging for patients, even those with insurance, to access timely healthcare outside of the ED. Families with limited English proficiency or limited health literacy may be at particular risk for communication challenges around coverage and treatment networks.
2. The clinical team felt that urinary testing was not needed at the original visit, but would be needed if she had persistent high fevers. Because they were concerned that she would not be able to be seen in clinic in the next 24–48 h, the testing was performed in the ED. The lack of access to follow-up care led to potentially unnecessary testing in the ED and a prolonged length of stay.
3. Current guidelines recommend renal and bladder ultrasound for young pediatric patients with their first urinary tract infection. If her urine testing had been positive, it would have been challenging for the team to obtain this important

outpatient test, and she probably would have had to remain in the ED for it—further increasing length of stay.

4. Efforts to reduce low-acuity visits are unlikely to be successfully resolved without attention to the personal, social, economic, and health system drivers of ED use and addressing the significant challenges to timely follow-up.

Discussion Questions

1. What are the best strategies for soliciting reasons why a patient or family chose to come to the ED (rather than other care locations) without sounding pejorative or judgmental?
2. What should the insurance provider/health system network change to prevent these situations happening in the future?
3. In your local institution, what access problems do you encounter for your patient population? What resources are available?

References

1. Penchansky R, Thomas JW. The concept of access: definition and relationship to consumer satisfaction. Med Care. 1981;19(2):127–40.
2. Saurman E. Improving access: modifying Penchansky and Thomas's theory of access. J Health Serv Res Policy. 2016;21(1):36–9.
3. Berchick ER, Barnett JC, Upton RD. Health insurance coverage in the United States: 2018. 2019. Available from: https://www.census.gov/library/publications/2019/demo/p60-267.html.
4. Medicaid Eligibility. Available from: https://www.medicaid.gov/medicaid/eligibility/index.html.
5. Health Insurance and Access to Care 2017. Available from: https://www.cdc.gov/nchs/data/factsheets/factsheet_hiac.pdf.
6. Sittig MP, Luu M, Yoshida EJ, Scher K, Mita A, Shiao SL, et al. Impact of insurance on survival in patients < 65 with head & neck cancer treated with radiotherapy. Clin Otolaryngol. 2020;45(1):63–72.
7. Rhodes KV, Bisgaier J, Lawson CC, Soglin D, Krug S, Van Haitsma M. "Patients who can't get an appointment go to the ER": access to specialty care for publicly insured children. Ann Emerg Med. 2012;61(4):394–403.
8. Emergency Medical Treatment & Labor Act (EMTALA) Centers for Medicare & Medicaid Services.
9. Hospital Survey and Construction Act of 1946 Title VI of the Public Health Service Act, Pub. L. No. 79-725, 60 Stat. 1040 (1946) (codified, as amended, at 42 U.S.C. §§ 291-291o (1976)).
10. Ansell DA, Schiff RL. Patient dumping. Status, implications, and policy recommendations. JAMA. 1987;257(11):1500–2.
11. Rosenbaum S. The enduring role of the emergency medical treatment and active labor act. Health Aff (Millwood). 2013;32(12):2075–81.
12. Rhodes KV, Smith KL. Short-term care with long-term costs: the unintended consequences of EMTALA. Ann Emerg Med. 2017;69(2):163–5.
13. Feero S, Hedges JR, Simmons E, Irwin L. Does out-of-hospital EMS time affect trauma survival? Am J Emerg Med. 1995;13(2):133–5.
14. Wei R, Clay Mann N, Dai M, Hsia RY. Injury-based geographic access to trauma centers. Acad Emerg Med. 2019;26(2):192–204.
15. Wandling M, Behrens J, Hsia R, Crandall M. Geographic disparities in access to urban trauma care: defining the problem and identifying a solution for gunshot wound victims in Chicago. Am J Surg. 2016;212(4):587–91.

16. Hsia RY, Shen YC. Rising closures of hospital trauma centers disproportionately burden vulnerable populations. Health Aff (Millwood). 2011;30(10):1912–20.
17. Hsia RY, Srebotnjak T, Kanzaria HK, McCulloch C, Auerbach AD. System-level health disparities in California emergency departments: minorities and Medicaid patients are at higher risk of losing their emergency departments. Ann Emerg Med. 2012;59(5):358–65.
18. Shen YC, Hsia RY. Association between emergency department closure and treatment, access, and health outcomes among patients with acute myocardial infarction. Circulation. 2016;134(20):1595–7.
19. Crandall M, Sharp D, Unger E, Straus D, Brasel K, Hsia R, et al. Trauma deserts: distance from a trauma center, transport times, and mortality from gunshot wounds in Chicago. Am J Public Health. 2013;103(6):1103–9.
20. Peng RB, Lee H, Ke ZT, Saunders MR. Racial disparities in kidney transplant waitlist appearance in Chicago: is it race or place? Clin Transpl. 2018;32(5):e13195. https://doi.org/10.1111/ctr.13195.
21. Malhotra J, Rotter D, Tsui J, Llanos AAM, Balasubramanian BA, Demissie K. Impact of patient-provider race, ethnicity, and gender concordance on cancer screening: findings from medical expenditure panel survey. Cancer Epidemiol Biomark Prev. 2017 Dec;26(12):1804–11. https://doi.org/10.1158/1055-9965.EPI-17-0660.
22. Fishman J, McLafferty S, Galanter W. Does spatial access to primary care affect emergency department utilization for nonemergent conditions? Health Serv Res. 2018;53(1):489–508.
23. Whittaker W, Anselmi L, Kristensen SR, Lau YS, Bailey S, Bower P, et al. Associations between extending access to primary care and emergency department visits: a difference-in-differences analysis. PLoS Med. 2016;13(9):e1002113.
24. Ismail SA, Gibbons DC, Gnani S. Reducing inappropriate accident and emergency department attendances: a systematic review of primary care service interventions. Br J Gen Pract. 2013;63(617):e813–20.
25. Solberg LI, Maciosek MV, Sperl-Hillen JM, Crain AL, Engebretson KI, Asplin BR, et al. Does improved access to care affect utilization and costs for patients with chronic conditions? Am J Manag Care. 2004;10(10):717–22.
26. Nath JB, Costigan S, Lin F, Vittinghoff E, Hsia RY. Access to federally qualified health centers and emergency department use among uninsured and medicaid-insured adults: California, 2005 to 2013. Acad Emerg Med. 2019;26(2):129–39.
27. Wallace NT, McConnell KJ, Gallia CA, Smith JA. How effective are copayments in reducing expenditures for low-income adult Medicaid beneficiaries? Experience from the Oregon health plan. Health Serv Res. 2008;43(2):515–30.
28. Lowe RA, McConnell KJ, Vogt ME, Smith JA. Impact of Medicaid cutbacks on emergency department use: the Oregon experience. Ann Emerg Med. 2008;52(6):626–34.
29. Baicker K, Taubman SL, Allen HL, Bernstein M, Gruber JH, Newhouse JP, et al. The Oregon experiment—effects of Medicaid on clinical outcomes. N Engl J Med. 2013;368(18):1713–22.
30. Finkelstein A, Taubman S, Wright B, Bernstein M, Gruber J, Newhouse JP, et al. The Oregon health insurance experiment: evidence from the first year. Q J Econ. 2012;127(3):1057–106.
31. Oregon health system transformation: CCO metrics 2017 final report: Oregon Health Authority; 2018 [6/20/19].
32. Basu S, Phillips RS. Reduced emergency department utilization after increased access to primary care. PLoS Med. 2016;13(9):e1002114.
33. Pourat N, Davis AC, Chen X, Vrungos S, Kominski GF. In California, primary care continuity was associated with reduced emergency department use and fewer hospitalizations. Health Aff (Millwood). 2015;34(7):1113–20.
34. Kim TY, Mortensen K, Eldridge B. Linking uninsured patients treated in the emergency department to primary care shows some promise in Maryland. Health Aff (Millwood). 2015;34(5):796–804.
35. Kangovi S, Barg FK, Carter T, Long JA, Shannon R, Grande D. Understanding why patients of low socioeconomic status prefer hospitals over ambulatory care. Health Aff (Millwood). 2013;32(7):1196–203.

36. Mathison DJ, Chamberlain JM, Cowan NM, Engstrom RN, Fu LY, Shoo A, et al. Primary care spatial density and nonurgent emergency department utilization: a new methodology for evaluating access to care. Acad Pediatr. 2013;13(3):278–85.
37. Taylor T, Salyakina D. Health care access barriers bring children to emergency rooms more frequently: a representative survey. Popul Health Manag. 2019;22(3):262–71.
38. Johnson WG, Rimsza ME. The effects of access to pediatric care and insurance coverage on emergency department utilization. Pediatrics. 2004;113(3 Pt 1):483–7.
39. Samuels-Kalow ME, Bryan MW, Shaw KN. Predicting subsequent high-frequency, low-acuity utilization of the pediatric emergency department. Acad Pediatr. 2017;17(3):256–60.
40. Piehl MD, Clemens CJ, Joines JD. "Narrowing the Gap": decreasing emergency department use by children enrolled in the Medicaid program by improving access to primary care. Arch Pediatr Adolesc Med. 2000;154(8):791–5.
41. Raven MC, Lowe RA, Maselli J, Hsia RY. Comparison of presenting complaint vs discharge diagnosis for identifying "nonemergency" emergency department visits. JAMA. 2013;309(11):1145–53.
42. Coffman JM, Rhodes KV, Fix M, Bindman AB. Testing the validity of primary care physicians' self-reported acceptance of new patients by insurance status. Health Serv Res. 2016;51(4):1515–32.
43. Rhodes KV, Miller FG. Simulated patient studies: an ethical analysis. Milbank Q. 2012;90(4):706–24.
44. Medicaid Access Study G. Access of Medicaid recipients to outpatient care. N Engl J Med. 1994;330(20):1426–30.
45. Asplin BR, Rhodes KV, Levy H, Lurie N, Crain AL, Carlin BP, et al. Insurance status and access to urgent ambulatory care follow-up appointments. JAMA. 2005;294(10):1248–54.
46. Rhodes KV, Vieth TL, Kushner H, Levy H, Asplin BR. Referral without access: for psychiatric services, wait for the beep. Ann Emerg Med. 2009;54(2):272–8.
47. Bisgaier J, Rhodes KV. Auditing access to specialty care for children with public insurance. N Engl J Med. 2011;364(24):2324–33.
48. Access to care: provider availability in medicaid managed care. Office of Inspector General; U.S. Department of Health and Human Services; 2014.
49. Samuels-Kalow ME, Rhodes KV, Henien M, Hardy E, Moore T, Wong F, et al. Development of a patient-centered outcome measure for emergency department asthma patients. Acad Emerg Med. 2017;24(5):511–22.
50. Sabbatini AK, Gallahue F, Newson J, White S, Gallagher TH. Capturing emergency department discharge quality with the care transitions measure: a pilot study. Acad Emerg Med. 2019;26(6):605–9. https://doi.org/10.1111/acem.13623. Epub 2019 May 23. PMID: 30256486.
51. Emergency department transitions of care: a quality measurement framework: national quality forum 2017.
52. Teach SJ, Crain EF, Quint DM, Hylan ML, Joseph JG. Improved asthma outcomes in a high-morbidity pediatric population: results of an emergency department-based randomized clinical trial. Arch Pediatr Adolesc Med. 2006;160(5):535–41.
53. Siu AL, Sonnenberg FA, Manning WG, Goldberg GA, Bloomfield ES, Newhouse JP, et al. Inappropriate use of hospitals in a randomized trial of health insurance plans. N Engl J Med. 1986;315(20):1259–66.
54. Roy S, Reyes F, Himmelrich S, Johnson L, Chokshi DA. Learnings from a large-scale emergency department care management program in New York City. NEJM Catalyst. 2018;4(1). https://catalyst.nejm.org/ed-care-management-program-nyc/.
55. Medford-Davis LN, Marcozzi D, Agrawal S, Carr BG, Carrier E. Value-based approaches for emergency care in a new era. Ann Emerg Med. 2017;69(6):675–83.
56. Medford-Davis LN, Lin F, Greenstein A, Rhodes KV. "I broke my ankle": access to orthopedic follow-up care by insurance status. Acad Emerg Med. 2017;24(1):98–105.

57. Medford-Davis LN, Prasad S, Rhodes KV. "What do people do if they don't have insurance?": ED-to-ED referrals. Acad Emerg Med. 2018;25(1):6–14.
58. American College of Emergency Physicians: Emergency Teleheath section webpage: https://www.acep.org/how-we-serve/sections/telehealth/. Accessed 15 Aug 2020.
59. Rhodes KV. Completing the play or dropping the ball?: The case for comprehensive patient-centered discharge planning. JAMA Intern Med. 2013;173(18):1723–4.
60. Emergency Triage, Treat, and Transport (ET3) Model: Centers for Medicare & Medicaid Services 2019.
61. Chaiyachati KH, Hom JK, Hubbard RA, Wong C, Grande D. Evaluating the association between the built environment and primary care access for new Medicaid enrollees in an urban environment using Walk and Transit Scores. Prev Med Rep. 2018;9:24–8.

9 Frequent Emergency Department Use: A Social Emergency Medicine Perspective

Maria C. Raven and Hemal K. Kanzaria

Key Points

- Frequent ED users are a heterogeneous population often experiencing unmet health and social needs.
- Rather than view ED visits in a negative light, health care systems can also consider the acute care system as a potential point of intervention – failing to address patients' underlying health and social needs may represent a missed opportunity to affect change.
- Coordination of care within and outside the health care system is crucial for frequent ED users.
- While case management and permanent supportive housing interventions have been found to be effective at reducing ED use, additional promising approaches leveraging information exchange technology and multi-disciplinary teams are currently being tested.

M. C. Raven
Department of Emergency Medicine, University of California, San Francisco, San Francisco, CA, USA
e-mail: Maria.raven@ucsf.edu

H. K. Kanzaria (✉)
Department of Emergency Medicine, University of California, San Francisco, San Francisco, CA, USA

Department of Care Coordination, Zuckerberg San Francisco General Hospital, San Francisco, CA, USA
e-mail: hemal.kanzaria@ucsf.edu

H. J. Alter et al. (eds.), *Social Emergency Medicine*,
https://doi.org/10.1007/978-3-030-65672-0_9

Foundations

Background

Frequent emergency department (ED) use is a critical area of interest for policy makers, payers, and clinicians. The topic has been covered widely in the lay press as problematic and is seen by many as representative of a broken health care system, a contributor to high health care costs, and evidence of gaps in the community and societal care. With few exceptions, frequent ED use is the result of underlying individual and population-level health and social needs that remain unmet. These needs are heterogeneous and can include distinct challenges for patients such as an inability to fill prescriptions, poor housing conditions causing exacerbation of underlying chronic disease, pain or malnutrition related to terminal illness, and untreated substance use disorder or mental illness.

Frequent ED use is variably defined. One review found 16 different definitions of frequent use among 31 studies [1]. Most studies defined a frequent ED user as an individual with at least 3–5 ED visits in a year. ED "super users" or "ultra-high" users have been defined as individuals with levels of annual use ranging from 15 to over 20 visits in a year [2]. And, recent research and policy has focused on a small subset of individuals with extreme levels of annual ED use that can extend for over a decade [3]. Regardless of the definition, frequent ED users account for a disproportionate number of ED visits overall [2].

Frequent ED user definitions are complicated by a few factors. First, many frequent ED users access care at more than one ED. Depending on the data source used to tally ED visits, all visits may not be accounted for, which can prevent providers and researchers from counting the full spectrum of visits by an individual patient [3]. Secondly, while the term "frequent ED user" focuses on ED use, most frequent ED users also access outpatient and inpatient hospital care, as well as other systems including substance use and mental health care, housing, and jails [2, 4]. Examining ED use in isolation does not address this larger picture. Moreover, while prior work has largely focused on ED reduction interventions for frequent ED users, it has rarely focused on understanding and addressing the underlying social and medical needs of these patients.

Evidence Basis

There is a large body of research characterizing frequent ED users, in part because their disproportionate use of care is of interest to policy makers and providers. While frequent ED users represent 4–8% of ED patients, they account for 21–28% of all ED visits [5], and can generate significant costs associated with the use of EDs, inpatient hospitalizations, and other types of acute care services. While researchers need to create categories and cut-offs to define frequent ED use for the purpose of data analyses [6], from a clinical standpoint most frequent ED use is a marker of the complexity of illness and/or unmet health and social needs [7] that can vary greatly from patient to patient. Compared to non-frequent ED users, individuals who use EDs frequently have been shown to have higher rates of underlying chronic medical conditions, mental health diagnoses, and substance use disorders [2, 8]. While there are some health and social characteristics that occur more commonly in frequent ED users compared to non-frequent users, the population is heterogeneous [9]. Whether an individual is a frequent ED user is associated with factors such as insurance type [10], access to outpatient care [11], and underlying medical conditions. For example, many studies have found that patients with public insurance have higher rates of frequent ED use than individuals who are privately insured [7]. Limited research has begun to focus on identifying social needs among frequent ED users and has found high rates of homelessness, a key social need that to date has not been well captured in administrative data but is often used for research [12, 13].

Increasingly, researchers recognize that frequent ED users often visit more than one ED [10]. While there has been only a limited amount of research in this area, technology platforms that allow data sharing across multiple EDs indicate that many frequent ED users travel to multiple sites and in some cases, geographic regions [3, 14]. Not all frequent ED users in 1 year will remain frequent ED users the next. Many studies of frequent ED users discuss the phenomenon of regression to the mean [3, 9], or the tendency in observational research for outliers to move to the center of the distribution over time. This is very important, and can also be difficult to measure accurately if data are limited to a single hospital or health system.

Because health care resources are limited, predicting who will become and remain a frequent ED user can help with developing and targeting interventions appropriately. Of all factors that have been examined to date, prior frequent use is the strongest predictor of future frequent use [2, 3]. However, there are differences among individuals with frequent ED use for a short period of time compared to those who are persistent frequent ED users for five to 10 consecutive years. For example, persistent frequent ED users are more likely to have a mental health diagnosis, make higher numbers of ED visits, visit more EDs, and be publicly insured [3]. Patterns of frequent ED use can also help to predict mortality: those with frequent ED use in the past year are more likely to die in both the short and long terms [15].

Given the heterogeneity and complexity of frequent ED users, there is no "off-the-shelf" intervention to improve or coordinate their care. Two recent comprehensive reviews of the literature on ED visit reduction programs [16, 17], which included programs with published data on interventions for frequent ED users, found that only case management interventions were effective at reducing ED use. This conclusion was based on a limited number of small studies, and most of the published literature on this topic has critical limitations. One recent study showed that short-term case management reduced ED visits and hospital admissions, and increased use of primary care among publicly insured frequent ED users, although individuals whose primary issues were substance use disorder and severe mental illness were excluded from the trial, and the investigators had access to data from only two EDs [18]. Multiple studies have examined the impact of permanent supportive housing (PSH) on health services use, yet few of these studies have focused specifically on frequent ED users. While some quasi-experimental studies have found that provision of PSH reduces the use of acute health services [19], other higher quality studies have not demonstrated net cost savings for PSH. Many individuals experiencing homelessness who are frequent ED users have serious underlying medical and behavioral health conditions and experience high mortality rates compared to similarly aged individuals who are not homeless. As a result, reductions in health services use and associated cost savings may not be a realistic outcome of PSH provision, and a focus on ED visit reductions as a primary outcome could undermine the opportunity to provide this much-needed intervention [20]. ED visit reduction may be a difficult outcome to achieve, especially in the short term, for frequent ED users, and additional outcomes including connection to primary care, sustained substance use and mental health treatment, service use outside of the health care system, and other measures of wellness must also be considered.

Emergency Department and Beyond

Bedside

Caring for patients who visit the ED frequently can feel challenging for emergency clinicians. As mentioned above, many have complex social needs that EDs are not currently well-designed to address [18]. In addition, patients who return for similar complaints again and again can create a sense for providers that preventing future visits is futile or, dangerously, that such patients are not ill. Frequent ED use has been found to be an independent predictor of short-term mortality [15], so the medical concerns of such patients should not be minimized.

Taking the time to identify and intervene around patients' social needs—often on top of the presenting medical or behavioral health complaint—may be challenging for busy EM clinicians due to time constraints. Partnering with supporting providers and staff including ED social workers and health care navigators may be needed to most effectively care for individuals with frequent ED use, and to provide care that address individuals' whole-person needs. While some ED clinicians may feel this is "not the ED's job," frequent ED users can be a captive audience during a long ED stay or hospitalization. This time can and should be used to identify social needs and engage patients with community-based resources.

In order to effectively care for frequent ED users, ED clinicians must be able to accurately identify them during their visits. Yet while many ED clinicians may think they "know" their frequent ED users, research reveals that frequent ED users travel to more than one ED, so that some individuals who are frequent ED users may not be identified as such if they visit multiple different EDs [3]. Levels of ED use can be severely underestimated if not accounting for a patient's entire universe of ED visits. Information technology solutions that allow emergency clinicians to see their patients' ED visits outside of their own hospital such as the Emergency Department Information Exchange (EDie) or EPIC Care Everywhere make this evident [21–23]. By linking patients across hospitals, a study of California state-wide data detected nearly 50% more frequent ED users than methodologies without record linkage would have found [3].

Why does this matter? Coordination of care within and outside of the health care system is crucial for frequent ED users. Few programs have proven effective at reducing ED visits among frequent ED users. However, the most promising programs are those that attempt to care for the whole individual over a more prolonged period of time, rather than intervening during only a single ED visit. These include ED or community-based case management programs that identify social needs and provide resources and ongoing contact with individuals both inside and outside the health care system. While most research on programs to reduce frequent ED use focus on reducing ED use as a primary outcome, this should not be the only goal. Many frequent ED users have multiple social needs including unstable housing, poor social support, and food insecurity. Attending to these needs, which influence health and health services utilization, is a worthy goal in and of itself. In addition to examining programs' impact on ED use, connections to sustaining services such as primary care, stable housing, and social supports must be included as key outcomes that help to determine a program's success.

Hospital/Healthcare System

ED visits in the US continue to increase every year, illustrating the large demand for acute services [24]. Many hospitals and health care systems view ED visits themselves as adverse outcomes. Rather than view ED visits in a negative light, hospitals and healthcare systems would be well served to look at the ED as a potential point of intervention [25]. Because frequent ED users are seen so often, failing to intervene to address their underlying health and social needs is a missed opportunity.

Some hospitals are investing in ED-based staff or programs to provide comprehensive non-traditional services. For example, in California, an increasing number of EDs are participating in state-funded "ED BRIDGE" programs to initiate medication-assisted treatment for opioid use disorder in EDs. A current bill under consideration in California would allocate state funds to provide EDs with substance use counselors who can intervene with high-risk patients during their visits and refer them to continued outpatient substance use treatment.

A growing number of EDs in California, Washington, Oregon, and over 20 other states are now on the EDie platform. This platform allows for real-time identification of frequent ED users and individuals with high-risk prescription substance use and facilitates the input of care guidelines that can be shared across all participating facilities in an effort to coordinate care and avoid duplication of services. Some health systems have invested in trained staff to work specifically with frequent ED users who are identified by EDie, most of whom have significant social needs that physicians may not have time to explore or address during the visit. Investing in these types of resources is critical for hospitals and health care systems.

While ED visits are often portrayed as "avoidable" by hospitals and health systems, as emergency clinicians at the bedside we understand that many circumstances underlying frequent ED visits are outside of our patients' control. Frequent ED users' high social needs often represent structural societal ills (e.g., lack of affordable housing, poverty), yet others result from the chronic illnesses which stem from these societal inequities [26]. It has been well documented that many frequent ED users are quite ill and often require hospital admission [5]. While ED providers and staff can take actions during the ED stay to begin to address social needs, coordination of care during the hospitalization and discharge planning from the inpatient setting are critically important and require staff who are knowledgeable about community-based resources and who have dedicated time to address patients' non-medical needs.

In addition, many frequent ED users also utilize primary care and specialty services including oncology, palliative care, nephrology, and other services. Research shows that increasingly, outpatient providers refer a large proportion of all ED visits nationwide [27]. Hospital systems can enable and encourage coordination of care that includes the ED in multiple ways. Outpatient providers can also help by documenting a plan and providing information around goals of care that are accessible to ED providers during the visit and can help guide emergency decision making. The ability for patients to have rapid post-ED visit follow-up in the outpatient setting is critical—health systems that allow for open access, next day, and after-hours appointments can meet this need. In addition, ED providers must be able to reach outpatient providers to assist with care coordination in real-time during an ED visit. Health systems that facilitate such opportunities for real-time consultation between outpatient and ED providers may also be able to avert hospital admissions.

It is also clear that frequent ED users access community-based services of which health system providers may not be aware [4, 5]. Multiple mechanisms exist to improve care for frequent ED users by allowing in-reach of community organizations and services into EDs. As an example, in San Francisco, a community paramedicine team, EMS-6, works with frequent 911 callers. EMS-6 can be called or paged from an ED to come to meet with a patient on their caseload whenever possible [28]. San Francisco is leveraging the EDie platform to enter data from city housing assessments that flag ED patients who are prioritized for scarce housing units, many of whom are also frequent ED users. When contacted, housing services staff will come to the ED to assess and coordinate care for patients around housing placement. Such partnerships were made possible by city and state commitments to improve care coordination (e.g., the San Francisco Health Plan [29], the SF General Fund, and the San Francisco Whole Person Care Pilot [30]). In addition, it required research combined with concerted efforts at stakeholder engagement over months and years to determine how to best direct efforts and educate direct care providers about the importance of in-reach into EDs.

Many hospital systems are starting to invest in or partner with community organizations to address social needs that contribute to frequent health system use. Recently, Kaiser Permanente invested $200 million in a program to prevent both eviction and homelessness [31], and also purchased a 41 unit building in Oakland for the purposes of providing permanent supportive housing to Oakland residents experiencing homelessness [32]. Other programs and research collaborations are developing to better understand social needs that are connected to frequent use of the health system. These initiatives leverage normally siloed data sources from housing, health care, the community, and jail to obtain a more comprehensive picture of frequent ED users' needs and provide empirical evidence that can support care coordination efforts [4]. In addition to the EMS-6 team, the public health system in San Francisco has invested in multiple programs to serve frequent ED users including a sobering center for frequent ED users with substance use disorders that has embedded case management [33, 34]. In addition, some hospitals are developing relationships with community-based providers of residential substance use services for which the hospital pays for a period of the patient's stay, allowing for a warm handoff to ongoing substance use treatment directly from the ED.

Societal Level

It is clear that numerous social needs underlie frequent ED use along with medical needs, many of which are themselves a result of social inequities and gaps in services outside of the health care system. Yet many parts of the health system are not responsive to the needs of frequent ED users, some of whom require very low barrier access to care. In the US, EDs provide the lowest barrier health care available. Emergency medicine is the only specialty mandated by the federal EMTALA law to provide care 24–7 for all comers. It should not be surprising, then, that the number of ED visitors increases every year as the supply of primary care and other services remains stable or decreases, despite societal needs [35].

A complete picture of frequent ED users will reveal that for many, their ED use alone does not define them, but is a symptom of other unmet needs and a fragmented health and social care system [4]. While much of the emphasis around frequent ED use has been focused on urban environments, it no doubt permeates rural and suburban environments as well [36], although the underlying social needs may vary from community to community. In many states, Medicaid agencies have realized that frequent ED users and frequent health system users in general will benefit from increased care coordination and management and have taken various measures that either mandate or incentivize innovation in this area.

The Washington State Medical Association, the Washington State Hospital Association, and the Washington State American College of Emergency Physician chapter collaborated on the "ER is for Emergencies" program to implement "seven best practices" to improve care coordination and curb costs associated with high ED use [37]. The best practices included developing patient care plans for frequent ED users, adopting interoperable health information exchange technology, and using feedback information to track data on frequent ED users and evaluate the efficacy of interventions.

These best practices were developed and legislated, in part, as an alternative to the misguided "Three Visit Rule" proposed by Washington State's Health Care Authority which would have denied reimbursement for "non-emergency" visits to EDs [38, 39]. Implementation of the best practices correlated with nearly $34 million of savings in 2013, improvements in coordination with primary care, an approximately 10% decrease in ED use (including by frequent ED users), and a 24% reduction in visits resulting in a scheduled drug prescription [39, 40].

Other interventions have focused more specifically on social needs. The Center for Medicare and Medicaid Services (CMS) has supported several initiatives at the state level to facilitate improvements in coordinated care for high-utilizing patients. In California, the state Medicaid program (Medi-Cal) received $1.5 billion to develop whole-person care (WPC) pilot projects as part of their renewed Medicaid 1115 waiver. The WPC pilots focus on reducing unplanned emergency medical care by better meeting the medical, behavioral health, and social needs of high-risk vulnerable populations. For example, San Francisco County's program is focused on improvements in health outcomes for patients experiencing chronic homelessness. Efforts to date have concentrated on a) developing and leveraging data integration across medical, mental health, substance use, and social services especially for frequent users; and b) enhancing care coordination efforts through the expansion of services like medical and psychiatric respite, multi-disciplinary "street medicine" teams, and improved inter-agency communication on high priority populations. Initiatives have fostered collaboration on the city/county level, between the Department of Public Health, Department of Homelessness and Supportive Housing, the Mayor's Office, the Human Services Agency, academic universities such as the University of California San Francisco, Medicaid managed care payors like the San Francisco Health Plan, and other entities like the San Francisco Fire Department (including EMS-6 as described above).

CMS is also currently supporting a 5-year $157 million Accountable Health Communities (AHC) program to examine how systematically screening and addressing patients' health-related social needs impacts health care utilization and costs among their beneficiaries. While robust literature on the health impact of social care interventions is still growing, there have been several recent well-done studies demonstrating improved health outcomes and reduced medical expenditures and costs [41–45]. For example, Hennepin Health—a county-based safety-net accountable care organization in Minneapolis, Minnesota created as a partnership between four organizations aimed to deliver integrated medical, behavioral, and social services—demonstrated a 9% decrease in ED visits over 1 year [46]. We expect to learn much more about the impact of social service screening and linkage on acute care utilization through the anticipated rigorous AHC evaluation program evaluation.

Recommendations for Emergency Medicine Practice

Basic

- Take every visit by a frequent ED user seriously. Recognize that frequent ED users are often medically ill, and that frequent ED use is an independent predictor of mortality [15]. At the same time, do not medicalize the social needs of frequent ED users.
- When possible, facilitate warm hand-offs (e.g., verbal communication with an outpatient care manager; provision of accompaniment or transportation to a referral site) for the highest-risk frequent ED users to facilitate improved care coordination. If social workers or care navigators are available in your ED, involve them in the care of frequent ED users.

Intermediate

- Take advantage of health information exchanges to gain a better understanding of the ED use patterns of frequent ED users and gain access to critical information such as prescription drug monitoring and care plans.
- Identify and, with appropriate inter-disciplinary support, address frequent ED users' medical *and* social needs with equal intensity and commitment. Partner with patients to understand and meet their self-identified needs, using their priorities as a guide. For example, if a patient presents to an ED reporting food insecurity as the highest priority, focus treatment efforts on addressing this concern.
- Develop and participate in a multi-disciplinary work group to identify the highest ED users in your health system and discuss how to better serve them in case conferences. These teams may include social workers, patient navigators, intensive care managers, and other staff from both the ambulatory care and inpatient care setting.

Advanced

- Work with partners across the health system (e.g., ambulatory care, pre-hospital care, intensive care management programs) to identify and advance opportunities that ensure frequent ED users get needed care in the least resource-intensive setting possible.
- Partner with agencies outside the health delivery system (e.g., housing and other social service agencies) to develop an integrated approach to caring for frequent ED users in a *whole-person* manner via data sharing and innovative payment methods.
- Work with your specialty society and advocacy organizations to highlight to policy makers the social needs and challenges that underpin frequent ED users.

Teaching Case

Clinical Case

X is a 59-year-old with more than 120 ED visits in a 1-year period, including 55 visits in a 2-month period. The majority of ED visits were to a single hospital, but she also had encounters with three other local hospitals. Based on X's social history in an integrated data system, X had been homeless for 10 years. X has a history of a traumatic brain injury (TBI) from a car accident, with resulting seizures and falls. X has no short-term memory whatsoever and does not recollect previous near daily visits to the ED, or any of the providers with whom she interacts. Because X lives on the street she is frequently brought to the ED by EMS after bystanders have witnessed a seizure. X often forgets to take prescribed medications due to both lack of stable housing and social support, as well as TBI. Providers treating X for the first time often incorrectly assume that she is under the influence of alcohol.

ED staff, including physicians, navigators, and social workers, coordinate with street medicine providers and the community paramedicine team to secure a bed for X at a Navigation Center (a service-intensive shelter). They also create a care plan within EDie with specific instructions that can be viewed by providers across the platform, should X present at other facilities. The care plan includes X's recent health and social history, including a description of her memory issues, contact numbers for her primary care provider and the paramedicine team, and instructions on how ED providers can redirect X back to the Navigation Center.

After several months in the Navigation Center, X still presents to the ED, but much less frequently than when she was unsheltered. ED providers hold a case conference with staff from Street Medicine and the Navigation Center to discuss how to best support X in the transition to permanent supportive housing, given X's propensity to wander, and the need for daily anti-seizure medications.

X moves into a small apartment, with in-home support services including assistance with meals, transportation, and medication administration. ED providers update the EDie care plan to include contact details for the front desk of X's apartment and how to redirect X there. X occasionally visits the ED, but had four visits in 3 months, compared to 75 visits in the same 3-month period during the preceding year. ED providers continue to work closely with X's community-based case manager to coordinate care.

Teaching Points

1. Recognize the social needs of a frequent ED-using patient and consider how these needs may impact their health.
2. Involve social workers, care navigators, and community-based staff in the care of these patients whenever possible. This may involve having these providers in-reach to their patients in the ED and having ED providers attend case conferences in the community.

3. Use health information exchanges to create care plans that are accessible to providers in other hospitals. Identify personnel to keep these care plans as up to date as possible, when key information changes (e.g., provider contact details, patient's status on waitlists for services).

Discussion Questions

1. In this instance, providers had access to an integrated data system to inform management and allow them a more comprehensive view of this patient's ED and other service use. Even when these systems are in place, not all providers take the time to look at them. How can we encourage the implementation of integrated data systems, and how might you get colleagues to actually use them to improve care for frequent ED users?
2. Frequent ED users are often unable to have their needs fully assessed or met in the current ED environment. How might you work to improve the ED visit so that the non-medical needs of frequent users are identified and intervened upon? How might you involve community-based organizations in ED "in-reach" (visiting patients when they are in the ED for interventions) and community-based care planning?
3. Many providers made assumptions about this patient—that she had an alcohol use disorder—and did not realize the documented traumatic brain injury might be the underlying cause of the seizure disorder and memory loss. How can ED providers avoid these types of biases in caring for frequent ED users?

Acknowledgments The authors are grateful to Caroline Cawley for her editorial assistance in preparation of this chapter.

Disclosure Author/Editor Dr. Maria Raven and author Dr. Hemal Kanzaria are unpaid clinical advisory board members for Collective Medical, the creator of the Emergency Department Information Exchange (EDie) platform discussed in this **chapter.**

References

1. Moe J, Kirkland SW, Rawe E, Ospina MB, Vandermeer B, Campbell S, et al. Effectiveness of interventions to decrease emergency department visits by adult frequent users: a systematic review. Acad Emerg Med. 2017;24(1):40–52.
2. Billings J, Raven MC. Dispelling an urban legend: frequent emergency department users have substantial burden of disease. Health Aff (Millwood). 2013;32(12):2099–108.
3. Kanzaria HK, Niedzwiecki MJ, Montoy JC, Raven MC, Hsia RY. Persistent frequent emergency department use: core group exhibits extreme levels of use for more than a decade. Health Aff (Millwood). 2017;36(10):1720–8.
4. Kanzaria HK, Niedzwiecki M, Cawley CL, Chapman C, Sabbagh SH, Riggs E, et al. Frequent emergency department users: focusing solely on medical utilization misses the whole person. Health Aff. 2019;38(11):1866–75.
5. LaCalle E, Rabin E. Frequent users of emergency departments: the myths, the data, and the policy implications. Ann Emerg Med. 2010;56(1):42–8.
6. Weber EJ. Defining frequent use: the numbers no longer count. Ann Emerg Med. 2012;60(1):33–4.
7. Zuckerman S, Shen YC. Characteristics of occasional and frequent emergency department users do insurance coverage and access to care matter? Med Care. 2004;42(2):176–82.

8. Doupe MB, Palatnick W, Day S, Chateau D, Soodeen R-A, Burchill C, et al. Frequent users of emergency departments: developing standard definitions and defining prominent risk factors. Ann Emerg Med. 2012;60(1):24–32.
9. Johnson TL, Rinehart DJ, Durfee J, Brewer D, Batal H, Blum J, et al. For many patients who use large amounts of health care services, the need is intense yet temporary. Health Aff (Millwood). 2015;34(8):1312–9.
10. Fuda KK, Immekus R. Frequent users of Massachusetts emergency departments: a statewide analysis. Ann Emerg Med. 2006;48(1):16.e1-.e8.
11. Fingar KR, Smith MW, Davies S, McDonald KM, Stocks C, Raven MC. Medicaid dental coverage alone may not lower rates of dental emergency department visits. Health Aff. 2015;34(8):1349–57.
12. Raven MC, Billings JC, Goldfrank LR, Manheimer ED, Gourevitch MN. Medicaid patients at high risk for frequent hospital admission: real-time identification and remediable risks. J Urban Health. 2009;86(2):230–41.
13. Raven MC, Tieu L, Lee CT, Ponath C, Guzman D, Kushel M. Emergency department use in a cohort of older homeless adults: results from the HOPE HOME study. Acad Emerg Med. 2017;24(1):63–74.
14. Horrocks D, Kinzer D, Afzal S, Alpern J, Sharfstein JM. The adequacy of individual hospital data to identify high utilizers and assess community health. JAMA Intern Med. 2016;176(6):856–8.
15. Niedzwiecki MJ, Kanzaria HK, Montoy JC, Hsia RY, Raven MC. Past frequent emergency department use predicts mortality. Health Aff. 2019;38(1):155–8.
16. Soril LJJ, Leggett LE, Lorenzetti DL, Noseworthy TW, Clement FM. Reducing frequent visits to the emergency department: a systematic review of interventions. PLoS One. 2015;10(4):e0123660.
17. Raven MC, Kushel M, Ko MJ, Penko J, Bindman AB. The effectiveness of emergency department visit reduction programs: a systematic review. Ann Emerg Med. 2016;68(4):467–83.e15.
18. Capp R, Kelley L, Ellis P, Carmona J, Lofton A, Cobbs-Lomax D, et al. Reasons for frequent emergency department use by medicaid enrollees: a qualitative study. Acad Emerg Med. 2016;23(4):476–81.
19. Larimer ME, Malone DK, Garner MD, Atkins DC, Burlingham B, Lonczak HS, et al. Health care and public service use and costs before and after provision of housing for chronically homeless persons with severe alcohol problems. JAMA. 2009;301(13):1349–57.
20. Kertesz SG, Baggett TP, O'Connell JJ, Buck DS, Kushel MB. Permanent supportive housing for homeless people — reframing the debate. N Engl J Med. 2016;375(22):2115–7.
21. Anderson S. Emergency department information exchange can help coordinate care for highest utilizers. ACEP Now. 2017.
22. Collective Medical. Collective Medical – Purpose 2019. Available from: http://collectivemedical.com/purpose/.
23. Epic. Organizations on the Care Everywhere Network 2019. Available from: https://www.epic.com/careeverywhere/.
24. Chow JL, Niedzwiecki MJ, Hsia RY. Trends in the supply of California's emergency departments and inpatient services, 2005–2014: a retrospective analysis. BMJ Open. 2017;7(5):e014721.
25. Raven MC. Policies that limit emergency department visits and reimbursements undermine the emergency care system: instead, let's optimize it. JAMA Netw Open. 2018;1(6):e183728-e.
26. Kushel MB, Perry S, Bangsberg D, Clark R, Moss AR. Emergency department use among the homeless and marginally housed: results from a community-based study. Am J Public Health. 2002;92(5):778–84.
27. Uscher-Pines L, Pines J, Kellermann A, Gillen E, Mehrotra A. Emergency department visits for nonurgent conditions: systematic literature review. Am J Manag Care. 2013;19(1):47–59.
28. Mayor Lee Launches Innovative Emergency Medical Response Team to Reduce 911 Calls & Better Serve Residents in Need [press release]. San Francisco, Thursday, February 11, 2016.
29. San Francisco Health Plan. San Francisco Health Plan – About Us San Francisco: SFHP; 2020. Available from: https://www.sfhp.org/about-us/.

30. San Francisco Department of Public Health. Whole Person Care San Francisco: City and County of San Francisco; 2020. Available from: https://www.sfdph.org/dph/comupg/oprograms/wpc/default.asp.
31. Announcing $200M impact investment to address housing crisis [press release]. Oakland: Kaiser Permanente; 2018.
32. Brinklow A. Healthcare giant Kaiser buys Oakland building for affordable housing. SF Curbed. 2019.
33. City and County of San Francisco. San Francisco Sobering Center San Francisco 2020. Available from: http://www.sfsoberingcenter.com/.
34. University of California San Francisco. Healthforce at UCSF, editor. San Francisco2016. [cited 2020]. Available from: https://healthforce.ucsf.edu/blog-article/san-francisco-sobering-center-model-nation.
35. Tang N, Stein J, Hsia RY, Maselli JH, Gonzales R. Trends and characteristics of US emergency department visits, 1997-2007. JAMA. 2010;304(6):664–70.
36. Doran KM, Raven MC, Rosenheck RA. What drives frequent emergency department use in an integrated health system? National data from the veterans health administration. Ann Emerg Med. 2013;62(2):151–9.
37. Washington State Hospital Association. ER is for emergencies. 2015.
38. Kellermann AL, Weinick RM. Emergency departments, medicaid costs, and access to primary care — understanding the link. N Engl J Med. 2012;366(23):2141–3.
39. Pines J, Schlicher N, Presser E, George M, McClellan M. Washington state medicaid: implementation and impact of "ER is for emergencies" program. The Brookings Institution. 2015.
40. Emergency Department Utilization: Update on Assumed Savings from Best Practices Implementation, Chapter 7, Laws of 2012, Washington State Legislature, Second Special Session (Partial Veto) Sess. March 20, 2014.
41. Berkowitz SA, Seligman HK, Rigdon J, Meigs JB, Basu S. Supplemental nutrition assistance program (SNAP) participation and health care expenditures among low-income adults. JAMA Intern Med. 2017;177(11):1642–9.
42. Gottlieb LM, Hessler D, Long D, Laves E, Burns AR, Amaya A, et al. Effects of social needs screening and in-person service navigation on child health: a randomized clinical trial. JAMA Pediatr. 2016;170(11):e162521-e.
43. Taylor LA, Tan AX, Coyle CE, Ndumele C, Rogan E, Canavan M, et al. Leveraging the social determinants of health: what works? PLoS One. 2016;11(8):e0160217.
44. Berkowitz SA, Delahanty LM, Terranova J, Steiner B, Ruazol MP, Singh R, et al. Medically tailored meal delivery for diabetes patients with food insecurity: a randomized cross-over trial. J Gen Intern Med. 2019;34(3):396–404.
45. Berkowitz SA, Terranova J, Hill C, Ajayi T, Linsky T, Tishler LW, et al. Meal delivery programs reduce the use of costly health care in dually eligible medicare and medicaid beneficiaries. Health Aff. 2018;37(4):535–42.
46. Sandberg SF, Erikson C, Owen R, Vickery KD, Shimotsu ST, Linzer M, et al. Hennepin health: a safety-net accountable care organization for the expanded medicaid population. Health Aff. 2014;33(11):1975–84.

Substance Use: A Social Emergency Medicine Perspective

10

Elizabeth A. Samuels, Ziming Xuan, and Edward Bernstein

Key Points

- Substance use disorders are complex, biopsychosocial diseases with high rates of morbidity and mortality.
- Addressing how underlying social determinants of health impact substance use will have the greatest and longest-lasting impact on improving population health.
- ED substance use initiatives include screening, motivational interviewing, initiation of and/or referral to treatment, and provision of harm reduction services.
- Patients with substance use disorders may need assistance with health-related social needs in order to achieve successful clinical outcomes and long-term well-being.
- Professional societies, departments of health, and lawmakers can be key partners to improving care within and across EDs and hospital systems.

E. A. Samuels (✉)
Department of Emergency Medicine, Alpert Medical School of Brown University, Providence, RI, USA
e-mail: elizabeth_samuels@brown.edu

Z. Xuan
Department of Community Health Sciences, Boston University School of Public Health, Boston, MA, USA
e-mail: zxuan@bu.edu

E. Bernstein
Department of Community Health Sciences, Boston University School of Public Health, Boston, MA, USA

Department of Emergency Medicine, Boston University School of Medicine, Boston, MA, USA
e-mail: ebernste@bu.edu

H. J. Alter et al. (eds.), *Social Emergency Medicine*,
https://doi.org/10.1007/978-3-030-65672-0_10

Foundations

Background

Emergency medicine providers care for patients with problems related to substance use on a daily basis. Compared to patients in other healthcare settings, adult emergency department (ED) patients have higher rates of substance use, including tobacco, alcohol, non-medical use of prescription medications, and illegal substances [1]. Over the last decade, substance use-related ED visits have significantly increased. From 2006 to 2014, US alcohol-related ED visits rose 61.6%, from 3,080,214 to 4,976,136 annual visits [2]. US ED drug-related visits doubled from 2005 to 2014, in large part due to increasing rates of opioid overdose [3] which increased by 29.7% from 2016 to 2017 alone [4].

Rising substance use-related ED visits are due to increased overall ED utilization, growing prevalence of substance use disorders (SUDs), increased diagnosis and detection of SUDs, and increased SUD severity and lethality of substances used. Both binge drinking and alcohol use disorder (AUD), for example, have increased nationally over the last two decades, especially among older males [5, 6]. There are approximately 88,000 alcohol-related deaths in the US each year, making alcohol the third leading preventable cause of death [7]. From 1999 to 2017, drug overdoses quadrupled and over 700,000 people died [8, 9]. Since 2013, drug overdose deaths due to synthetic opioids other than methadone increased exponentially among people from all races, ethnicities, sexes, and age groups [10]. Most of these deaths involved potent synthetic opioids, specifically fentanyl [11, 12]. While crude overdose death rates have been highest among White people, the highest percent increase in mortality due to synthetic opioids other than methadone has been among non-Hispanic/Latinx Black people. From 2013 to 2017, non-Hispanic/Latinx Black people had an 18-fold increase in overdose mortality, people who identified as Hispanic/Latinx had a 12.3-fold increase, and non-Hispanic/Latinx White people had a 9.2-fold mortality increase [12].

Despite the increasing national prevalence and severity of SUDs, engagement in formal addiction treatment remains low [2, 3, 13–15]. In 2018, there were an estimated 21.2 million people over the age of 12—7.8% of the US population—who needed treatment for a SUD, but only about 11.3% of people with a SUD (2.4 million people) received specialty treatment [16]. This treatment gap is due to multiple factors, including stigma and discrimination, limited treatment availability, cost, un- and under-insurance, failure to perceive a treatment need, and unmet health-related social needs [17–20]. In this chapter, we describe how, as a key interface between the community and the healthcare system, emergency medicine providers have an important opportunity to help close the SUD treatment gap and improve public health by advocating for policies and programs to provide early treatment initiation, reduce treatment barriers, and address health-related social needs.

Evidence Basis

Historically, SUDs have been mischaracterized as moral failings resulting from bad "choices." Neuroscience research on addiction has demonstrated the neuronal circuitry changes from exposure to alcohol or drugs that result in compulsive substance use despite adverse personal consequences [21]. Advancements in our neurobiological understanding of SUDs now support the treatment of SUDs as chronic diseases. However, SUDs also have complex psychosocial components; they often co-occur with psychiatric conditions and are deeply intertwined with social determinants of health (SDOH). These factors influence not only the development SUDs, but also use-related harm and treatment initiation and retention.

The social determinants of SUDs are complex and vary by individual, geography, and substance, but there are some common associated factors. Housing status, adverse childhood experiences, and exposure to trauma and violence are all associated with substance use and the development of SUDs [22–25]. ED patients experiencing homelessness, for example, have increased prevalence and severity of substance use compared to ED patients who are stably housed [26]. Other individual-level factors, such as lower level of educational attainment, having low income, and being unemployed, underemployed, or employed in a job with a high risk of injury, such as manual labor occupations, have also been associated with increased rates of substance use and opioid overdose [27, 28]. The rise in deaths among middle-aged, less-educated, working-class White people due to alcoholism, overdose, and suicide—the so-called "deaths of despair"—have been connected to high unemployment, poor job prospects, and widening social inequalities [29–31]. The relationship between SUDs and social determinants of SUD is bidirectional and cyclical. For example, just as substance use can result in loss of income, housing, or employment, lack of housing or employment can impair an individual's ability to successfully engage in SUD treatment and sustained recovery. Due to structural racism, economic opportunities and SUD treatment access and maintenance are even more limited for Black people [32, 33].

Interventions to address SUD in the medical setting have typically focused on individual medical treatment and/or behavior changes. These are important, but addressing the social determinants that underlie substance use will have a greater impact on improving overall population health [34]. A comprehensive approach to addressing SUDs takes into account not only the medical treatment of SUDs, but also health-related social needs and "upstream" SDOH. The socioecological model in Fig. 10.1 outlines key substance use-related factors and multilevel interventions ranging from the individual and intrapersonal levels to organizational, community, and policy spheres. Using an upstream approach to understand SUDs can help inform ED strategies and practices at the bedside as well as broader system-level and community-engaged interventions.

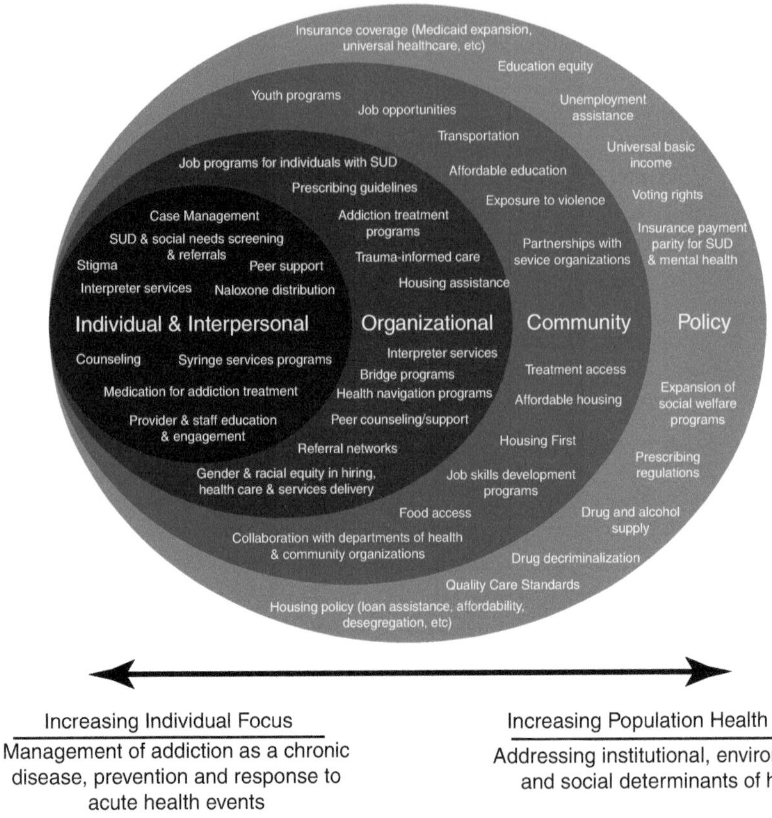

Fig. 10.1 Socioecological model and public health impact of interventions for substance use disorders [34, 173, 217]

Emergency Department and Beyond

Bedside

While addressing the social determinants of SUDs may have the greatest impact on reducing population-level SUD morbidity and mortality, ED clinicians can have a significant and long-term impact on the health and wellbeing of individuals with SUD even in a brief clinical encounter. A social emergency medicine approach to ED treatment of SUDs entails four basic components: (1) identification of harmful substance use; (2) treatment initiation and harm reduction; (3) referral to treatment; and (4) identifying and addressing health-related social needs.

Many people with SUDs have a prior history of trauma, including physical, emotional, and sexual assault and/or abuse [35–37]. Using a trauma-informed approach during the healthcare encounter is key to identifying patients with SUDs and successful patient engagement. This includes awareness of the widespread impact of trauma, recognizing the signs and symptoms of trauma, integrating knowledge about trauma into patient care, and preventing re-traumatization [38, 39]. Providers can avoid re-traumatizing patients by communicating with and caring for patients in a respectful, non-judgmental way that prioritizes a patient's sense of safety; uses person-centered, de-stigmatizing language; demonstrates provider trustworthiness and transparency; encourages peer support, mutual self-help, and patient empowerment; and incorporates considerations about cultural and historical context, gender, race, sexuality, and economics [38, 40].

Identification of Substance Use Disorders

Many substance use-related ED visits are due to intoxication, acute overdose, injury, suicidality, or injection-related injuries or infections. Other reasons for seeking care include acute exacerbations of chronic conditions related to substance use, such as liver failure from alcoholic cirrhosis. For many ED patients, however, harmful substance use or the presence of a SUD may not be obvious. Given the high prevalence of tobacco, alcohol, and drug use among ED patients overall, many EDs have implemented either universal or targeted substance use screening [41].

Substance use screening, brief intervention, and referral to treatment (SBIRT) interventions have been widely studied for tobacco, alcohol, and other drugs [42]. ED SBIRT interventions can be done by ED clinicians, social work, or drug and alcohol counselors and typically take 5-30 minutes, depending on intervention scope and content [43]. While studies have yielded variable results [41], SBIRT interventions in ED and non-ED settings overall show efficacy for not only reducing consumption of alcohol, tobacco, marijuana, and other drugs (e.g., cocaine, methamphetamine, and heroin) [44], but also substance use-related risk behavior, such as driving while intoxicated [45]. More information about SBIRT can be found through the SAMHSA-HRSA Center for Integrated Health Solutions (https://www.integration.samhsa.gov/clinical-practice/sbirt).

Screening can be targeted or universal and there are several short screening tools that have been adapted for ED use, such as the AUDIT-C [46–48], CAGE [49], and NIAAA single-use question [50–52]. Early detection and intervention for harmful substance use can prevent SUD development. Screening is especially relevant for adolescents, who may have early or undetected harmful substance use [53]. SBIRT has been widely applied and recommended in pediatrics using the CRAFFT screening tool, which is validated, brief, and developmentally appropriate [54, 55].

Screening paired with motivational interviewing, a patient-centered approach that incorporates a patient's preferences and choice [56], will inform the type of ED treatment and/or harm reduction services provided and subsequent treatment referral. A positive screen or identification of unhealthy alcohol and/or drug use can be followed by a brief conversation to elicit the patient's perspective and any steps they might take to be safer and healthier.

Treatment Initiation and Harm Reduction

Medical treatment of toxidromes and withdrawal are routinely taught during residency training to meet basic emergency medicine clinical competencies [57]. As a result, ED clinicians have considerable expertise in the management of intoxication, alcohol withdrawal, and overdose. ED provision of medications for SUD includes nicotine replacement for tobacco cessation and buprenorphine for the treatment of opioid use disorder (OUD) [58–61]. Naltrexone and acamprosate are effective medications to treat AUD, but their initiation in the ED has not yet been described [62]. After alcohol, tobacco is the next most common substance used among ED patients. ED-provided motivational interviewing and nicotine replacement have been shown to be effective in reducing tobacco use, including among low-income ED patients [63]. In just a few minutes, ED providers can assess for tobacco use, provide smoking cessation counseling, and prescribe nicotine replacement while a person is awaiting admission or upon discharge with referral to outpatient resources [58, 64, 65].

Key ED OUD initiatives include naloxone distribution for overdose rescue, ED initiation of buprenorphine, and behavioral counseling with referral to treatment [66]. Treatment with agonist medications for OUD (MOUD)—specifically buprenorphine and methadone—decreases mortality by more than 50% [67] and has also been shown to reduce overdose and acute care utilization [68]. Currently, however, only a minority of people with OUD are treated with these medications [69]. Receipt of MOUD in the year following an ED visit for an opioid overdose is associated with reductions in all-cause and opioid-related mortality [70]. A single site study of ED-initiated buprenorphine demonstrated a greater likelihood of follow-up at 30 days, decreased self-reported opioid use, and cost-effectiveness [61, 71]. Any prescriber with a DEA license can order buprenorphine for administration in the ED. However, completion of required training and receipt of a Drug Addiction and Treatment Act of 2000 (DATA 2000) waiver is needed to provide patients with a discharge prescription [72]. If a provider does not have a waiver, patients may return to the ED for up to 72 hours to get additional doses of buprenorphine [73, 74]. Providing patients with a prescription for a short course of buprenorphine until they can be seen for outpatient follow-up is preferred. Some EDs use hospital-based bridge programs or low barrier access clinics that stabilize patients on MOUD and link patients to outpatient office-based or opioid treatment programs (OTPs) for ongoing treatment. Other EDs provide referral directly to community-based OTPs or office-based treatment providers [75–77]. Success of ED-initiated buprenorphine is dependent on the availability of outpatient office-based buprenorphine providers and OTPs. There is significant variability in access to providers, with large access gaps by geography [17, 78, 79] and race [32]. Telemedicine is one strategy used to try to address gaps in outpatient treatment access [80–85].

Harm reduction, an approach which seeks to reduce drug-related harm while respecting individuals' autonomy, is an essential component of comprehensive SUD care. Intervention cornerstones include syringe services programs and overdose education and naloxone distribution programs. Recommended by the World Health Organization and the Centers for Disease Control and Prevention (CDC), syringe services programs provide sterile injection supplies and teach safe injection practices to individuals who inject drugs, including opioids and methamphetamine.

These programs have been shown to reduce HIV and hepatitis C transmission without increasing substance use [86–91]. Syringe services programs also provide additional services including HIV and hepatitis C testing, naloxone distribution, personalized support, continuity, and linkage to care. Given these benefits, patients should be referred to syringe services programs for continued harm reduction and support services, if available.

Community overdose education and naloxone distribution programs have been shown to reduce overdose mortality [92–95]. Naloxone is an opioid antagonist that reverses the effects of an opioid overdose. Naloxone distribution has increased substantially in recent years through community-based harm reduction organizations, public health departments, pharmacies, and other medical and community settings. EDs are increasingly distributing naloxone to patients at risk of opioid overdose [96–99]. While the liberal distribution of naloxone in EDs is recommended, some particularly high-risk groups include people recently released from prison [100, 101], people who inject opioids or concurrently use opioids and other sedating substances (e.g., benzodiazepines, alcohol) [102], and people with significant co-morbidities such as liver and lung disease [103]. Naloxone can be prescribed by any ED prescriber. Some EDs distribute take-home naloxone kits that include a mouth barrier for rescue breathing and educational materials. All patients at risk of overdose, or who are around individuals at risk of overdose, should be offered naloxone. In states that allow for third party prescribing of naloxone, naloxone can also be provided to family and friends of individuals at risk for opioid overdose [104].

Referral to Treatment

ED SUD screening, motivational interviewing, and referral to treatment can be done by any ED provider. Training in these practices is short and can be easily incorporated into emergency medicine residency training [105–107]. Depending on availability, Health Promotion Advocates, peer recovery specialists, substance use navigators, health coaches, or social workers can provide not only treatment referral, but can also help address issues like transportation, insurance coverage, and food and housing insecurity, which are key factors in treatment initiation, engagement, and retention.

One of the earliest examples of these types of programs, Project ASSERT (Alcohol & Substance Use Services, Education, and Referral to Treatment), uses Health Promotion Advocates, who are licensed peer alcohol and drug counselors, to provide substance use assessments and linkage to care. The first Project ASSERT program was started at Boston Medical Center (BMC) ED in 1994. Similar programs have since been implemented at EDs across the US [108–110]. Project ASSERT Health Promotion Advocates provide bedside psychosocial assessments, determine the appropriate treatment level of care, and arrange for placement in a continuum of treatment services as well as referrals to community support services. Support services provided include placement in shelters, social peer support referrals, overdose education, and naloxone rescue kits, primary care linkage, transportation assistance, and referrals to the BMC's food and clothing banks.

Patients interested in MOUD or AUD treatment receive same or next-day appointments at the affiliated BMC Faster Paths low-barrier bridge clinic [108, 111].

Some EDs use community-based peer recovery specialists, also known as recovery coaches, to assist with treatment linkage. Peer recovery specialists complete training (and, in some states, a certification exam) about addiction, SUD treatment, trauma-informed care, and motivational interviewing. They use their personal experience to provide individualized peer support to help individuals with SUD identify and remove barriers to treatment, including health-related social needs such as housing, lack of insurance, and lack of sufficient employment or education. Peer specialists can be hired by community-based organizations, departments of health [112], or be hospital-affiliated or employed [113–115]. For example, the AnchorED program in Rhode Island [96, 113] is a partnership between hospitals and a community-based organization, Anchor Recovery Community Center. After a patient presents to an ED, consulted AnchorED peer recovery coaches assess patients for readiness for change and provide motivational interviewing, naloxone and overdose prevention education, treatment navigation, and follow-up after an ED visit. Similar models are being increasingly applied in other municipalities and states [112, 116], but effectiveness has yet to be fully assessed [97, 112, 117].

Referral to treatment is best provided by a multidisciplinary team but for EDs that do not have these resources, clinicians can still provide screening, counseling, and referral to treatment. Community treatment services can be found using SAMHSA's National Helpline (1-800-622-HELP [4357]) or their online treatment locator: https://findtreatment.samhsa.gov/. Local peer support groups can also be found through Alcoholic Anonymous (www.aa.org), Narcotic Anonymous (www.na.org), or SMART Recovery (Self-Management and Recovery Training at www.smart.org)

Identifying and Addressing Health-Related Social Needs
Because social needs impact substance use and access to SUD treatment, comprehensive ED initiatives to address SUDs must simultaneously address unmet social needs. At the time of an ED visit, individual-level assistance with housing, food insecurity, transportation, or health insurance can be provided. As these may be key barriers to accessing addiction treatment, assistance with these factors from the ED can help facilitate successful outpatient treatment engagement and retention. This may include the provision of services from the ED or referral to community resources. In 2019, the American College of Emergency Physicians (ACEP) recognized the importance of addressing social needs in the ED by passing a resolution endorsing staffing EDs with social workers [118]. Furthermore, hospitals and healthcare systems can develop infrastructure to assist patients with insurance enrollment, healthcare, and social services navigation, providing any needed clothing, jackets, or shoes, and provision of transportation vouchers and/or assistance to help patients attend outpatient follow-up appointments [119, 120]. As is discussed in greater detail below, housing is a key area of need for many patients with SUDs. For individuals with SUDs, referrals to Housing First programs—housing for

individuals experiencing homelessness without prerequisites around sobriety or completion of SUD treatment [121–123]—may be preferred to programs that require abstinence.

Hospital/Healthcare System

Health and hospital systems as well as professional societies can encourage departments of health and state and local governments to make systemic and policy changes to that would ultimately improve care for ED patients with SUDs and address their unmet social needs. Emergency physicians have led such efforts across the US. Parity for SUD treatment under the Affordable Care Act has improved access to care, but significant barriers remain, largely related to gaps in treatment availability, challenges in addressing concurrent unmet social needs, and pervasive stigma [124, 125]. Black, Indigenous, and people of color (BIPOC) [33, 126–128], women [129], residents of rural areas [17, 78, 130, 131], and low-income individuals [125, 132] have lower initiation and maintenance of addiction treatment. In areas with limited addiction medicine provider availability [32, 79], telemedicine is being used to fill treatment gaps by providing either direct services or behavioral interventions [81], patient to provider communications, or enhancing the capacity of local providers to provide treatment for SUDs [133, 134]. However, these services can require ongoing access to the internet and/or smartphones, which may be cost-prohibitive or unavailable for some low-income individuals and people living in rural areas, who may be in most need of these services [135–138].

Reimbursement metrics can be used to reward care quality and value rather than quantity. Some of these initiatives are driven by, or done in partnership with, payers. MassHealth, Massachusetts' Medicaid program, for example, has developed a network of Accountable Care Organizations to improve care for patients with SUDs [139]. The growing emphasis on care quality has spurred some hospital systems to develop quality and population health initiatives [140] including housing programs, local workforce development initiatives, food pantries and gardens, and healthcare managers and navigation programs for patients with SUDs. Given the high prevalence of co-occurring substance use and unmet social needs among individuals who frequently use the ED [141–143], some initiatives designed to address frequent ED utilization have incorporated SUD treatment along with addressing health-related social needs [144]. Some programs use community health workers or offer intensive case management to provide individualized support and healthcare navigation [144–147].

Some municipalities have established respite programs where people without housing can stay while receiving treatment for medical problems that do not require hospitalization, but could not be safely addressed at a shelter or while living on the street [148]. Expanding indications to include beginning someone on SUD treatment is one potential strategy that could help improve SUD treatment initiation among patients experiencing homelessness.

Some municipalities and hospitals have invested in alternate sites of care for people who are intoxicated. One example is sobering centers, locations where uninjured and medically stable intoxicated individuals can safely regain short-term sobriety as an alternative to being brought to an ED or jail [149, 150]. Sobering centers have been implemented around the US and vary in their organizational affiliation, triage mechanism, and funding. Currently, there is no standardized scope of work or certification process for sobering centers. While they typically do not deliver onsite treatment for substance use, they can provide peer support and linkage to treatment.

To improve the quality of care within and across health systems, professional societies, public health departments, and state and federal governments have released treatment recommendations, guidelines, standards, and requirements for the treatment of SUDs in the ED. All trauma centers are required to screen, at a minimum, for alcohol use to maintain their trauma center certification [151]. Level I and Level II certified trauma centers must not only screen, but also provide interventions for all individuals who screen positive. The Colorado [152] American College of Emergency Physicians (ACEP) chapter has written treatment guidelines and California ACEP [59] has collaborated on and endorsed guidelines for ED buprenorphine initiation. Nationally, ACEP has not only supported ED naloxone distribution, but also the use of opioid prescribing guidelines and ED buprenorphine initiation. ACEP is also working to improve ED care of OUD at academic and community EDs. Through its Emergency Quality Network (E-QUAL) Opioid Initiative, ACEP provides online education and training resources and quality metric reporting and measurement for participating EDs nationwide [153].

Some states have begun to regulate and/or incentivize OUD treatment standards. After the passage of a substance use discharge planning law in 2016, the Rhode Island Department of Health released hospital treatment standards for the treatment of OUD and opioid overdose. These standards require EDs to provide standardized SUD screening; use peer recovery services; refer patients to treatment; and provide naloxone for overdose rescue, among other requirements [154]. These standards also include additional care recommendations for hospitals that can initiate and maintain patients on MOUD. Similarly, the city of Baltimore also released hospital standards of care [155], all New York City hospitals have agreed to treatment guidance put forward by the NYC Department of Health and Mental Hygiene [156], and last year the Massachusetts legislature passed a law requiring that EDs have protocols and capacity to possess, dispense, administer, and prescribe buprenorphine. To assist EDs with implementation, the Massachusetts Health and Hospital Association and the Massachusetts ACEP chapter released treatment guidelines [157]. Using financial incentives rather than regulation, the Pennsylvania Department of Human Services added an opioid-specific incentive to their Hospital Quality Improvement Program [158]. This program provides financial incentives for hospitals to develop clinical pathways to link Medicaid patients with OUD into treatment within 7 days of an ED visit [158, 159]. Moving beyond incentives, in February 2019, California passed legislation funding EDs to hire and pilot substance use disorder peer navigators and behavioral health peer navigators [160].

Societal Level

While substance use and SUDs are present among all genders, races, geographic regions, and social strata, they are not evenly distributed or equitably treated. Socioeconomic conditions have a significant impact on the development, progression, and treatment of SUDs. These conditions include, but are not limited to, housing conditions, healthcare access, availability of community resources, poverty, immigration status, employment, exposure to violence, neighborhood stressors, and discrimination [30, 161–163]. Many of the neighborhood-based inequalities we observe today are the direct result of historical housing policies known as "redlining" which not only resulted in the segregation of US metropolitan areas, but also systematically limited Black people's opportunities for economic advancement [164]. Widening income inequality, weakening social welfare and healthcare safety net programs, and diminishing social cohesion over the last 50 years all contribute to social distress that is associated with higher substance use severity and mortality [30, 31, 165, 166]. Areas with the lowest levels of social capital and socioeconomic status, for example, tend to have more liquor stores [169] and higher overdose rates [167, 168]. Conversely, wealthier counties and communities have fewer liquor retail outlets [170], lower levels of tobacco and alcohol use [171], and lower rates of opioid overdose [168]. Where people live and local demographics also impact what treatment options are available to them. For example, counties that are majority White are more likely to have access to buprenorphine, as compared to predominantly Black counties which are more likely to have access to methadone [32].

To support broader structural change and improvements in population health, emergency physicians can leverage their frontline experience, data, and knowledge to identify and advocate for programs, initiatives, policies, and legislation that address social determinants of SUDs. Societal level changes that will impact the development and treatment of SUDs can be divided into three general categories: (1) individual social factors; (2) public policy; and (3) the drug supply environment [172]. Each of these factors impacts not only substance use and addiction treatment, but also individual and population health and social outcomes.

Individual Social Factors

As previously discussed, structural inequities, SDOH, neighborhood exposures, and one's living environment have an important impact on substance use and treatment engagement. In addition to addressing individual patient social needs as described earlier, ED providers and health systems can work on a larger community or structural level to improve the social determinants of SUDs. For example, stigma and discrimination toward drug use, people who use drugs, and people with SUDs present barriers to treatment access [173, 174]. To address this barrier, the Grayken Center for Addiction at Boston Medical Center has launched an anti-stigma campaign, which includes training and advocacy for use of non-stigmatizing language when talking about people with SUDs. This also includes employment initiatives that build workforce capacity for people with SUDs, public housing strategies to support people with SUDs, and educational programs and services for people with SUDs to get re-enrolled in school and gain higher education to help them improve their future job opportunities [175].

Housing is another key determinant of substance use that health systems can help address by screening patients for housing needs, referring patients to existing housing resources, and even providing funding or land for housing-related initiatives in conjunction with community and governmental stakeholders. There are two basic housing models to assist people with SUDs: Housing First permanent supportive housing and Recovery Housing (aka Sober Living Houses) [176]. Housing First is a model of permanent supportive housing not limited to people with SUDs. Residents are provided with permanent housing that uses a harm reduction approach to provide individuals with voluntary support and resources for their substance use [121]. These programs have demonstrated good efficacy in housing retention for people with SUDs [121–123]. Recovery Housing can range from independent, resident-run homes to staff-managed residences that provide clinical services [177]. Such programs are heterogeneous, and many use a model of peer-support with abstinence-based, Alcoholics Anonymous/Narcotics Anonymous type teachings. Some evaluations of Recovery Housing programs have shown decreased substance use, increased monthly income, and decreased incarceration [177]. While housing is essential for successful recovery, given the variability of Recovery Housing programs, some programs may not be well suited for patients with OUD if they do not accept individuals on MOUD.

Public Policy

Addressing socioeconomic determinants of SUDs requires changes in current public policy including expansion of social programs. Available transportation, education, health, addiction treatment, housing, social and legal advocacy services, and interactions with the criminal justice system, have significant impacts on SUD development and treatment. Medicaid expansion has improved treatment access in expansion states, but has not fully addressed overall treatment gaps or racial and gender disparities in care [32, 33, 70, 125, 127, 132, 178]. Addressing racial and gender disparities in treatment engagement and retention will require identifying and addressing systematic bias in treatment programs and the development of culturally competent, linguistically accessible, and/or specific treatment programs, including for pregnant people with SUDs [178–180]. Restrictive immigration policies, detention, deportations, anti-immigrant sentiments, and limited interpreter services are additional barriers faced by immigrants and refugees accessing mental health and substance use services [181].

Emergency physicians can be important advocates for the establishment or removal of laws to reduce drug use-related harm. Historically, the US has used a criminal justice policy strategy to address SUDs [182]. This has had important implications for SUD prevalence, morbidity, mortality, and public health. Rather than decreasing substance use, this approach has largely pushed substance use underground, resulting in increased use-related harm. In the case of witnessed opioid overdose, for example, many report fear of arrest as a primary cause of not calling 911 [183]. The passage of Good Samaritan Legislation offers some protection against prosecution to encourage utilization of emergency medical services in case of an overdose. Other state-specific legislation, specifically naloxone and syringe access laws, offer important opportunities to provide harm reduction services that can prevent opioid overdose as well as injection-related complications such as cellulitis, endocarditis, HIV, and hepatitis C [184–186]. The effectiveness of these laws is undermined, however, by the presence of paraphernalia laws, which criminalize possession of equipment for drug consumption, including sterile syringes, even if acquired from a syringe services program or purchased at a pharmacy [187]. Despite their public health utility, many states severely restrict syringe access at pharmacies or community-based syringe services programs, especially in rural communities and the Southeast and Midwest regions [187]. For more information about laws related to opioid prescribing, MOUD, Good Samaritan Legislation, naloxone access, and paraphernalia, see the Prescription Drug Abuse Policy System (PDAPS) at: http://pdaps.org.

Although all people who use illegal drugs face increased policing and scrutiny from the criminal justice system, communities of color and low-income communities are disproportionately impacted [188, 189]. Longer sentences for drug possession, including three-strikes laws and mandatory minimums, have not reduced the prevalence of substance use or rates of overdose [190] but have resulted in the mass incarceration of BIPOC communities [191]. These policy failures are related to the recurrent relapsing nature of SUDs, a high prevalence of trauma and post-traumatic stress disorder among incarcerated people, limited-to-no economic opportunities for those who have been incarcerated, and concurrent defunding of social welfare programs [190, 191, 192].

Furthermore, the use of evidence-based treatment in prison is limited and, where available, racial disparities persist, with decreased provision to Black people who are incarcerated [193]. Following release from prison, 77% of individuals with OUD return to opioid use and the risk of death increases threefold [194]. A minority of prisons and jails offer people with OUD evidence-based treatment with MOUD. Many prisons and jails even force individuals to stop their methadone and/or buprenorphine upon entry. Offering or continuing MOUD in prison can have a significant impact on overdose deaths. Studies from the US and internationally exploring prison buprenorphine programs have observed reductions in post-release opioid overdose death of up to 85% [195–197].

Punishing substance use with incarceration has not reduced substance use, nor has it decreased substance-use related morbidity and mortality. Furthermore, criminal justice involvement has a detrimental impact on a person's ability to obtain housing, employment, and education and a significant negative impact on the health and wellbeing of the families and communities of people who are incarcerated—all of which can undermine successful engagement in SUD treatment after incarceration [198]. Outside of drug legalization, policy changes to mitigate the impact of criminal justice drug policies would include job and education programs for people with criminal justice involvement, defelonization of drug possession, equitable access to SUD treatment, removal of screening for prior convictions on employment applications [199], and decoupling felony convictions from eligibility for housing and occupational licenses [200, 201].

Drug or diversion courts are one strategy used to reduce incarceration for drug-related charges and to improve access to addiction treatment. Some have shown effectiveness in reducing substance use and recidivism, but they vary in form, use, and provision of evidence-based treatment [202–204]. One study showed that only 53% of drug courts allowed MOUD to be part of an individual's treatment plan [194]. Patient advocates have voiced concern about the role of coercion in acceptance of treatment, the efficacy of coerced treatment, and the persistence of racial discrimination in the courts, as treatment has been observed to be preferentially provided to White people facing criminal charges [205]. If implemented effectively, however, these programs could have the potential to reduce incarceration and improve treatment engagement.

A public health policy approach to SUD would entail a fundamental and radical change in drug policy. Drug and drug paraphernalia decriminalization and expansion of evidence-based addiction treatment could not only reduce the health consequences of drug use and incarceration but could also lower barriers to treatment and harm reduction services. In 2001, Portugal decriminalized drug consumption and expanded access to treatment and harm reduction services. Portugal had previously been considered the "heroin capital of Europe," with an estimated 1% of its population using heroin. Since decriminalization, from 1999 to 2013, overdose deaths have decreased by 80%, treatment engagement increased over 60%, new diagnoses of HIV decreased by 94%, per capita social cost of drug misuse decreased by 18%, and the percentage of people in prison for drug law violations decreased by 45% [206–208].

Drug and Alcohol Supply

Many efforts to reduce substance use and substance use-related harm have focused on reducing and restricting the drug and alcohol supply. These efforts have ranged from total prohibition to legalization with government regulation and enforcement. Alcohol was previously prohibited and is now regulated through taxes and other restrictions by age, time, location, and types of beverage. Tobacco is similarly restricted through taxes, and restrictions on age and locations of consumption, which have contributed to reductions in tobacco use and in mortality from smoking-related illnesses [209, 210]. States are increasingly passing legislation to legalize and regulate recreational and medical marijuana consumption, despite federal restrictions [211, 212].

Drug supply strategies to address opioid overdose have focused on reducing opioid prescribing, use of prescription drug monitoring programs, and Drug Enforcement Administration responses to the global drug market. Reducing prescription supply without increasing the availability of SUD treatment and harm reduction services has shifted demand toward illegal drug use. Since the release of the CDC's opioid prescribing guidelines in 2016, there has been a decline in opioid prescribing [213, 214]. However, despite decreasing prescriptions, overall overdose deaths remain high with many overdoses involving illegal opioid use alone and in combination with prescription opioids [9].

Successfully addressing the opioid epidemic, and substance use disorders more broadly, requires looking beyond biological and medical approaches of prevention, diagnosis, and treatment to strategies that address broader global structural issues that drive the drug trade, such as underdevelopment and poverty. As clinicians, we start at home, one patient at a time. But, by also advocating for hospital, health system, and societal changes in partnership with local communities and policy makers, ED providers can have a meaningful systemic impact to reduce SUD morbidity and mortality.

Recommendations for Emergency Medicine Practice

Basic

- Identify patients with substance use disorders based on their clinical presentation or use a one-question or brief substance use disorder screening tool (https://www.integration.samhsa.gov/clinical-practice/screening-tools) [41].
- Provide take-home naloxone for all patients who have presented after an opioid overdose or who are at risk for an overdose.
- Treat opioid *withdrawal* in the ED with opioid agonists; methadone if the patient is on methadone, buprenorphine if the patient is not on long-acting opioids and has no contraindications [215].
- Counsel patients on treatment options and refer patients requesting help to appropriate inpatient or outpatient treatment programs that offer medications for SUDs.
- Offer rapid HIV testing to individuals who inject drugs.
- Refer patients who use injection drugs to syringe services programs and provide them with education about sterile injection techniques to reduce soft tissue infections and transmission of HIV and hepatitis C.

Intermediate

- For patients with moderate to severe OUD who are not in treatment or taking long-acting opioids (e.g., methadone), prescribe buprenorphine and provide referral to outpatient follow-up. While an "X-waiver" is required to write a discharge prescription for buprenorphine, any DEA-licensed ED provider can order a dose of buprenorphine to be administered in the ED (see ACEP's Buprenorphine Use in the ED Tool [60] https://www.acep.org/patient-care/bupe/ or the ED Bridge Guide [216]: https://ed-bridge.org/guide).
- Employ ED health navigators (peers or Health Promotion Advocates) to conduct substance use assessments, behavioral counseling, and linkage to treatment.
- Identify and assess unmet social needs among patients with substance use disorders and refer to community resources as appropriate.

Advanced

- Establish hospital-based programs or community partnerships to address patients' unmet social needs as a key part of addressing SUDs.

- Establish hospital workforce development programs for individuals with SUDs.
- Advocate for changes in public policy that expand access to addiction treatment and harm reduction rather than punishment.

Teaching Case

Clinical Case

Ms. B is a 30-year-old Latina female with limited English proficiency and a past medical history of chronic low back pain, post-traumatic stress disorder secondary to sexual assault, and bipolar disease. She reports a history of heroin use and more recent use of opioid pills that she buys on the street. She presents to the ED experiencing opioid withdrawal symptoms and requesting Percocet. She receives one dose of 4mg buprenorphine/1mg naloxone by the ED attending and is referred to the ED's Faster Paths to Treatment's Bridge Clinic for further evaluation the next day.

At Faster Paths, she reports that one year prior to her visit her oxycodone prescription was abruptly discontinued by her primary care physician, who was concerned about the new state prescribing regulations and monitoring system. She began buying "Percocet 30mg" and tramadol on the street to prevent withdrawal and relieve her low back pain. The Faster Paths bridge clinic physician prescribes buprenorphine 4mg/naloxone 1mg twice daily, and a 4mg Naloxone HCL nasal spray to carry on her person. After a month of monitoring her cravings, negative urine drug testing for opioids except for buprenorphine, and dosage adjustments with frequent visits, she is transferred to the hospital's Office-Based Addiction Treatment (OBAT) Maintenance Clinic.

During her last visit to Faster Paths, the patient reports extreme anxiety after receiving a letter notifying her to vacate her apartment by the end of the month. She reports that the landlord is selling the building in her gentrifying neighborhood and the new landlord intends to modernize the property and vastly increase rent. Ms. B voices worry about losing her low-rent apartment, which is within walking distance from her part-time job and addiction, medical, and psychiatric care at the same hospital. She reports poor credit, though she could afford first and last month's rent for a new apartment. She is scared of becoming homeless, which would threaten her stable work, safety for herself, and the security of her medications. She was most concerned about triggering a relapse to heroin, which she has not used in more than 5 years. An emergency physician working in the Faster Paths low barrier bridge clinic wrote a letter to her landlord requesting an extension on her lease. The clinic coordinator consulted BMC THRIVE, an online resource guide developed for direct referrals, and provided Ms. B with information on the BMC medical-legal partnership services and a state program that offers financial assistance for moving costs. The physician's letter helped Ms. B get a month's extension. She applied to Boston Public Housing Authority for Section 8 housing without success; however, prior to eviction, because of her resilience, support system, and resources, she found low-cost housing in a suburb outside of Boston. One year later, Ms. B continues to take her prescribed buprenorphine and regularly attend her medical, addiction, and psychiatric appointments.

Teaching Points

1. The relationship between substance use disorder and social needs is bidirectional. Addressing housing instability is a critical component for patient engagement and retention in substance use disorder treatment.
2. Screening for substance use disorder, homelessness, and housing insecurity should be integrated into ED practice, documented in the electronic health record, and linked to a referral and/or consult for an ED social worker or peer advocate, as available.
3. Emergency providers can write letters to landlords and government agencies and refer patients to local eviction prevention or legal services like the medical-legal partnership.
4. ED providers can provide treatment for opioid withdrawal and opioid use disorder using medication for opioid use disorder. Beginning treatment in the ED and providing a warm handoff/referrals are especially important for patients with concurrent social needs, who may not have the resources or ability to access outpatient treatment options without assistance.
5. EDs work within a system of care and cannot provide all needed services themselves. EDs should develop relationships with a referral network of substance use disorder treatment services and housing and other resources to help patients meet their social needs in their community. The SAMHSA practitioner locator guide is helpful to start the process of identifying referral sites for substance use disorder (https://www.samhsa.gov/medication-assisted-treatment/practitioner-program-data/treatment-practitioner-locator).

Discussion Questions

1. The patient presented to the ED in opioid withdrawal. The ED physician effectively treated the patient with 4 mg of SL buprenorphine following ED protocol and state law. How does this compare to your ED's current practice? What are some benefits, barriers, and enablers for providing buprenorphine in the ED? What mechanisms for outpatient follow-up are available in your local community?
2. What is the impact of homelessness and housing insecurity on the behavioral and physical health of people with substance use disorders? How do substance use disorders contribute to homelessness and housing insecurity? How can ED physicians identify housing insecurity and homelessness? How can providers direct patients to appropriate community resources for unmet health-related social needs? How is this currently done at your institution? What are the ways this could be improved?
3. The ED physician provided a warm handoff to colleagues in the ED low-barrier bridge clinic that provides buprenorphine prescriptions and case management. What are your thoughts on the role of ED physicians to identify and refer patients with substance use disorder? What are the best ways to do this?

4. What institutional or other supports would be needed for the ED to provide comprehensive care for patients with substance use disorders who are unstably housed? What resources are available at your ED/hospital? How can ED physicians advocate for their hospital to engage with and support community agencies already involved with housing issues?

References

1. Bernstein SL, D'Onofrio G. A promising approach for emergency departments to care for patients with substance use and behavioral disorders. Health Aff (Millwood). 2013;32(12):2122–8.
2. White AM, Slater ME, Ng G, Hingson R, Breslow R. Trends in alcohol-related emergency department visits in the United States: results from the nationwide emergency department sample, 2006 to 2014. Alcohol Clin Exp Res. 2018;42(2):352–9.
3. Weiss A, Bailey M, O'Malley L, Barrett M, Elixhauser A, Steiner C. Patient characteristics of opioid-related inpatient stays and emergency department visits nationally and by state, 2014. Agency for Healthcare Research and Quality: Rockville; 2017.
4. Vivolo-Kantor AM, Seth P, Gladden RM, Mattson CL, Baldwin GT, Kite-Powell A, et al. Vital signs: trends in emergency department visits for suspected opioid overdoses – United States, July 2016–September 2017. MMWR Morb Mortal Wkly Rep. 2018;67(9):279–85.
5. Han BH, Moore AA, Sherman S, Keyes KM, Palamar JJ. Demographic trends of binge alcohol use and alcohol use disorders among older adults in the United States, 2005–2014. Drug Alcohol Depend. 2017;170:198–207.
6. Grucza RA, Sher KJ, Kerr WC, Krauss MJ, Lui CK, McDowell YE, et al. Trends in adult alcohol use and binge drinking in the early 21st-century United States: a meta-analysis of 6 national survey series. Alcohol Clin Exp Res. 2018;42(10):1939–50.
7. Centers for Disease Control and Prevention. Average for United States 2006–2010 alcohol-attributable deaths due to excessive alcohol use 2013. Available from: www.cdc.gov/ARDI.
8. Centers for Disease Control and Prevention. Multiple cause of death 1999–2017 on CDC WONDER online database: National Center for Health Statistics.; December 2018. Available from: https://wonder.cdc.gov/mcd.html.
9. NIDA. Overdose Death Rates 2019 [updated January 2019]. Available from: https://www.drugabuse.gov/related-topics/trends-statistics/overdose-death-rates.
10. Spencer MR, Warner M, Bastian BA, Trinidad JP, Hedegaard H. Drug overdose deaths involving fentanyl, 2011–2016. Hyattsville: National Center for Health Statistics; 2019. Contract No.: 3.
11. Rudd RA, Seth P, David F, Scholl L. Increases in drug and opioid-involved overdose deaths – United States, 2010–2015. MMWR Morb Mortal Wkly Rep. 2016;65(5051):1445–52.
12. Scholl L, Seth P, Kariisa M, Wilson N, Baldwin G. Drug and opioid-involved overdose deaths – United States, 2013–2017. MMWR Morb Mortal Wkly Rep. 2018;67(5152):1419–27.
13. NIDA. Nationwide trends June 25, 2015. Available from: https://www.drugabuse.gov/publications/drugfacts/nationwide-trends.
14. Mack KA, Jones CM, Ballesteros MF. Illicit drug use, illicit drug use disorders, and drug overdose deaths in metropolitan and nonmetropolitan areas – United States. MMWR Surveill Summ. 2017;66(19):1–12.
15. National Survey on Drug Use and Health 2018 [Internet]. 2020. Available from: https://www.samhsa.gov/data/data-we-collect/nsduh-national-survey-drug-use-and-health.
16. Substance Abuse and Mental Health Services Administration. Key substance use and mental health indicators in the United States: results from the 2018 National Survey on Drug Use and Health (HHS Publication No. PEP19-5068, NSDUH Series H-54). Center for Behavioral

Health Statistics and Quality, Substance Abuse and Mental Health Services Administration: Rockville; 2019.
17. Andrilla CHA, Moore TE, Patterson DG, Larson EH. Geographic distribution of providers with a DEA waiver to prescribe buprenorphine for the treatment of opioid use disorder: a 5-year update. J Rural Health. 2019;35(1):108–12.
18. Haffajee RL, Lin LA, Bohnert ASB, Goldstick JE. Characteristics of us counties with high opioid overdose mortality and low capacity to deliver medications for opioid use disorder. JAMA Netw Open. 2019;2(6):e196373.
19. Lagisetty PA, Ross R, Bohnert A, Clay M, Maust DT. Buprenorphine treatment divide by race/ethnicity and payment. JAMA Psychiatry. 2019;76(9):979–81.
20. Kulesza M, Matsuda M, Ramirez JJ, Werntz AJ, Teachman BA, Lindgren KP. Towards greater understanding of addiction stigma: intersectionality with race/ethnicity and gender. Drug Alcohol Depend. 2016;169:85–91.
21. American Psychiatric Association. Substance-related and addictive disorders. Diagnostic and statistical manual of mental disorders. 5th ed. Washington, D.C.: American Psychiatric Association; 2013.
22. Shin SH, McDonald SE, Conley D. Patterns of adverse childhood experiences and substance use among young adults: a latent class analysis. Addict Behav. 2018;78:187–92.
23. Svingen L, Dykstra RE, Simpson JL, Jaffe AE, Bevins RA, Carlo G, et al. Associations between family history of substance use, childhood trauma, and age of first drug use in persons with methamphetamine dependence. J Addict Med. 2016;10(4):269–73.
24. Hughes K, Bellis MA, Hardcastle KA, Sethi D, Butchart A, Mikton C, et al. The effect of multiple adverse childhood experiences on health: a systematic review and meta-analysis. The Lancet Public Health. 2017;2(8):e356–e66.
25. Dube SR, Felitti VJ, Dong M, Chapman DP, Giles WH, Anda RF. Childhood abuse, neglect, and household dysfunction and the risk of illicit drug use: the adverse childhood experiences study. Pediatrics. 2003;111(3):564–72.
26. Doran KM, Rahai N, McCormack RP, Milian J, Shelley D, Rotrosen J, et al. Substance use and homelessness among emergency department patients. Drug Alcohol Depend. 2018;188:328–33.
27. Compton WM, Jones CM, Baldwin GT. Relationship between nonmedical prescription-opioid use and heroin use. N Engl J Med. 2016;374(2):154–63.
28. Rigg KK, Monnat SM. Urban vs. rural differences in prescription opioid misuse among adults in the United States: informing region specific drug policies and interventions. Int J Drug Policy. 2015;26(5):484–91.
29. Case A, Deaton A. Mortality and morbidity in the 21st century: Washington, DC. Brookings Institute; 2017.
30. Ulirsch JC, Weaver MA, Bortsov AV, Soward AC, Swor RA, Peak DA, et al. No man is an island: Living in a disadvantaged neighborhood influences chronic pain development after motor vehicle collision. Pain. 2014;155(10):2116–23.
31. Knapp EA, Bilal U, Dean LT, Lazo M, Celentano DD. Economic insecurity and deaths of despair in US Counties. Am J Epidemiol. 2019;188(12):2131–9.
32. Goedel WC, Shapiro A, Cerda M, Tsai JW, Hadland SE, Marshall BDL. Association of racial/ethnic segregation with treatment capacity for opioid use disorder in counties in the United States. JAMA Netw Open. 2020;3(4):e203711.
33. Saloner B, Cook BL. Blacks and hispanics are less likely than whites to complete addiction treatment, largely due to socioeconomic factors. Health Affairs. 2013;32(1):135–45.
34. Frieden TR. Shattuck lecture: the future of public health. N Engl J Med. 2015;373(18):1748–54.
35. Finkelhor D, Turner HA, Shattuck A, Hamby SL. Violence, crime, and abuse exposure in a national sample of children and youth: an update. JAMA Pediatr. 2013;167(7):614–21.
36. Finkelhor D, Turner HA, Shattuck A, Hamby SL. Prevalence of childhood exposure to violence, crime, and abuse: results from the national survey of children's exposure to violence. JAMA Pediatr. 2015;169(8):746–54.

37. Smith SG, Zhang X, Basile KC, Merrick MT, Wang J, Kresnow M, et al. The National Intimate Partner and Sexual Violence Survey (NISVS): 2015 data brief – updated release. National Center for Injury Prevention and Control, Centers for Disease Control and Prevention: Atlanta; 2018.
38. Center for Substance Abuse Treatment. Trauma-informed care in behavioral health services. Rockville: Substance Abuse and Mental Health Services Administration; 2014.
39. Substance Abuse and Mental Health Services Administration. SAMHSA's concept of trauma and guidance for a trauma-informed approach. HHS Publication No. (SMA) 14-4884. Substance Abuse and Mental Health Services Administration: Rockville; 2014.
40. Corbin TJ, Purtle J, Rich LJ, Rich JA, Adams EJ, Yee G, et al. The prevalence of trauma and childhood adversity in an urban, hospital-based violence intervention program. J Health Care Poor Underserved. 2013;24(3):1021–30.
41. Hawk K, D'Onofrio G. Emergency department screening and interventions for substance use disorders. Addict Sci Clin Pract. 2018;13(1):18.
42. Bernstein E, Topp D, Shaw E. A preliminary report of knowledge translation: lessons from taking screening and brief intervention techniques from the research setting into regional systems of care. Acad Emerg Med. 2009;16:1225.
43. SBIRT: Screening, Brief Intervention, and Referral to Treatment Washington, D.C.: SAMHSA-HRSA Center for Integrated Health Solutions. Available from: https://www.integration.samhsa.gov/clinical-practice/sbirt.
44. Madras BK, Compton WM, Avula D, Stegbauer T, Stein JB, Clark HW. Screening, brief interventions, referral to treatment (SBIRT) for illicit drug and alcohol use at multiple healthcare sites: comparison at intake and 6 months later. Drug Alcohol Depend. 2009;99(1-3):280–95.
45. Barata IA, Shandro JR, Montgomery M, Polansky R, Sachs CJ, Duber HC, et al. Effectiveness of SBIRT for alcohol use disorders in the emergency department: a systematic review. West J Emerg Med. 2017;18(6):1143–52.
46. Smith PC, Schmidt SM, Allensworth-Davies D, Saitz R. Primary care validation of a single-question alcohol screening test. J Gen Intern Med. 2009;24(7):783–8.
47. Smith PC, Schmidt SM, Allensworth-Davies D, Saitz R. A single-question screening test for drug use in primary care. Arch Intern Med. 2010;170(13):1155–60.
48. AUDIT-C Overview. Available from: https://www.integration.samhsa.gov/images/res/tool_auditc.pdf.
49. O'Brien CP. The CAGE questionnaire for detection of alcoholism: a remarkably useful but simple tool. JAMA. 2008;300(17):2054–6.
50. NIAAA. Helping patients who drink too much: a clinician's guide. National Institute on Alcohol Abuse and Alcoholism; Washington, DC. 2005.
51. Bernstein E, Topp D, Shaw E, Girard C, Pressman K, Woolcock E, et al. A preliminary report of knowledge translation: lessons from taking screening and brief intervention techniques from the research setting into regional systems of care. Acad Emerg Med. 2009;16(11):1225–33.
52. Smith PC, Cheng DM, Allensworth-Davies D, Winter MR, Saitz R. Use of a single alcohol screening question to identify other drug use. Drug Alcohol Depend. 2014;139:178–80. https://doi.org/10.1016/j.drugalcdep.2014.03.027. Epub 2014 Apr 5. PMID: 24768061; PMCID: PMC4085274.
53. Committee on Substance A, Levy SJ, Kokotailo PK. Substance use screening, brief intervention, and referral to treatment for pediatricians. Pediatrics. 2011;128(5):e1330–40.
54. Knight JR, Sherritt L, Shrier LA, Harris SK, Chang G. Validity of the CRAFFT substance abuse screening test among adolescent clinic patients. Arch Pediatr Adoles Med. 2002;156(6):607.
55. The Center for Adolescent Substance Use Research. The CRAFFT 2.1 manual. Boston: The Center for Adolescent Substance Use Research; 2018.
56. Miller WR, Rollnick S. Motivational interviewing: preparing people to change addictive behavior. New York: The Guilford Press; 2013.
57. Counselman FL, Babu K, Edens MA, Gorgas DL, Hobgood C, Marco CA, et al. The 2016 model of the clinical practice of emergency medicine. J Emerg Med. 2017;52(6):846–9.

58. Lemhoefer C, Rabe GL, Wellmann J, Bernstein SL, Cheung KW, McCarthy WJ, et al. Emergency department-initiated tobacco control: update of a systematic review and meta-analysis of randomized controlled trials. Prev Chronic Dis. 2017;14:E89.
59. CA BRIDGE. Resources. https://cabridge.org/tools/resources/. Accessed 26 Feb 2021.
60. Ketcham E, Ryan R. BUPE: buprenorphine use in the Emergency Department Tool Dallas, TX: American College of Emergency Physicians; 2018. Available from: https://www.acep.org/patient-care/bupe/.
61. D'Onofrio G, O'Connor PG, Pantalon MV, Chawarski MC, Busch SH, Owens PH, et al. Emergency department-initiated buprenorphine/naloxone treatment for opioid dependence: a randomized clinical trial. Jama. 2015;313(16):1636–44.
62. Mark T, Montejano L, Kranzler H, Chalk M, Gastfriend D. Comparison of healthcare utilization among patients treated with alcoholism medications. Am J Manag Care. 2010;16(12):879–88.
63. Bernstein SL, D'Onofrio G, Rosner J, O'Malley S, Makuch R, Busch S, et al. Successful tobacco dependence treatment in low-income emergency department patients: a randomized trial. Ann Emerg Med. 2015;66(2):140–7.
64. Bernstein SL, Boudreaux ED, Cydulka RK, Rhodes KV, Lettman NA, Almeida SL, et al. Tobacco control interventions in the emergency department: a joint statement of emergency medicine organizations. J Emerg Nurs. 2006;32(5):370–81.
65. Katz DA, Vander Weg MW, Holman J, Nugent A, Baker L, Johnson S, et al. The Emergency Department Action in Smoking Cessation (EDASC) trial: impact on delivery of smoking cessation counseling. Acad Emerg Med. 2012;19(4):409–20.
66. Duber HC, Barata IA, Cioe-Pena E, Liang SY, Ketcham E, Macias-Konstantopoulos W, et al. Identification, management, and transition of care for patients with opioid use disorder in the emergency department. Ann Emerg Med. 2018;18:S0196–644.
67. Sordo L, Barrio G, Bravo MJ, Indave BI, Degenhardt L, Wiessing L, et al. Mortality risk during and after opioid substitution treatment: systematic review and meta-analysis of cohort studies. BMJ. 2017;357:j1550.
68. Wakeman SE, Larochelle MR, Ameli O, Chaisson CE, McPheeters JT, Crown WH, et al. Comparative effectiveness of different treatment pathways for opioid use disorder. JAMA Netw Open. 2020;3(2):e1920622.
69. Substance Abuse and Mental Health Services Administration. Key substance use and mental health indicators in the United States: results from the 2018 National Survey on Drug Use and Health. Center for Behavioral Health Statistics and Quality, Substance Abuse and Mental Health Services Administration: Rockville; 2019.
70. Larochelle MR, Bernson D, Land T, Stopka TJ, Wang N, Xuan Z, et al. Medication for opioid use disorder after nonfatal opioid overdose and association with mortality: a Cohort Study. Ann Intern Med. 2018;169(3):137–45.
71. Busch SH, Fiellin DA, Chawarski MC, Owens PH, Pantalon MV, Hawk K, et al. Cost-effectiveness of emergency department-initiated treatment for opioid dependence. Addiction. 2017;112(11):2002–10.
72. Kampman K, Jarvis M. American Society of Addiction Medicine (ASAM) national practice guideline for the use of medications in the treatment of addiction involving opioid use. J Addict Med. 2015;9(5):358–67.
73. Administering or dispensing of narcotic drugs., 21 CFR § 1306.07(b) (1998).
74. Davis CS, Carr DH. The law and policy of opioids for pain management, addiction treatment, and overdose reversal. Indiana Health Law Rev. 2017;14(1):1.
75. D'Onofrio G, McCormack RP, Hawk K. Emergency departments – A 24/7/365 option for combating the opioid crisis. N Engl J Med. 2018;379(26):2487–90.
76. Substance Abuse and Mental Health Services Administration. Use of medications for opioid use disorder in emergency departments. Rockville: National Mental Health and Substance Use Policy Laboratory; Bethesda, MD, 2019.
77. CA BRIDGE. Buprenorphine (Bup) Hospital Quick Start. Updated Sept 2020. https://cabridge.org/resource/buprenorphine-bup-hospital-quick-start/. Accessed 26 Feb 2021.

78. Joudrey PJ, Edelman EJ, Wang EA. Drive times to opioid treatment programs in urban and rural counties in 5 US states. JAMA. 2019;322(13):1310–2.
79. Grimm C. Geographic disparities affect access to buprenorphine services for opioid use disorder. Washington, D.C.: U.S. Department of Health and Human Services. Office of Inspector General; 2020. Contract No.: OEI-12-17-00240.
80. Eibl JK, Gauthier G, Pellegrini D, Daiter J, Varenbut M, Hogenbirk JC, et al. The effectiveness of telemedicine-delivered opioid agonist therapy in a supervised clinical setting. Drug Alcohol Depend. 2017;176:133–8.
81. Huskamp HA, Busch AB, Souza J, Uscher-Pines L, Rose S, Wilcock A, et al. How is telemedicine being used in opioid and other substance use disorder treatment? Health Aff (Millwood). 2018;37(12):1940–7.
82. Uscher-Pines L, Huskamp HA, Mehrotra A. Treating patients with opioid use disorder in their homes: an emerging treatment model. JAMA. 2020;324:39.
83. Yang YT, Weintraub E, Haffajee RL. Telemedicine's role in addressing the opioid epidemic. Mayo Clin Proc. 2018;93(9):1177–80.
84. Zheng W, Nickasch M, Lander L, Wen S, Xiao M, Marshalek P, et al. Treatment outcome comparison between telepsychiatry and face-to-face buprenorphine medication-assisted treatment for opioid use disorder: a 2-year retrospective data analysis. J Addict Med. 2017;11(2):138–44.
85. Samuels EA, Clark SA, Wunsch C, Keeler LAJ, Reddy N, Vanjani R, et al. Innovation during COVID-19: improving addiction treatment access. J Addict Med. 2020;14:e8.
86. Palmateer N, Kimber J, Hickman M, Hutchinson S, Rhodes T, Goldberg D. Evidence for the effectiveness of sterile injecting equipment provision in preventing hepatitis C and human immunodeficiency virus transmission among injecting drug users: a review of reviews. Addiction. 2010;105(5):844–59.
87. MacArthur GJ, van Velzen E, Palmateer N, Kimber J, Pharris A, Hope V, et al. Interventions to prevent HIV and Hepatitis C in people who inject drugs: a review of reviews to assess evidence of effectiveness. Int J Drug Policy. 2014;25(1):34–52.
88. Aspinall EJ, Nambiar D, Goldberg DJ, Hickman M, Weir A, Van Velzen E, et al. Are needle and syringe programmes associated with a reduction in HIV transmission among people who inject drugs: a systematic review and meta-analysis. Int J Epidemiol. 2014;43(1):235–48.
89. Platt L, Minozzi S, Reed J, Vickerman P, Hagan H, French C, et al. Needle and syringe programmes and opioid substitution therapy for preventing HCV transmission among people who inject drugs: findings from a Cochrane Review and meta-analysis. Addiction. 2018;113(3):545–63.
90. Des Jarlais D, Perlis T, Arasteh K, Torian L, Hagan H, Beatrice S, et al. Reductions in hepatitis C virus and HIV infections among injecting drug users in New York City, 1990–2001. AIDS. 2005;19(Suppl 3):S20–5.
91. Nguyen TQ, Weir BW, Des Jarlais DC, Pinkerton SD, Holtgrave DR. Syringe exchange in the United States: a national level economic evaluation of hypothetical increases in investment. AIDS Behav. 2014;18(11):2144–55.
92. Walley AY, Xuan Z, Hackman HH, Quinn E, Doe-Simkins M, Sorensen-Alawad A, et al. Opioid overdose rates and implementation of overdose education and nasal naloxone distribution in Massachusetts: interrupted time series analysis. BMJ. 2013;346:f174.
93. Piper TM, Stancliff S, Rudenstine S, Sherman S, Nandi V, Clear A, et al. Evaluation of a naloxone distribution and administration program in New York City. Subst Use Misuse. 2008;43(7):858–70.
94. Doe-Simkins M, Walley AY, Epstein A, Moyer P. Saved by the nose: bystander-administered intranasal naloxone hydrochloride for opioid overdose. Am J Public Health. 2009;99(5):788–91.
95. Maxwell S, Bigg D, Stanczykiewicz K, Carlberg-Racich S. Prescribing naloxone to actively injecting heroin users: a program to reduce heroin overdose deaths. J Addict Dis. 2006;25(3):89–96.
96. Samuels E. Emergency department naloxone distribution: a Rhode Island department of health, recovery community, and emergency department partnership to reduce opioid overdose deaths. R I Med J. 2014;97(10):38–9.

97. Samuels EA, Baird J, Yang ES, Mello MJ. Adoption and utilization of an emergency department naloxone distribution and peer recovery coach consultation program. Acad Emerg Med. 2019;26(2):160–73.
98. Samuels EA, Hoppe J, Papp J, Whiteside L, Raja AS, Bernstein E. Naloxone distribution strategies needed in emergency departments. ACEPNow. American College of Emergency Physicians. March 16, 2016. https://www.acepnow.com/article/naloxone-distribution-strategies-needed-in-emergency-departments/.
99. Dwyer K, Walley AY, Langlois BK, Mitchell PM, Nelson KP, Cromwell J, et al. Opioid education and nasal naloxone rescue kits in the emergency department. West J Emerg Med. 2015;16(3):381–4.
100. Binswanger IA, Blatchford PJ, Mueller SR, Stern MF. Mortality after prison release: opioid overdose and other causes of death, risk factors, and time trends from 1999 to 2009. Ann Intern Med. 2013;159(9):592–600.
101. Bukten A, Stavseth MR, Skurtveit S, Tverdal A, Strang J, Clausen T. High risk of overdose death following release from prison: variations in mortality during a 15-year observation period. Addiction. 2017;112(8):1432–9.
102. Garg RK, Fulton-Kehoe D, Franklin GM. Patterns of opioid use and risk of opioid overdose death among medicaid patients. Med Care. 2017;55(7):661–8.
103. Nadpara PA, Joyce AR, Murrelle EL, Carroll NW, Carroll NV, Barnard M, et al. Risk factors for serious prescription opioid-induced respiratory depression or overdose: comparison of commercially insured and veterans health affairs populations. Pain Med. 2018;19(1):79–96.
104. Naloxone Overdose Prevention Laws: prescription drug abuse policy system; 2017. Available from: http://pdaps.org/datasets/laws-regulating-administration-of-naloxone-1501695139.
105. Ryan S, Pantalon MV, Camenga D, Martel S, D'Onofrio G. Evaluation of a pediatric resident skills-based screening, brief intervention and referral to treatment (SBIRT) curriculum for substance use. J Adolesc Health. 2018;63(3):327–34.
106. Duong DK, O'Sullivan PS, Satre DD, Soskin P, Satterfield J. Social workers as workplace-based instructors of alcohol and drug screening, brief intervention, and referral to treatment (SBIRT) for emergency medicine residents. Teach Learn Med. 2016;28(3):303–13.
107. Tetrault JM, Green ML, Martino S, Thung SF, Degutis LC, Ryan SA, et al. Developing and implementing a multispecialty graduate medical education curriculum on Screening, Brief Intervention, and Referral to Treatment (SBIRT). Subst Abus. 2012;33(2):168–81.
108. Improving the hospital and emergency department response to substance use disorders: a Project ASSERT Case Study: American Hospital Association; [cited 2019]. Available from: https://www.aha.org/system/files/content/17/project-assert-case-study.pdf.
109. Project ASSERT: Boston Medical Center. Available from: https://www.bmc.org/programs/project-assert.
110. The BNI ART Institute. Project ASSERT: SBIRT in Emergency Care: The Boston University School of Public Health. Available from: https://www.bu.edu/bniart/sbirt-experience/sbirt-programs/sbirt-project-assert/.
111. Faster Paths to Treatment: Boston Medical Center. Available from: https://www.bmc.org/programs/faster-paths-to-treatment.
112. Welch AE, Jeffers A, Allen B, Paone D, Kunins HV. Relay: a peer-delivered emergency department-based response to nonfatal opioid overdose. Am J Public Health. 2019;109(10):1392–5.
113. Addiction Policy Forum. Anchor ED. Washington, D.C. February 2017. www.addictionpolicy.org/post/anchored-rhode-island.
114. Waye KM, Goyer J, Dettor D, Mahoney L, Samuels EA, Yedinak JL, et al. Implementing peer recovery services for overdose prevention in Rhode Island: an examination of two outreach-based approaches. Addict Behav. Washington, D.C. 2019;89:85–91.
115. McGuire AB, Powell KG, Treitler PC, Wagner KD, Smith KP, Cooperman N, et al. Emergency department-based peer support for opioid use disorder: emergent functions and forms. J Subst Abuse Treat. 2020;108:82–7.

116. NYC Department of Health and Mental Hygiene. Information for Peers. Available from: https://www1.nyc.gov/site/doh/health/health-topics/alcohol-and-drug-use-information-for-peers.page.
117. Samuels EA, Bernstein SL, Marshall BDL, Krieger M, Baird J, Mello MJ. Peer navigation and take-home naloxone for opioid overdose emergency department patients: preliminary patient outcomes. J Subst Abuse Treat. 2018;94:29–34.
118. Summary of 2019 council resolutions: resolution 50 social work in the Emergency Department American College of Emergency Physicians; 2019. Available from: https://washingtonacep.org/wp-content/uploads/2019-Resolutions-Adopted-by-the-Council-Board.pdf.
119. Boston Medical Center. The preventive food pantry Boston, MA. Available from: https://www.bmc.org/nourishing-our-community/preventive-food-pantry.
120. Bridle-Fitzpatrick S. Hospital – food bank partnerships: a recipe for community health: hunger and health; April 25, 2018. Available from: https://hungerandhealth.feedingamerica.org/2018/04/hospital-food-bank-partnerships-recipe-community-health/.
121. Tsemberis S, Gulcur L, Nakae M. Housing First, consumer choice, and harm reduction for homeless individuals with a dual diagnosis. Am J Public Health. 2004;94(4):651–6.
122. Urbanoski K, Veldhuizen S, Krausz M, Schutz C, Somers JM, Kirst M, et al. Effects of comorbid substance use disorders on outcomes in a Housing First intervention for homeless people with mental illness. Addiction. 2018;113(1):137–45.
123. Palepu A, Patterson ML, Moniruzzaman A, Frankish CJ, Somers J. Housing first improves residential stability in homeless adults with concurrent substance dependence and mental disorders. Am J Public Health. 2013;103(Suppl 2):e30–6.
124. Vuolo L, Oster R, Weber E. Evaluating the promise and potential of the parity act on its tenth anniversary, Health Affairs Blog October 10, 2018. Available from: https://www.healthaffairs.org/do/10.1377/hblog20181009.356245/full/.
125. Meinhofer A, Witman AE. The role of health insurance on treatment for opioid use disorders: evidence from the Affordable Care Act Medicaid expansion. J Health Econ. 2018;60:177–97.
126. Lagisetty PA, Ross R, Bohnert A, Clay M, Maust DT. Buprenorphine treatment divide by race/ethnicity and payment. JAMA Psychiatry. 2019;76:979.
127. Nowotny KM. Race/ethnic disparities in the utilization of treatment for drug dependent inmates in U.S. state correctional facilities. Addict Behav. 2015;40:148–53.
128. Guerrero EG, Marsh JC, Duan L, Oh C, Perron B, Lee B. Disparities in completion of substance abuse treatment between and within racial and ethnic groups. Health Serv Res. 2013;48(4):1450–67.
129. Gilbert PA, Pro G, Zemore SE, Mulia N, Brown G. Gender differences in use of alcohol treatment services and reasons for nonuse in a national sample. Alcohol Clin Exp Res. 2019;43(4):722–31.
130. Cerda M, Gaidus A, Keyes KM, Ponicki W, Martins S, Galea S, et al. Prescription opioid poisoning across urban and rural areas: identifying vulnerable groups and geographic areas. Addiction. 2017;112(1):103–12.
131. Amiri S, Lutz R, Socias ME, McDonell MG, Roll JM, Amram O. Increased distance was associated with lower daily attendance to an opioid treatment program in Spokane County Washington. J Subst Abuse Treat. 2018;93:26–30.
132. Guerrero EG, Garner BR, Cook B, Kong Y, Vega WA, Gelberg L. Identifying and reducing disparities in successful addiction treatment completion: testing the role of Medicaid payment acceptance. Subst Abuse Treat Prev Policy. 2017;12(1):27.
133. Komaromy M, Duhigg D, Metcalf A, Carlson C, Kalishman S, Hayes L, et al. Project ECHO (Extension for Community Healthcare Outcomes): a new model for educating primary care providers about treatment of substance use disorders. Subst Abus. 2016;37(1):20–4.
134. Tofighi B, Abrantes A, Stein MD. The role of technology-based interventions for substance use disorders in primary care: a review of the literature. Med Clin North Am. 2018;102(4):715–31.
135. Lord S, Moore SK, Ramsey A, Dinauer S, Johnson K. Implementation of a substance use recovery support mobile phone app in community settings: qualitative study of clinician and staff perspectives of facilitators and barriers. JMIR Ment Health. 2016;3(2):e24.

136. Anderson M, Kumar M. Digital divide persists even as lower-income Americans make gains in tech adoption: Pew Research Center; 2019. Available from: https://www.pewresearch.org/fact-tank/2019/05/07/digital-divide-persists-even-as-lower-income-americans-make-gains-in-tech-adoption/.
137. Federal Communications Commission. 2019 Broadband Deployment Report. Federal Communications Commission. https://www.fcc.gov/reports-research/reports/broadband-progress-reports/2019-broadband-deployment-report. Published May 29, 2019. Accessed August, 2020.
138. Perrin A. Digital gap between rural and nonrural America persists: Pew Research Center; [updated May 31, 2019]. Available from: https://www.pewresearch.org/fact-tank/2019/05/31/digital-gap-between-rural-and-nonrural-america-persists/.
139. MassHealth Launches Restructuring To Improve Health Outcomes for 1.2 Million Members [press release]. Boston: Executive Office of Health and Human Services; 2018.
140. Centers for Medicare and Medicaid Services Innovation Center. Accountable Health Communities Model [updated Nov 13, 2019]. Available from: https://innovation.cms.gov/initiatives/ahcm.
141. Hunt KA, Weber EJ, Showstack JA, Colby DC, Callaham ML. Characteristics of frequent users of emergency departments. Ann Emerg Med. 2006;48(1):1–8.
142. LaCalle E, Rabin E. Frequent users of emergency departments: the myths, the data, and the policy implications. Ann Emerg Med. 2010;56(1):42–8.
143. Curran G, Sullivan G, Williams K, Han X, Allee E, Kotrla K. The association of psychiatric comorbidity and use of the emergency department among persons with substance use disorders: an observational cohort study. BMC Emerg Med J. 2008;8:17.
144. Moe J, Kirkland SW, Rawe E, Ospina MB, Vandermeer B, Campbell S, et al. Effectiveness of interventions to decrease emergency department visits by adult frequent users: a systematic review. Acad Emerg Med. 2017;24(1):40–52.
145. Kwan BM, Rockwood A, Bandle B, Fernald D, Hamer MK, Capp R. Community health workers: addressing client objectives among frequent emergency department users. J Public Health Manag Pract. 2018;24(2):146–54.
146. Bodenmann P, Velonaki VS, Ruggeri O, Hugli O, Burnand B, Wasserfallen JB, et al. Case management for frequent users of the emergency department: study protocol of a randomised controlled trial. BMC Health Serv Res. 2014;14:264.
147. Sadowski LS, Kee RA, VanderWeele TJ, Buchanan D. Effect of a housing and case management program on emergency department visits and hospitalizations among chronically ill homeless adults: a randomized trial. JAMA. 2009;301(17):1771–8.
148. Doran KM, Ragins KT, Gross CP, Zerger S. Medical respite programs for homeless patients: a systematic review. J Health Care Poor Underserved. 2013;24(2):499–524.
149. Warren O, Smith-Bernardin S, Jamieson K, Zaller N, Liferidge A. Identification and practice patterns of sobering centers in the United States. J Health Care Poor Underserved. 2016;27(4):1843–57.
150. Jarvis SV, Kincaid L, Weltge AF, Lee M, Basinger SF. Public intoxication: sobering centers as an alternative to incarceration, Houston, 2010–2017. Am J Public Health. 2019;109(4):597–9.
151. American College of Surgeons Committee on Trauma. Resources for the optimal care of the injured patient. Chicago: American College of Surgeons; 2014.
152. Colorado ACEP. Opioid prescribing & treatment guidelines. Colorado ACEP: Northglenn; 2017.
153. American College of Emergency Physicians. E-QUAL Network Opioid Initiative 2019. Available from: https://www.acep.org/administration/quality/equal/emergency-quality-network-e-qual/e-qual-opioid-initiative/.
154. Rhode Island Department of Behavioral Healthcare Developmental Disabilities, and Hospitals, Rhode Island Department of Health. Levels of Care for Rhode Island Emergency Departments and Hospitals for Treating Overdose and Opioid Use Disorder. 2017 March.

155. Baltimore City Health Department. Levels of care for Baltimore City Hospitals responding to the opioid epidemic guide for Hospitals Baltimore, MD August 2018. Available from: https://health.baltimorecity.gov/sites/default/files/Levels%20of%20Care%20-%20Guide.pdf.
156. NYC Department of Health and Mental Hygiene. Guidance for the care of patients presenting to New York City Emergency Departments following a non-fatal opioid overdose New York City: NYC Department of Health and Mental Hygiene; May 31, 2019. Available from: https://www1.nyc.gov/assets/doh/downloads/pdf/basas/non-fatal-overdose-providers.pdf.
157. Massachusetts Health and Hospital Association. Guidelines for medication for addiction treatment for opioid use disorder within the emergency department: Massachusetts Health and Hospital Association; January 2019. Available from: https://www.mhalink.org/MHADocs/MondayReport/2019/18-01-04MATguidelinesNEWFINAL.pdf.
158. Kilaru AS, Perrone J, Kelley D, Siegel S, Lubitz SF, Mitra N, et al. Participation in a hospital incentive program for follow-up treatment for opioid use disorder. JAMA Netw Open. 2020;3(1):e1918511.
159. Pennsylvania Department of Human Services. Follow-up treatment after ED visit for opioid use disorder June 6, 2018. Available from: https://www.pamedsoc.org/docs/librariesprovider2/pamed-documents/new-opioid-measures-hqip.pdf?sfvrsn=8524485b_2.
160. Santiago. Health and Safety Code Section 11757: substance use disorder treatment: peer navigators: California State Legislature; Feb 5 2019. Available from: https://leginfo.legislature.ca.gov/faces/billTextClient.xhtml?bill_id=201920200AB389.
161. Boardman JD, Finch BK, Ellison CG, Williams DR, Jackson JS. Neighborhood disadvantage, stress, and drug use among adults. J Health Soc Behav. 2001;42(2)
162. Galea S, Nandi A, Vlahov D. The social epidemiology of substance use. Epidemiol Rev. 2004;26:36–52.
163. Galea S, Rudenstine S, Vlahov D. Drug use, misuse, and the urban environment. Drug Alcohol Rev. 2005;24(2):127–36.
164. Rothstein R. The Color of Law: a forgotten history of how our government segregated America. Liveright: New York; 2017.
165. US Census Bureau. 2018 American Community Survey February 13, 2020. Available from: https://www.census.gov/programs-surveys/acs/news/data-releases/2018/release.html#par_textimage_copy.
166. Schaeffer K. 6 facts about economic inequality in the U.S. Washington, D.C.: Pew Research Center; February 7, 2020. Available from: https://www.pewresearch.org/fact-tank/2020/02/07/6-facts-about-economic-inequality-in-the-u-s/.
167. Zoorob MJ, Salemi JL. Bowling alone, dying together: the role of social capital in mitigating the drug overdose epidemic in the United States. Drug Alcohol Depend. 2017;173:1–9.
168. Pear VA, Ponicki WR, Gaidus A, Keyes KM, Martins SS, Fink DS, et al. Urban-rural variation in the socioeconomic determinants of opioid overdose. Drug Alcohol Depend. 2019;195:66–73.
169. LaVeist TA, Wallace JM. Health risk and inequitable distribution of liquor stores in African American neighborhood. Soc Sci Med. 2000;51(4):613–7.
170. Bluthenthal RN, Cohen DA, Farley TA, Scribner R, Beighley C, Schonlau M, et al. Alcohol availability and neighborhood characteristics in Los Angeles, California and southern Louisiana. J Urban Health. 2008;85(2):191–205.
171. Hamad R, Brown DM, Basu S. The association of county-level socioeconomic factors with individual tobacco and alcohol use: a longitudinal study of U.S. adults. BMC Public Health. 2019;19(1):390.
172. Saloner B, McGinty EE, Beletsky L, Bluthenthal R, Beyrer C, Botticelli M, et al. a public health strategy for the opioid crisis. Public Health Rep. 2018;133(1_suppl):24S–34S.
173. Ashford RD, Brown AM, Ryding R, Curtis B. Building recovery ready communities: the recovery ready ecosystem model and community framework. Addict Res Theory. 2019;28:1–11.
174. Cloud W, Granfield R. Conceptualizing recovery capital: expansion of a theoretical construct. Subst Use Misuse. 2008;43(12-13):1971–86.

175. Matthew DB. Un-burying the lead: public health tools are the key to beating the opioid epidemic. USC-Brookings Schaeffer Initiative for Health Policy; 2018.
176. Wittman FD, Polcin DL, Sheridan D. The architecture of recovery: two kinds of housing assistance for chronic homeless persons with substance use disorders. Drugs Alcohol Today. 2017;17(3):157–67.
177. Jason LA, Olson BD, Ferrari JR, Lo Sasso AT. Communal housing settings enhance substance abuse recovery. Am J Public Health. 2006;96(10):1727–9.
178. Schiff DM, Nielsen T, Hoeppner BB, Terplan M, Hansen H, Bernson D, et al. Assessment of racial and ethnic disparities in the use of medication to treat opioid use disorder among pregnant women in Massachusetts. JAMA Netw Open. 2020;3(5):e205734.
179. Guerrero EG, Marsh JC, Khachikian T, Amaro H, Vega WA. Disparities in Latino substance use, service use, and treatment: implications for culturally and evidence-based interventions under health care reform. Drug Alcohol Depend. 2013;133(3):805–13.
180. Valdez LA, Flores M, Ruiz J, Oren E, Carvajal S, Garcia DO. Gender and cultural adaptations for diversity: a systematic review of alcohol and substance abuse interventions for latino males. Subst Use Misuse. 2018;53(10):1608–23.
181. Perreira KM, Pedroza JM. Policies of exclusion: implications for the health of immigrants and their children. Annu Rev Public Health. 2019;40:147–66.
182. Alexander M. The New Jim Crow: mass incarceration in the age of colorblindness. New York: Jackson, Tenn New Press; 2010.
183. Koester S, Mueller SR, Raville L, Langegger S, Binswanger IA. Why are some people who have received overdose education and naloxone reticent to call Emergency Medical Services in the event of overdose? Int J Drug Policy. 2017;48:115–24.
184. Gonsalves GS, Crawford FW. Dynamics of the HIV outbreak and response in Scott County, IN, USA, 2011–15: a modelling study. The Lancet HIV. 2018;5(10):e569–e77.
185. Flath N, Tobin K, King K, Lee A, Latkin C. enduring consequences from the war on drugs: how policing practices impact HIV risk among people who inject drugs in Baltimore city. Subst Use Misuse. 2017;52(8):1003–10.
186. Goedel WC, King MRF, Lurie MN, Galea S, Townsend JP, Galvani AP, et al. Implementation of syringe services programs to prevent rapid human immunodeficiency virus transmission in rural counties in the United States: a modeling study. Clin Infect Dis. 2019;
187. Davis CS, Carr DH, Samuels EA. Paraphernalia laws, criminalizing possession and distribution of items used to consume illicit drugs, and injection-related harm. Am J Public Health. 2019;109(11):1564–7.
188. Fazel S, Yoon IA, Hayes AJ. Substance use disorders in prisoners: an updated systematic review and meta-regression analysis in recently incarcerated men and women. Addiction. 2017;112(10):1725–39.
189. The Drug Policy Alliance. The drug war, mass incarceration and race. The Drug Policy Alliance: New York; 2018.
190. Pew Charitable Trusts. More imprisonment does not reduce state drug problems: data show no relationship between prison terms and drug misuse. Washington, D.C.: Pew Charitable Trusts; 2018.
191. Wolff N, Huening J, Shi J, Frueh BC. Trauma exposure and posttraumatic stress disorder among incarcerated men. J Urban Health. 2014;91(4):707–19.
192. Hinton E. From the war on poverty to the war on crime: the making of mass incarceration in America. Cambridge, MA: Harvard University Press; 2016.
193. Rosenberg A, Groves AK, Blankenship KM. Comparing black and white drug offenders: implications for racial disparities in criminal justice and reentry policy and programming. J Drug Issues. 2017;47(1):132–42.
194. Substance Abuse and Mental Health Services Administration. Medication-Assisted Treatment (MAT) in the criminal justice system: brief guidance to the states: Substance Abuse and Mental Health Services Administration; March 2019. Available from: https://store.samhsa.gov/system/files/pep19-matbriefcjs.pdf.

195. Green TC, Clarke J, Brinkley-Rubinstein L, Marshall BDL, Alexander-Scott N, Boss R, et al. Postincarceration Fatal Overdoses After Implementing Medications for Addiction Treatment in a Statewide Correctional System. JAMA Psychiatry. 2018;75(4):405–7.
196. Marsden J, Stillwell G, Jones H, Cooper A, Eastwood B, Farrell M, et al. Does Exposure to Opioid Substitution Treatment in Prison Reduce the Risk of Death After Release? A National Prospective Observational Study in England. Addiction. 2017;112(8):1408–18.
197. Degenhardt L, Larney S, Kimber J, Gisev N, Farrell M, Dobbins T, et al. The impact of opioid substitution therapy on mortality post-release from prison: retrospective data linkage study. Addiction. 2014;109(8):1306–17.
198. Wildeman C, Wang EA. Mass incarceration, public health, and widening inequality in the USA. The Lancet. 2017;389(10077):1464–74.
199. Ban the Box Campaign. Available from: https://bantheboxcampaign.org.
200. Matthew DB. Un-burying the lead: public health tools are the key to beating the opioid epidemic. USC-Brookings Schaeffer Initiative for Health Policy; 2018.
201. Ordinance 125393: fair chance housing ordinance August 23, 2017. Available from: http://seattle.legistar.com/View.ashx?M=F&ID=5387389&GUID=6AA5DDAE-8BAE-4444-8C17-62C2B3533CA3.
202. Peters RH, Young MS, Rojas EC, Gorey CM. Evidence-based treatment and supervision practices for co-occurring mental and substance use disorders in the criminal justice system. Am J Drug Alcohol Abuse. 2017;43(4):475–88.
203. Tsai J, Flatley B, Kasprow WJ, Clark S, Finlay A. Diversion of veterans with criminal justice involvement to treatment courts: participant characteristics and outcomes. Psychiatr Serv. 2017;68(4):375–83.
204. The Center for Health and Justice at TASC. No entry: a national survey of criminal justice diversion programs and initiatives. Chicago: The Center for Health and Justice at TASC; 2013.
205. Nicosia N, Macdonald JM, Arkes J. Disparities in criminal court referrals to drug treatment and prison for minority men. Am J Public Health. 2013;103(6):e77–84.
206. Goncalves R, Lourenco A, Silva SN. A social cost perspective in the wake of the Portuguese strategy for the fight against drugs. Int J Drug Policy. 2015;26(2):199–209.
207. Ferreira S. Portugal's radical drugs policy is working. Why hasn't the world copied it? The Guardian. 2017 Dec 5.
208. Bajekal N. Want to win the war on drugs? Portugal might have the answer. Time. 2018 Aug 1.
209. Sharbaugh MS, Althouse AD, Thoma FW, Lee JS, Figueredo VM, Mulukutla SR. Impact of cigarette taxes on smoking prevalence from 2001-2015: a report using the Behavioral and Risk Factor Surveillance Survey (BRFSS). PLoS One. 2018;13(9):e0204416.
210. Frazer K, Callinan JE, McHugh J, van Baarsel S, Clarke A, Doherty K, et al. Legislative smoking bans for reducing harms from secondhand smoke exposure, smoking prevalence and tobacco consumption. Cochrane Database Syst Rev. 2016;(2):CD005992.
211. National Conference of State Legislatures. State Medical Marijuana Laws June 5, 2019. Available from: http://www.ncsl.org/research/health/state-medical-marijuana-laws.aspx.
212. National Conference of State Legislatures. Marijuana Overview May 28, 2019. Available from: http://www.ncsl.org/research/civil-and-criminal-justice/marijuana-overview.aspx.
213. Kuehn B. Declining opioid prescriptions. JAMA. 2019;321(8):736.
214. Dowell D, Haegerich T, Chou R. CDC guideline for prescribing opioids for chronic pain – United States, 2016. MMWR Recomm Rep. 2016;65(1):1–49.
215. Herring AA, Perrone J, Nelson LS. Managing opioid withdrawal in the emergency department with buprenorphine. Ann Emerg Med. 2019;73(5):481–7.
216. Herring A, Snyder H, Moulin A, Luftig J, Sampson A, Trozky R, et al. Buprenorphine Guide: ED-BRIDGE; 2018. Available from: https://ed-bridge.org/guide.
217. Sallis JF, Owen N, Fisher EB. Ecological models of health behavior. In: Glanz KRB, Viswanath K, editors. Health behavior and health education. 4th ed. San Francisco: Wiley; 2008. p. 465–85.

Part IV

Social Needs

Education and Employment

11

Erik S. Anderson and Jennifer Avegno

Key Points

- Employment status, and the working conditions of one's employment, are social determinants of health (SDOH) that should be considered in emergency department (ED) clinical decision making.
- Emergency physicians can leverage legal services such as medical-legal partnerships to help address employment issues affecting health, and advocate for policies that improve access to quality, safe, and stable employment for patients in their communities.
- Low educational attainment is associated with risky behaviors that result in ED visits; screening with linkage to community resources to improve educational attainment, particularly in younger patients, may be a reasonable strategy for ED patients.
- Low educational attainment can be addressed as a social need in the ED through patient-provider communication strategies and partnerships with community-based organizations.

E. S. Anderson (✉)
Department of Emergency Medicine, Highland Hospital – Alameda Health System, Oakland, CA, USA
e-mail: esanderson@alamedahealthsystem.org

J. Avegno
Department of Emergency Medicine, University Medical Center New Orleans, New Orleans, LA, USA
e-mail: javegn@lsuhsc.edu

H. J. Alter et al. (eds.), *Social Emergency Medicine*,
https://doi.org/10.1007/978-3-030-65672-0_11

Foundations

Background

The idea that conditions such as employment, work status, and wages influence health contributed substantially to the conceptualization of social determinants of health (SDOH). The formative Whitehall study, initiated in 1967, found a strong inverse correlation between social class, as defined by the type of employment and wages, and mortality from a wide range of diseases among British civil servants. They noted that while overall mortality fell from 1970 to 1980, the mortality rates in manual labor occupations declined more slowly than in non-manual labor occupations. Because manual occupations were generally lower-earning, the difference in existing health disparities between social classes widened [1]. In 1991, Marmot and colleagues published a follow-up to this study, the Whitehall II study, showing that these differences in mortality rates persisted over the subsequent two decades [2]. These landmark studies provide empirical evidence that employment and wages are associated with disparities in mortality. As the wealth gap in the US widens, recent evidence suggests such disparities will only become more marked [3].

Recent research on inequity in mortality between lower and higher income earners points to income as a marker for both employment and education, where median income rises with each tier of educational attainment [4, 5]. Chetty et al. reported in 2014 that there was a 15-year difference in life expectancy for men and a 10-year difference in life expectancy for women when comparing the richest 1% of individuals and the poorest 1% in the US [3]. This is a similar differential in life expectancy attributable to a lifetime of smoking. Given that wealth is even more unequally distributed than income, health disparities may persist or worsen as wealth is passed from generation to generation [3]. In addition to differences in mortality, low-income individuals with chronic diseases are more than three times as likely as affluent individuals with chronic disease to have impairments in activities of daily living [5].

Evidence Basis

Education, employment, and income are distinct but related social determinants that individually and together can affect the need for ED care, which can serve as a safety net both for high social needs and acute medical problems [6, 7]. Individuals living in poverty and those with Medicaid insurance are more likely to use the ED for routine health needs, at the same time as those needs are concentrated with poverty [6–8], compounding the effect. For example, in a study of more than 60,000 outpatients, educational attainment was independently associated with the prevalence of chronic disease [9]. Further, among participants who developed chronic kidney disease, those who completed college had a 24% lower mortality rate over the 7-year observation period than those who only completed some high school [9].

Un- and under-employment was a powerful covariate in a study of frequent ED use, where patients in this predicament were more than twice as likely to use the ED five or more times in a year than those employed full time [10]. Although the relationship between employment and ED use is complex, much of it is driven by the persistence of employment-linked insurance in the US, and the primacy of insurance in access to quality care [11, 12].

Patients with lower educational attainment are more likely to be prescribed opioids, suggesting that this population may be disproportionately impacted by iatrogenic consequences of opioid prescribing [13]. The influence of education on health outcomes can also be seen in individual diseases. Patients with sickle cell disease who do not have a high school diploma are more likely to visit the ED when compared to those who have graduated from high school, after adjustment for disease severity, psychosocial functioning, and employment status [14]. The impact, of course, for patients with illness early in life is bidirectional, as their underlying sickle cell disease directly inhibits some patients' ability to receive a quality education [15]. Addressing and optimizing care for young patients who are chronically ill becomes especially critical as impacts on their education can have compounding effects on their health outcomes.

Low educational attainment and unemployment are social needs that ED providers are not accustomed to addressing. Although these needs may seem disconnected from the use of health services and health outcomes, research shows that education and employment are deeply rooted social risk factors, and the effects of inadequate education and low employment are generational and bidirectional [16]. Specifically, improving an individual's education level seems to have the most health benefit for people whose parents are poorly educated, beyond the observation that personal educational attainment seems to counteract the effects of poorly educated parents [17]. In a study analyzing 37 years of follow-up data from an early childhood intervention program in a low-income African American community in Michigan, Muenning et al. found that subsequent educational attainment, insurance status, and income were higher over time in the intervention group [18]. Improving education through dedicated interventions in vulnerable communities is a means for addressing health inequities and overcoming generational cycles of poverty.

There is a significant and consistent association with high school grades and health risk behaviors [19]. Students who have lower grades in high school are more likely to carry a weapon, drink alcohol regularly, smoke cigarettes, be currently sexually active, and are more likely to be sedentary and less physically active. Students with lower grades also are more likely to misuse prescription drugs, take part in a physical fight, be sexually assaulted, and not use a seatbelt in a car [19]. Understanding the effects of educational attainment early in life can lead to clear risk factors for acute care visits, and advocating for policies that support low-achieving students may mitigate some of these health behaviors and thus, possibly, downstream ED use.

Despite the gradient in education and earnings, the cutoff of having earned a high school diploma is a clear binary discriminator in wages and in health outcomes. A 25-year-old woman with a high school degree can expect to live to an age of 86 years, as opposed to a similar aged woman without a degree whose life expectancy is shorter by 11 years; the gap for men is 15 years [20]. Another estimate finds annual attributable deaths from lack of a high school education to be comparable to the number of deaths that could be averted if all current smokers had the mortality rate of former smokers—potentially an apples-to-apples assessment when considering the role of socioeconomic status in both exposures [21]. It is important to be aware of the intersectionality of educational disparities with inequities in race and gender, all of which can work together to compound differential health outcomes [22, 23].

Finally, in the setting of an overburdened emergency system where throughput and acute care delivery are paramount, interventions will need to hold to the tenets of social EM and consider individual patient-provider interactions, EDs, and hospital systems, and at the same time take a societal perspective. This is not to say that it is within the purview of emergency physicians to improve a person's wages, directly improve access to high-quality education and opportunity in their community, or find patients suitable employment during a routine ED visit; but if the health effects of poverty are similar to a lifetime of smoking, we ought to consider it in our practice just as we consider other more traditional risk factors. In this chapter, we outline pragmatic approaches to address inequities rooted in education and employment, and how practicing emergency physicians can consider these factors when providing clinical care in the ED. We outline an approach to care that considers these needs, tailors disposition planning and bedside teaching appropriately, and recognizes the value in ED care for visits that some may view as "non-urgent."

Emergency Department and Beyond

Bedside

The most apparent impact of a patient's employment in the ED setting may be from dangerous working conditions or inherently dangerous professions. Construction workers, for example, have a work-related injury rate of up to four times the general population, and agricultural injuries are quite common in rural emergency departments [24, 25]. Additionally, certain occupations, most famously firefighters, have an alarming risk of both on-duty occupational mortality as well as marked chronic disease risks associated with their professions [26].

Low educational attainment may be among the most prevalent and readily recognizable SDOH for patients in the clinical setting when it manifests as poor health literacy (see elsewhere, *Language and Literacy*). We can, however, adopt the principles of communication aimed at improving care for patients with low levels of educational attainment, which include: (1) universal precautions and assuming a lower level of patient educational attainment; (2) developing visual aids for patient education; (3) simplifying written materials; and (4) engaging the use of patient navigators, and specifically engaging the target audience [27]. While these approaches may seem like common sense, their impact for patients can be profound.

Recognizing education as a critically important social determinant, investigators in one pediatric ED used the ED visit as an opportunity to enroll children in a pilot project that aimed to increase reading in the home. This intervention, which consisted of distributing a brochure that stressed the importance of reading at home and a take-home children's book, did not have a meaningful impact on parent-child reading [28]. Despite the observed outcome, it is one of the few studies in the literature to attempt an ED-based educational intervention, suggesting that more work is needed to investigate whether or not there are feasible ways to promote improved educational performance in patients in such a context, or whether efforts may be better spent advocating for educational reforms in a community setting.

While EDs may not clearly be poised to improve patient educational attainment during a single encounter, we can modify practices to mitigate the impact of education on health outcomes. There are a few practical strategies that emergency providers can take to address some of the impacts of educational attainment and employment social risk factors, and these primarily are accomplished through linkage to resources in the community. Providers can begin to collect information on educational resources and opportunities for job training in their area. Access to quality medical care is a potential mediator of the relationship between educational attainment and employment and health. Linking patients to available comprehensive population health-focused community programs—programs that can engage in patient's social and medical risks longitudinally [29, 30] —is a unique opportunity in the ED. Screening strategies for social needs in the ED are a focus of ongoing research, but it does seem that both patients and community partners feel this is an important part of ED care [31]. In one qualitative study by Samuels-Kalow et al., patients did not necessarily expect that the ED would be able to fully address complex SDOH immediately from an ED visit, but viewed any effort as a positive step forward. There remains significant work to be done regarding implementation strategies for screening that can be integrated into ED care [31].

Hospital/Healthcare System

Hospitals and health systems have a key role in creating an environment and culture that engages patients at all levels of educational attainment. This can take the form of medical-legal help desks, investment by hospitals in struggling communities, and specifically focusing on improving care for common problems facing communities with low educational attainment and high rates of under- or unemployment.

The American Academy of Medical Colleges (AAMC) has developed a set of tools that address social determinants of health, and as many as a quarter of the implementation strategies address education with a focus on the pediatric and adolescent population [32]. Strategies outlined by the AAMC include programs that contribute to kindergarten and college readiness, as well as initiatives that use telemedicine to decrease school absences for school-aged children.

Medical-legal partnerships (MLPs) present another opportunity for hospitals and EDs to directly offer patients new educational and employment opportunities while they interact with the healthcare system. A study by Losonczy et al. examined the impact of a help desk for health-related social needs, integrated with a medical-legal partnership and based in an urban ED [33]. They found that employment was identified by almost 25% of all patients as their top social need when interacting with the help desk. This particular help desk aimed to help the patient navigate social services, including unemployment benefits, as well as job training opportunities. The provision of resources that aim to address employment and other SDOH while patients are in the hospital or ED through the help desk model offers an opportunity to expand our scope of services and engage in upstream interventions without the need for clinicians to engage in-depth at the bedside during a patient's acute care encounter.

In most areas, hospitals and healthcare systems are among the largest employers, which offers another avenue to improve education and employment at the community level. A variety of jobs at different skill levels are required to deliver comprehensive healthcare, and some hospitals are leveraging this demand into investments in their communities, with a focus on justice and equity, directing funds already slated for employment opportunities. These institutions then position themselves as anchor institutions, defined as hospitals and EDs that develop and maintain deep roots in their communities through jobs and job training [34]. The concept of "outside-in"—direct hiring from the neighborhood into the institution—and "inside-up"—training and promoting locally hired talent to management positions—employment practices are bedrock to the anchor institution concept.

Pipeline programs focus on hiring, educating, and mentoring young people in health professions with the goal of increasing the proportion of underrepresented communities in health professions. Such initiatives are a way for hospitals to not only employ members of the communities they serve, but also to invest in their futures by facilitating experiences in the health professions in underserved communities. One such program in Oakland, California—Mentoring in Medicine and Science (MIMS)—employs local youth from communities that are underrepresented in medicine as health coaches [35]. Their jobs are augmented with mentoring and shadowing opportunities and reflect an upstream approach to education and employment that is based within the existing healthcare system. MIMS also conducts science enrichment and mentoring programs in struggling high schools and middle schools. Kaiser Permanente began to experiment with vocational training as a community health intervention in the 1980s, though long-term data on outcomes from this approach have not been published. The dual mission of their School of Allied Health Sciences, "to meet the demands of technologist shortages and provide community outreach and vocational training," was first deployed in the struggling community of Richmond, California [36].

Hospital-based violence intervention programs, such as the Violence Intervention Advocacy Program at Boston Medical Center and Cure Violence at University Medical Center in New Orleans, are often staffed by survivors of violent injury, and also engage other survivors of violence through advocacy, education, and employment opportunities. Engaging victims of violence has led to some early positive outcomes for patients, who have found the education and job training initiatives embedded in these more comprehensive programs to be particularly beneficial [37]. Johns Hopkins Medical Center also has a formal plan and approach to investing in their local communities through utilizing minority- and women-owned companies, increasing hiring of community residents, and providing job opportunities to neighborhoods with high rates of unemployment [38]. Such investments in the community by hospitals and healthcare institutions represent systems-level interventions that address SDOH with the aim of lifting up communities economically, and improving long-term health outcomes.

Societal Level

Education and employment are some of the largest issues facing society from a political and economic perspective, and are a focus of local, regional, and national policy discussions. Many experts have argued that educational policy specifically falls within the domain of public health action and advocacy, but the clear connection between education and employment and health outcomes is less commonly made at the policy level. Clearly articulating the connection between educational attainment and health outcomes leverages the public health community—and the evidence base—to make a more robust argument for addressing SDOH to improve health equity. The literature supporting this argument has been strong for decades.

While there are many specific policies that address equitable educational attainment and employment conditions for communities, one approach that leverages our position as healthcare providers is expanding the role of Medicaid in SDOH. Medicaid specifically aids low-income and vulnerable communities, making it an appropriate avenue for addressing social risk factors aimed at improving population health. In *Health Affairs*, DeSalvo and Leavitt estimate that 40 states have Medicaid programs that address SDOH through contracts and waivers with the Centers for Medicare and Medicaid Services (CMS) [39]. Advocating for support, expansion, and dissemination of Medicaid-supported partnerships with managed care organizations and other stakeholders may create a sustainable, integrated approach to integrating SDOH into community health.

Recommendations for Emergency Medicine Practice

Basic

- Emergency physicians should recognize that a patient's educational attainment and employment status are key drivers of health outcomes. These outcomes are manifest in underlying chronic disease as well as acute presentations in the ED.
- A focus on populations at high risk of suffering adverse outcomes related to low education, such as children with chronic diseases, adolescents with substance

use disorders, or patients who are being considered for opioid prescriptions, may help mitigate the impact of education as a social determinant.
- Caring for patients who are under- or unemployed is a routine part of emergency medicine, and understanding risk factors associated with certain occupations, as well as understanding limitations to discharge plans related to employment issues, can help us care for this population.

Intermediate

- A medical-legal partnership or help desk for health-related social needs, where patients are referred after a brief screening in the ED, is a reasonable parallel-process intervention that has some evidence to support its practice.
- Hospitals can take on the role of anchor institutions by investing in their local communities through stable employment and educational opportunities.

Advanced

- Emergency physicians should consider supporting policy proposals that aim to improve educational quality, specifically for vulnerable and poor communities.
- Support the sustainable expansion of services aimed at addressing SDOH through Medicaid and public-private partnerships.

Teaching Case

Clinical Case

Ms. V. is a 22-year-old woman presenting to the ED with an acute asthma exacerbation. She reports using her budesonide inhaler every 2 hours and her albuterol inhaler every morning as she reports being instructed. She has been progressively worsening over the last several weeks, and has been unable to be seen in primary care as her sliding scale clinic has not had any space for appointments and she does not have health insurance.

Ms. V is in obvious respiratory distress with poor air entry and wheezing in bilateral lung fields and she is initially placed on non-invasive positive pressure ventilation with continuous nebulized albuterol. She also receives intravenous steroids and magnesium. While she improves over the next several hours and awaits admission to the hospital, you have the opportunity to discuss with the patient her medication management with a culturally appropriate color-coded sheet showing the brown color on the "controller" medication and the red color on the albuterol "rescue" medication.

She also discusses her difficulty finding health insurance and regular preventative care access due to unsteady "gig" employment. She did not graduate from high school, and work in the area is hard to come by without a high school diploma. You ask the ED Help Desk team to speak with her and she is connected with a local community program that provides free classes toward a GED. This community center also works with the hospital to place young people in health career internships and employment opportunities at the medical center.

Teaching Points

1. Culturally appropriate medication instructions in a simple format are a straightforward way to mitigate the impact of education on short-term health outcomes. This may be particularly useful for chronic disease management and prevention of acute exacerbations requiring ED visits.
2. A knowledge of local community resources is helpful to connect patients in education and employment resources, and "help desks" can offload this task when available.

Discussion Questions

1. What resources are available at your hospital for simple patient-oriented handouts with health information?
2. You live in the same urban area as Ms. V. What community interventions or initiatives could you support to help address some of the upstream educational barriers she faces?

References

1. Fuller JH, Shipley MJ, Rose G, Jarrett RJ, Keen H. Coronary-heart-disease risk and impaired glucose tolerance. The Whitehall study. The Lancet. 1980;1(8183):1373–6.
2. Marmot MG, Stansfeld S, Patel C, North F, Head J. Health inequalities among British civil servants: the Whitehall II study. Lancet. 1991;337(8754):1387–93.
3. Weller CE, Hanks A. The widening racial wealth gap in the United States after the great recession. InForum Soc Econ. 2018;47(2):237–52. Routledge.
4. Chetty R, Stepner M, Abraham S, Lin S, Scuderi B, Turner N, et al. The association between income and life expectancy in the United States, 2001–2014. JAMA. 2016;315(16):1750.
5. Chokshi DA. Income, poverty, and health inequality. JAMA. 2018;319(13):1312–3.
6. Rodriguez RM, Fortman J, Chee C, Ng V, Poon D. Food, shelter and safety needs motivating homeless persons' visits to an urban emergency department. Ann Emerg Med. 2009; 53(5):598–602.e1.
7. Kushel MB, Perry S, Bangsberg D, Clark R, Moss AR. Emergency department use among the homeless and marginally housed: results from a community-based study. Am J Public Health. 2002;92(5):778–84.
8. Tang N, Stein J, Hsia RY, Maselli JH, Gonzales R. Trends and characteristics of US emergency department visits, 1997–2007. JAMA. 2010;304(6):664.
9. Choi AI, Weekley CC, Chen S-C, Li S, Kurella Tamura M, Norris KC, et al. Association of educational attainment with chronic disease and mortality: the Kidney Early Evaluation Program (KEEP). Am J Kidney Dis. 2011;58(2):228–34.

10. Behr JG, Diaz R. Emergency department frequent utilization for non-emergent presentments: results from a regional urban trauma center study. PLoS One. 2016;11(1):e0147116.
11. Giannouchos TV, Kum HC, Foster MJ, Ohsfeldt RL. Characteristics and predictors of adult frequent emergency department users in the United States: a systematic literature review. J Eval Clin Pract. 2019;25(3):420–33.
12. Bisgaier J, Rhodes KV. Auditing access to specialty care for children with public insurance. N Engl J Med. 2011;364(24):2324–33.
13. Platts-Mills TF, Hunold KM, Bortsov AV, Soward AC, Peak DA, Jones JS, et al. More educated emergency department patients are less likely to receive opioids for acute pain. Pain. 2012;153(5):967–73.
14. Jonassaint CR, Beach MC, Haythornthwaite JA, Bediako SM, Diener-West M, Strouse JJ, et al. The association between educational attainment and patterns of emergency department utilization among adults with sickle cell disease. Int J Behav Med. 2016;23(3):300–9.
15. Schwartz LA, Radcliffe J, Barakat LP. Associates of school absenteeism in adolescents with sickle cell disease. Pediatr Blood Cancer. 2009;52(1):92–6.
16. Gugushvili A, McKee M, Murphy M, Azarova A, Irdam D, Doniec K, King L. Intergenerational mobility in relative educational attainment and health-related behaviours. Soc Indic Res. 2019;141(1):413–41.
17. Miech R, Pampel F, Kim J, Rogers RG. The enduring association between education and mortality: the role of widening and narrowing disparities. Am Sociol Rev. 2011;76(6):913–34.
18. Muennig P, Schweinhart L, Montie J, Neidell M. Effects of a prekindergarten educational intervention on adult health: 37-year follow-up results of a randomized controlled trial. Am J Public Health. 2009;99(8):1431–7.
19. Health Risk Behaviors and Academic Achievement. Center for Disease Control. https://www.cdc.gov/healthyyouth/health_and_academics/pdf/health_risk_behaviors.pdf Accessed 10/1/2019.
20. Sasson I. Trends in life expectancy and lifespan variation by educational attainment: United States, 1990-2010. Demography. 2016;53(2):269–93.
21. Krueger PM, Tran MK, Hummer RA, Chang VW. Mortality attributable to low levels of education in the United States. PLoS One. 2015;10(7):e0131809.
22. Harris A, Leonardo Z. Intersectionality, race-gender subordination, and education. Rev Res Educ. 2018;42(1):1–27.
23. Sasson I, Hayward MD. Association between educational attainment and causes of death among white and black US adults, 2010–2017. JAMA. 2019;322(8):756.
24. Zwerling C, Miller ER, Lynch CF, Torner J. Injuries among construction workers in rural Iowa: emergency department surveillance. J Occup Environ Med. 1996;38(7):698–704.
25. Stueland D, Zoch T, Stamas P, Krieg G, Boulet W. The spectrum of emergency care of agricultural trauma in central Wisconsin. Am J Emerg Med. 1990;8(6):528–30.
26. Kales SN, Soteriades ES, Christophi CA, Christiani DC. Emergency duties and deaths from heart disease among firefighters in the United States. N Engl J Med. 2007;356(12):1207–15.
27. US Department of Health and Human Services, Office of Disease Prevention and Health Promotion. National Action Plan to improve health literacy. Washington, D.C.: Author; 2010.
28. Nagamine WH, Ishida JT, Williams DR, Yamamoto RI, Yamamoto LG. Child literacy promotion in the emergency department. Pediatr Emerg Care. 2001;17(1):19–21.
29. Levesque JF, Harris MF, Russell G. Patient-centred access to health care: conceptualising access at the interface of health systems and populations. Int J Equity Health. 2013;12(1):18.
30. Wesson D, Kitzman H, Halloran KH, Tecson K. Innovative population health model associated with reduced emergency department use and inpatient hospitalizations. Health Aff. 2018;37(4):543–50.
31. Samuels-Kalow ME, Molina MF, Ciccolo GE, Curt A, Manchanda EC, de Paz NC, Camargo CA Jr. Patient and community organization perspectives on accessing social resources from the emergency department: a qualitative study. Western J Emerg Med. 2020;21(4):964.

32. Teaching Hospitals' Commitment to Addressing the Social Determinants of Health. American Academy of Medical Colleges. https://www.aamc.org/system/files/c/2/480618-aamc-teaching-hospitals-addressing-sdoh.pdf. Accessed 8/1/2020.
33. Losonczy L Ilona, Hsieh D, Hahn C, Fahimi J. More than just meds: National survey of providers' perceptions of patients' social, economic, environmental, and legal needs and their effect on emergency department\ldots. Soc Ldots. 2014.
34. Democracy Collaborative. https://democracycollaborative.org/. Accessed 8/1/2020.
35. Freeman Garrick J. Commentary: screening, education, and advocacy on SDOH: the complete package for treating the whole patient. Ann Emerg Med. 2019;74(5 Supplement):S25–7.
36. Kaiser Foundation Hospital – Northern California Region. https://oshpd.ca.gov/ml/v1/resources/document?rs:path=%5CData-And-Reports%5CCommunity-Benefit-Plans%5C2017%5C106074093_NS_CBP_2017.pdf. Accessed 5/1/2020.
37. James TL, Bibi S, Langlois BK, Dugan E, Mitchell PM. Boston violence intervention advocacy program: a qualitative study of client experiences and perceived effect. Acad Emerg Med. 2014;21(7):742–51.
38. Simmons M. Johns Hopkins pledges more jobs and investment for Baltimore. Baltimore Business Journal. https://www.bizjournals.com/baltimore/news/2020/01/29/johnshopkins-pledges-more-jobs-and-investment-for.html#:~:text=Investing%20at%20least%20%2475%20million,more%20citizens%20with%20criminal%20records.&text=Increasing%20local%20business%20spending%20by,owned%20and%20veteran%2Downed%20companies. Accessed 5/1/2020.
39. DeSalvo K, Leavitt MO. For an option to address social determinants of health, look to medicaid. Health Affairs Blog. https://www.healthaffairs.org/do/10.1377/hblog20190701.764626/full/. Accessed 8/1/2020.

Financial Insecurity

12

Stephen B. Brown and Karen D'Angelo

Key Points

- Due to the widening gap in American income and wealth, financial insecurity is a substantial and growing problem in the US.
- Financial insecurity directly and indirectly affects other social determinants of health. It adversely impacts patients' access to social and economic resources that contribute to health (e.g., access to healthy food, secure housing, utilities, and heat), as well as their ability to adhere to medical treatment recommendations.
- Emergency medicine staff can assess financial insecurity with screening questions and direct patients into programs that can address their financial stress. Additionally, the department and/or hospital should seek to integrate technology that identifies and refers patients to community-based resources.
- Some healthcare systems have begun to address financial insecurity beyond the individual patient level by adopting an anchor institution mission, recognizing that hospitals can play a vital economic role in the communities that they serve and can help build prosperous and healthier neighborhoods.
- Structural competency, defined as the need for healthcare systems and healthcare providers to understand and address the structural roots of poor health, should be incorporated into medical school and continuing education curricula.

S. B. Brown (✉)
Department of Emergency Medicine, University of Illinois Hospital & Health Sciences System, Chicago, IL, USA
e-mail: sbbrown9@uic.edu

K. D'Angelo
Department of Social Work, Southern Connecticut State University, New Haven, CT, USA
e-mail: dangelok1@southernct.edu

H. J. Alter et al. (eds.), *Social Emergency Medicine*,
https://doi.org/10.1007/978-3-030-65672-0_12

Foundations

Background

> A dramatic divergence between data and experience is confounding America's policy debates. The data seem to show that households have attained unprecedented prosperity, and wages have (at worst) held their own against inflation, or (at best) risen much faster than prices. By conventional measures, material living standards everywhere in the income distribution are at all-time highs, and technological progress continues to improve them. Yet many jobs able to support a family in the past no longer do. Millennials are in worse financial shape than were those of Generation X at the same age, who themselves had fallen behind the baby boomers. The stories appear irreconcilable [1].
>
> *–Oren Cass, Executive Director, American Compass*

Financial insecurity is a growing problem in the US. According to a national survey conducted by the Center for Financial Services Innovation, only 70 million people, or about 28% of US adults, think they are "financially healthy" [2].

Despite optimistic economic perceptions after the recovery from the Great Recession of 2008, data shows that the country has seen slow growth in the living standards of low- and middle-income Americans and rising rates of inequality [3]. In 2017, 55% of US adults, or 138 million Americans, reported struggling with some financial concerns, and 42 million (17%) were struggling with all, or nearly all, aspects of their financial lives [2]. While very high wage workers have seen increases of 41% since 1980, middle class wages have increased just 6%, and low wage workers have seen a −5% cumulative change, contributing to an ever widening wage stagnation in the lower and middle class [4]. There is also a profound Black-White wealth gap revealing the accumulated effects of inequality and discrimination, and which has resulted in decreased intergenerational wealth transfer within Black families. An examination of US family wealth in 2016 found a staggering tenfold difference between White and Black families ($171,000 versus $17,650, respectively) [5].

A generation ago, a typical male worker could cover a family's expenses of four major expenditures (housing, healthcare, transportation, and education) on 30 weeks of salary, leaving 22 weeks of pay for everything else a family wants and needs, such as food, clothing, entertainment, and savings. By 2018, it took 53 weeks just for the four major expenditures [1].

"Financial insecurity" is used broadly to describe living paycheck to paycheck and/or concerns about making ends meet. People who are financially insecure are economically vulnerable; they have little savings, often spend as much as or more than they make, and are frequently crippled by unmanageable debt [6]. The Federal Reserve Board found that 40% of adults, "if faced with an unexpected expense of $400, would either not be able to cover it or would cover it by selling something or borrowing money" [7]. Increased basic living expenses, including rising food, housing, and healthcare costs, as well as the rising cost of higher education, have the potential to lead to economic instability. Certain groups are disproportionately affected by financial insecurity, including, for example, women, Black and Hispanic individuals and families, younger people, people with lower educational attainment, people with unstable work schedules, people living in the southern and western US, and people who grew up poor [2].

More recently, the crisis caused by COVID-19 has resulted in significant job losses and financial insecurity for many Americans [8, 9]. There are signs this is getting worse, with 32% of Americans missing house payments in July 2020 [10], mortgage delinquencies hitting a record high in April 2020, exceeding those seen during the Great Recession [11] and estimates that if the crisis persists, 28 million could become homeless [12].

The pandemic has exposed the persistent structural racial disparities related to financial security, with 73% of Black Americans and 70% of Hispanic Americans stating they do not have financial reserves to cover emergency expenses, compared to 47% of White Americans [8, 9]. Without the economic stability that comes with well-compensated and stable employment, people in the US often have limited access to vital resources [13] such as affordable housing, nutritious food, quality childcare and education, reliable transportation, safe neighborhoods for exercise and/or play, and comprehensive healthcare. For the most vulnerable in our society, the struggles to afford basic necessities are compounded by the exorbitant cost of healthcare, estimated to be two times per capita compared to other countries, coupled with the chronic underfunding of social services [14, 15]. Economic security directly and indirectly influences the social conditions in which we live and influences people's ability to maintain healthy lives and adhere to their healthcare providers' recommendations.

The federal government provides assistance to low-income families to improve overall health and decrease inequities [16]. Examples of government assistance programs are SNAP (Supplemental Nutrition Assistance Program), commonly known as food stamps, and the Housing Choice Voucher Program, also known as Section 8 [17]. Many of the federal government's programs use means-testing as a qualifying criterion. In order to be eligible, the recipient's income and assets must be below an identified threshold, often in relation to the Federal Poverty Level (FPL, currently $26,200 for a family of 4) [18]. Yet the FPL has not risen over time to account for inflation and the actual costs of living [19]. Thus, obsolete qualifying guidelines may make government programs inaccessible to those who are functionally impoverished yet considered to be too well off for aid.

Program criteria may be overly restrictive and counterintuitive and create disincentives to seek employment or self-sufficiency. For those with a medical condition(s) that renders them unable to work, even if they do not have a work history, Supplemental Security Income (SSI) provides limited income that is 30% below the FPL (approximately $750 per month) [20]. It can be further reduced by an array of disqualifying events, such as going back to work and earning more than $1220 per month or $14,640 a year. A person cannot have assets over $2000, so that someone who may own a car, no matter how old or in disrepair it may be, may see their benefits reduced or eliminated [21]. Thus, while government programs help mitigate financial strain, requirements to receive and maintain benefits are complicated. Consequently, patients may be unsure of what economic supports they may continue to receive. This may further exacerbate economic stress for these patients and potentially confound concerns about affordability of care.

Evidence Basis

Impact of Financial Insecurity on Health

In addition to providing timely, high-quality emergent care for life-threatening illness and injury, the emergency department (ED) functions as a crucial element in the social safety net that serves vulnerable populations with high rates of material and financial needs [22]. Since health in the US follows a linear socioeconomic gradient [23] but the value of a dollar is fixed, those most deprived are disproportionately affected by financial insecurity and its consequent outcomes [24, 25]. In other words, potentially preventable repeat ED visits and health crises manifesting from the social determinants of health such as food insecurity, transportation issues, difficulties paying for utilities, housing instability, and other health-related social needs may be rooted in financial insecurity [26, 27].

Financial insecurity is associated with poor health outcomes, including both physical and mental illness [28–32]. Chest pain is a common presenting concern in EDs, accounting for approximately 7 million visits [33]. Between 30–50% of patients with non-specific chest pain are found to be suffering from a panic or anxiety disorder [34]. One study of chest pain attributed to panic disorder found that financial insecurity contributed to the experience of pain [35]. Financial insecurity is not just associated with anxiety, panic, and pain however; a study published in February 2019 demonstrated that income volatility, a particular type of financial strain defined as an unexpected drop of earnings of 25% or more, was associated with a twofold risk of heart attack, heart failure, stroke, and early death. As previous research on heart health and income has shown, low-income individuals have a higher risk for heart disease than high-income earners [36].

Stress related to financial insecurity appears to exacerbate pain. A study by Chou et al. found a significant relationship between unemployment and the use of pain medication. Participants were asked to recall a time in their lives when they felt financially vulnerable. Those that reported vulnerability described almost double the amount of physical pain when compared to those who were economically stable, after controlling for age, gender, negative affect, and employment status [35].

The number of suicides in the US rose 30.4% between 1999 and 2015, now ranking as one of the top 10 leading causes of death [37], while it fell in most western European countries except the Netherlands [38]. Research has generally found that the higher the level of income inequality in the US states, the higher the probability of death by suicide. When there is a large gap between the rich and poor, those at or near the bottom struggle more, making them more susceptible to addiction, mental illness, and criminality. Controlling for other variables, states with higher per capita spending on social services had lower rates of suicide [37]. Evidence suggests that policies that improve financial security, such as increasing the minimum wage or the earned income tax credit (EITC), may reduce the suicide rate. It is estimated that raising the EITC by 10% would prevent 1230 suicides annually, according to the National Bureau of Economic Research [39].

Financial insecurity also often impacts victims of intimate partner violence (IPV). Victims of IPV have four times the odds of experiencing housing instability as the result of economic abuse, where the perpetrator controls a person's ability to acquire, use, and maintain economic resources [40]. It is found in almost all battering relationships [41]. Thus, financial insecurity and financial dependence may reinforce IPV victims' decisions to remain with abusive partners.

Impact on Healthcare Utilization

Financial strain can be an important factor in making healthcare decisions for many low-income individuals, who often forgo medical care in favor of essential basic needs like food and rent. They are sometimes forced to make trade-offs between household and individual needs. In a 2017 survey of US adults, 27% of adults went without some form of recommended medical care. For families making $40,000 or less per year, that figure was closer to 40%. Moreover, 20% had significant, unexpected medical bills to pay, 37% of whom incurred unpaid debt from those bills [7]. This follows an income gradient, with 65% of respondents earning $50,000 or less putting off timely care [42].

The lack of flexible medical payment options hurt families with children. They are less likely to be able to pay their out-of-pocket costs in full and, compared to individuals, are twice as likely to have their medical bills sent to collections [43]. The burden of unexpected, expensive medical bills has been attributed to almost two-thirds of bankruptcies and 57% of mortgage foreclosures [44, 45]. Foreclosures have been associated with an increased probability of Child Protective Services (CPS) involvement, and increases in ED utilization [46].

Cost-related nonadherence (CRN) is patient behavior that seeks to reduce or avoid the cost of care. This is most often described in studies about prescription medications. Those individuals affected by CRN report more comorbidities. Two national studies noted significant associations between multiple chronic diseases and CRN after controlling for income and sociodemographic factors [47].

Out-of-pocket medication costs are increasing for many Americans, and as the adult population ages, the number of prescriptions may increase. Increased dependency on medications often occurs during a point in the life course when people's incomes may be decreasing, or people are living on a fixed income. Even with Medicare Part D, the prescription drug benefit for seniors, out-of-pocket expenses may be substantial. Specialty tier drugs – which Medicare defines as those costing more than $670 per month – accounted for over 20% of all Part D spending in 2019, up from 6% to 7% before 2010. Commercial Part D plans are permitted to charge coinsurance premiums that could exceed $5100 a year – unaffordable for many seniors on a fixed income [48].

Compared to patients with commercial insurance, patients with Medicaid encounter more barriers to primary care, such as lack of transportation, inability to connect on the telephone, long waits in the physician's office, and inconvenient office hours, and the presence of these are associated with higher ED utilization [49]. Yet those that seek ED care come with a wide variety of material needs related to financial insecurity [22]. Asked about their choice to use EDs, low-income patients in a Pennsylvania study (both the insured and uninsured) explained that they consciously chose the ED over nonhospital settings because they perceived that "the care was cheaper, the quality of care was seemingly better, transportation options were more readily accessible, and, in some cases, the hospital offered more respite than a physician's office" [27].

Emergency Department and Beyond

Bedside

> *Social determinants of health is an abstract term, but for millions of Americans, it is a very tangible, frightening challenge: How can someone manage diabetes if they are constantly worrying about how they're going to afford their meals each week? How can a mother with an asthmatic son really improve his health if it's their living environment that's driving his condition? This can feel like a frustrating, almost fruitless position for a healthcare provider, who understands what is driving the health conditions they're trying to treat, who wants to help, but can't simply write a prescription for healthy meals, a new home, or clean air.* [50]
>
> –Alex M. Azar II, US Secretary of Health and Human Services

The conundrum described by Secretary Azar is one that emergency medicine providers face every shift. The challenges of addressing the excess burden of acute medical needs in communities where the health effects of deep poverty contribute to ED use and poorer outcomes can, at times, be overwhelming [38]. It can be difficult to uncover the causes of repeated utilization and poorly managed chronic disease with the unending flow of patients and time critical diagnoses in the ED. The time and effort needed to identify social needs and to intervene is not something one provider can do. Addressing social determinants of health and financial insecurity calls for a coordinated, systematic, team-based approach that includes nursing, social work, care coordinators, peer recovery specialists, and community health workers.

Because patients may feel shame, embarrassment, or guilt if directly asked about financial insecurity, they may not disclose this information. However, clinicians should remain cognizant that many unmet social needs are related to underlying financial concerns. For example, patients may struggle to pay for electricity, which is needed not only for refrigeration, cooking, heating, and cooling but also to run, for example, a nebulizer. Barriers to electricity affect compliance with medical devices and can lead to medical crises requiring emergency care [51].

Early in the course of the clinical encounter, ED providers should consider financial barriers, especially when devising a patient's discharge treatment plan. Facilitating access to no- or low-cost prescription medication is one way to do this. Prior to prescribing medications, have a conversation and ask questions to determine whether a person can afford the medications: "I need to prescribe you some medications. How do you normally pay for medicines? Do you have insurance?" Most of the major retail pharmacy chains – CVS Caremark, Walgreens, Target, Walmart – offer $4–10 formularies of commonly prescribed generic medications for 30–90 days [52, 53]. ED physicians can adjust their prescribing to align with these formularies. As an added benefit, EDs can purchase $4, $5, or $10 gift cards to be distributed to patients unable to afford their medications and provide a list of 24-hour pharmacies near the patient's home or the hospital. The website needymeds.org offers lower-cost alternatives for some specialty care medications [54]. By working with the patient to address social care challenges, the ED provider and their team may be able to mitigate potential complications and facilitate a better outcome.

Additionally, documenting social determinants in clinical notes will begin to quantify the scope of the issue. The International Statistical Classification of Diseases and Related Health Problems taxonomy (ICD-10) of medical diagnoses and procedures has a section for factors influencing health status. These "Z codes" are intended to document social needs that impact health. These include low income (Z59.6), insufficient social insurance and welfare support (Z59.7), and problems related to housing and economic circumstances, unspecified (Z59.9) [55].

Hospital/Healthcare System

Social Determinant of Health Screening

Healthcare systems can put into place programs that screen for and mediate individual patients' health-related social needs while also more broadly addressing the upstream factors – the social determinants – that impact the health of their patients from the surrounding communities [56]. A 2017 survey by Deloitte found that 88% of hospitals are beginning to incorporate screening for social needs but that many are ad hoc or intermittent, with 40% reporting no current capability to measure outcomes [57, 58]. The largest multi-site project to date, the Centers for Medicare and Medicaid Services' (CMS) Accountable Health Communities (AHC) model, has been designed to integrate the recognition of social determinants in order to bridge the gap between healthcare and human service providers in a hospital's primary service area. The AHC supports 31 clinical-community linkages throughout the US with a goal to improve health outcomes and reduce costs by identifying and addressing health-related social needs and working within communities to increase social services. These AHCs must screen patients for five social conditions: housing instability, food insecurity, transportation needs, interpersonal violence, and utility needs. AHCs must develop an inventory of social services in their communities of service providers, identify shortages in those services, and work with community members to develop a plan to ameliorate the gap in services [59].

There are numerous screening instruments that can aid providers in identifying unmet social needs, but two are widely used. The AHC Health-Related Social Needs Screening Tool has a core module of ten questions regarding living situation, food insecurity, transportation, utilities, and safety. There are also 16 supplemental questions that cover financial strain, employment, family and community support, education, physical activity, substance use, mental health, and disabilities [60]. The National Association of Community Health Centers Protocol for Responding to and Assessing Patients' Assets, Risks and Experiences tool (PRAPARE) includes 20 questions that can be directly uploaded into many electronic health records as structured data [61]. Both instruments screen for social needs related to financial insecurity such as housing instability, food insecurity, and utility needs. The AHC screening tool also contains a direct question about financial strain which EM providers or staff can use if a system is in place (i.e., social or case worker in the ED or referrals) to address responses indicating hardship: "How hard is it for you to pay for the very basics like food, housing, medical care, and heating?" with response options of not hard at all, somewhat hard, or very hard [57].

Hospitals and clinics commonly seek to refer their patients to community-based social service organizations that assist with non-medical needs. In the past, they generated informal lists of community-based organizations (CBO) on paper or in electronic lists that were not regularly updated. Community resource referral platforms are online web-based tools that catalog community-based social and healthcare services, provide search capabilities, and have the ability to send referrals to CBOs. A 2019 report by Social Interventions Research & Evaluation Network (SIREN) compared features of nine vendor platforms. A recent innovation is the addition of a "close-the-loop" communication which notifies healthcare providers that a patient accessed a referred agency [62].

Clinical Programs

For patients having difficulty affording their medications, pharmaceutical manufacturers have developed over 200 pharmaceutical assistance programs (PAP) that provide free to low-cost subsidies for name-brand medications [63]. There are also copay assistance programs for expensive, lifelong drugs such as transplant anti-rejection formularies. Burley et al. found that in the 12 months after PAP enrollment, patients experienced a 51% reduction in the likelihood of visiting an ED or hospital [64]. There are inconsistent eligibility requirements among these programs, so the authors of the study recommend embedding a patient prescription coordinator in the ED. Alternatively, pharmacy departments, social workers, or community-based pharmacies can assist patients to apply for benefits.

Financial Assistance

Most hospitals have financial case management departments that will apply for Medicaid or Medicare coverage on their patient's behalf. It is in the financial self-interest of hospitals to do so. Although reimbursement may be a fraction of commercial rates, hospitals recover at least a portion of what would become unrecoverable debt [65]. This service benefits patients since they otherwise would have to pay out of pocket.

A requirement of the Affordable Care Act (ACA) is that non-profit hospitals must offer charity care and the assistance to apply for it. It also encourages hospitals to add complimentary financial services [66]. Hospitals may be able to assist with Medicaid re-determination of benefits when the patient's coverage period lapses, usually after 12 months. Additional on-site or off-site medical-financial partnerships for low-income patients can provide services such as financial counseling (focusing on credit, debt reduction, savings, and budgeting), free tax preparation, job assistance, and public benefits referral [67].

The Social Security Administration (SSA) has two programs that can provide financial assistance. Supplemental Security Income (SSI) is a needs-based program for individuals who are blind, disabled, or elderly, with low income/resources. Social Security Disability Insurance (SSDI) is for blind or disabled individuals who have a work history and are insured through employee and employer contributions to the Social Security Trust Fund [20]. A sister government agency, the US Substance Abuse and Mental Health Services Administration (SAMHSA), has a program called SSI/SSDI Outreach, Access, and Recovery (SOAR), an expedited service to help qualified applicants to secure benefits [68]. The use of SOAR has been linked to an increased first-time acceptance rate compared to those applicants who do not use SOAR [68]. Having a source of income greatly improves the chances of being able to obtain housing, making SOAR a particularly valuable service to homeless patients. Some hospitals have begun training their financial case managers to be SOAR counselors.

The Anchor Mission

Today, universities and health systems play an increasingly important economic role in communities. There is a growing awareness that these institutions have considerable untapped resources to leverage in their local communities, in order to address long-standing structural deficits and poor economic vitality. This *Anchor Mission* [69] acknowledges the anchor hospital's "institutional priority to improve community health and well-being by leveraging all [its] assets, including hiring, purchasing, and investment for equitable economic impact… [This] can powerfully impact the upstream determinants of health and help build inclusive and sustainable local economies" [70]. The healthcare sector contributes $800 billion in annual expenditures to the US economy and has accumulated $500 billion in investment dollars. As "Anchor Institutions," hospitals are called to realign their traditional business practices to "consciously apply the long-term, place-based economic power of the institution, in combination with its human and intellectual resources, to better the long-term welfare of the community in which it is located" [69]. Westside United, a partnership of seven hospitals, including two university hospitals (Rush University Medical Center and University of Illinois Hospital and Health Sciences system), is an example of the Anchor Mission in Chicago. Collectively, the hospitals have agreed to support the local economy by purchasing $100 million annually of locally sourced laundry, food service, and supplies; to place a priority of hiring locally; to provide wealth management and financial guidance to increase home ownership for Chicago's Westside residents and employees who live there; and to provide paid summer internships for local high school students [71].

Societal Level

The American Medical Association and United Healthcare, the largest health insurance carrier in the US, are collaborating to standardize how social determinant data is collected, processed, and integrated. This partnership will create over 20 new

ICD-10 codes related to the social determinants of health, with an intention that the codes will trigger more systematic referrals to social and government services, connecting them to local resources [72].

As a result of a disjointed, unaffordable, and sometimes inaccessible healthcare system, ED providers in the US are confronted with a variety of unmet social and economic needs, which cause and exacerbate many illnesses. Since EDs are open 24/7 and treat all patients in need, ED providers work with many patients' who are financially insecure [73, 74]. Traditionally ED physicians have received limited education on how to address these needs, which generally fall outside the scope of clinical practice and perceived physician role.

Literature suggests this lack of agency is a source of frustration for the ED physician and may contribute to burnout as well [73, 75]. ED providers have the highest rates of burnout out of all physician specialties [76], and burnout "is directly correlated to a personal sense of disempowerment to effect change in the work environment" [73]. Marked by emotional exhaustion, depersonalization, and reduced sense of personal accomplishment that results in decreased work effectiveness, burnout may also result in suboptimal care [77]. Thus, improved medical and residency education to explicitly prepare physicians to better understand and address patients' unmet financial needs (and other social determinants of health) may empower physicians working in the ED [73, 77].

Healthcare is delivered within the overarching context and history of our surrounding communities where our patients live. Illness and injury are often complications of long-standing structural violence and the inadequate and often inequitable application of public policy (e.g., zoning regulations, food systems, housing infrastructure, labor laws, tax rules, criminal justice sentencing guidelines, public education and social programming, etc.) [73, 78].

Medical education will need to recognize that "social and economic forces produce symptoms" and facilitate gene expression, and consequently we need "medical models for structural change" [79]. Named "structural competency," this paradigm incorporates the impact of systemic and institutionalized social and economic marginalization. Training on structural competency for medical trainees may be impactful. Medical residents who attended a training on structural competency reported that this framework was particularly helpful to better understand their patients and thereby "build a partnership" with them [80]. Additionally, it helped trainees reframe thinking toward patients, away from misconceptions that lead providers to "inadvertently blame patients for harm caused by structural violence" [81].

Health systems and EM providers can also become community advocates and powerful constituents by engaging with community-based organizations that address the structural determinants that lead to financial strain [67]. EM providers can advocate for policies, practice changes, and/or community projects that aim to improve the economic security of the neighborhood through workforce development, economic development, small business loans development, and/or affordable housing. For example, the US Department of Housing & Urban Development (HUD) supports the Continuum of Care (CoC) program. It designates a lead agency to promote communitywide commitment to the goal of ending homelessness. "Community" can be an entire state, a county, or a city [82]. Through its H2 program, HUD is encouraging systems integration between healthcare and housing.

In order to maintain tax-exempt status, nonprofit hospitals must demonstrate that the institution serves the health interests of the surrounding community. In this context, community benefit refers to "the initiatives and activities undertaken by nonprofit hospitals to improve health in the communities they serve" [83]. Many hospitals do this via a community benefits officer or a department. For example, Trinity Health Care has a Community Benefits Ministry for each of its hospitals across the US and outlines the specific ways in which it reinvests profits into the local community to improve health outcomes and improve access to healthcare [84]. These departments coordinate activities to maintain nonprofit status, such as the triennial Community Health Needs Assessment (CHNA). It supports activities by providing community action grants that promote community health, like CPR training and health fairs.

There is a large corpus of research examining the link between poverty, inequality, and their resultant impact on financial strain and poor health outcomes [22, 25, 28]. EM providers can contribute to this body of work by doing research that lends insight into actionable interventions that meaningfully reduce the disparities in health outcomes which are rooted in financial insecurity.

Recommendations for Emergency Medicine Practice

Basic

- Ask questions to determine if patients have financial insecurity and how this is related to their health, healthcare use, or other health-related social needs.
- Consider evaluating and purchasing an online community resource referral platform. Several software companies offer curated online human services directories with listings for services that address financial insecurity such as rental assistance, job training and placement, financial counseling, assistance with application to government assistance programs, rent-to-own, and free tax preparation.
- Write generic prescriptions that align with local big box retailers' low-cost formularies.
- Refer to departmental resources if available, early in the patient's ED presentation, such as a social worker or case worker/care management within your emergency department. If these resources are lacking within the ED, advocate for hiring personnel to help address these complex patient social needs.

Intermediate

- Financial insecurity is a base from which many other social needs stem. It can emerge from difficulty seeking meaningful, sustainable employment in jobs that

provide a living wage and adequate benefits or the increased risk of being fired after a workplace injury [85]. Create referral relationships with community-based organizations (CBO) that offer integrated job training, legal aid, employment services, and financial counseling.
- Training institutions can incorporate a structural competency framework to better equip the next generation of clinicians to effectively identify and address patients' unmet social and economic needs [65]. This is often best done by working collaboratively with public health departments, allied health professionals, and community health workers with indigenous knowledge of the grassroots community.

Advanced

- Engage your hospital's leadership to explore creating or joining an Anchor Mission. The process for engaging your community and setting shared priorities is explained on the Democracy Collaborative website (https://healthcareanchor.network/2020/02/anchor-mission-communications-toolkit/) [86].
- Working with hospital leadership, take a population health approach by creating the workforce necessary to identify, screen, and refer at-risk patients. Leadership should convene healthcare system stakeholders and community service providers to determine what determinants are most important to patients in the community and create referral relationships to agencies that provide those services.
- Form alliances with affordable housing, mental health, food insecurity, and other advocacy agencies to influence local, state, and federal officials to increase access to services for affordable and supportive housing, homelessness, and mental illness, among others. National agencies that have a national presence with offices in many large cities include the Corporation for Supportive Housing (CSH), National Alliance to End Homelessness (NAEH), Enterprise Community Partners, and National Alliance on Mental Illness (NAMI) [87–89]. If these agencies do not have a presence in your area, learn who the active local agencies are and seek to find shared priorities.

Teaching Case

Clinical Case

A 48-year-old male presents to the ED for shortness of breath related to asthma. This is his fifteenth ED visit in the past 3 months for a multitude of complaints. He has had three asthma exacerbations, the most recent resulting in intubation. He has also been worked up for an ankle injury, knee pain, and diarrhea and was brought in by the police twice after being assaulted. He looks unkempt with poor hygiene and soiled clothing. He arrives at 10:30 at night. During this visit, he reports that he is having shortness of breath and chest pain. The weather outside is clear with a temperature of 22 degrees Fahrenheit. His vital signs are stable with a pulse oximetry of 96%, and he has only mild wheezing on lung auscultation.

The patient is given an albuterol nebulizer treatment that appears to resolve his symptoms, but he insists he needs to be admitted, that he is not feeling well. While continuing the conversation, you ask where the patient has been living and he admits he was staying with family but was kicked out 2 weeks ago. With the help of social work, the patient agrees to be discharged to a crisis shelter.

An off-service intern who is responsible for the patient's care is discharging the patient. You walk in during the intern's instructions for follow-up care. He has written a prescription for a rescue inhaler and is handing it to the patient. The patient becomes visibly agitated, reaches over to his coat on the chair, and pulls out a handful of scripts. He says to the intern, "I told the other docs, I don't have any money - I couldn't get my meds before - what makes you think I can get them now!?"

Teaching Points

1. The patient's care is complicated by his homeless status and financial insecurity. There are many layers to untangle with the patient.
2. Although there may be a direct benefit to the patient's health by helping him find housing, this may be a solution that is out of immediate reach for the hospital.
3. The patient may qualify for Medicaid or local charity care. Verify if your hospital has a financial case manager or an embedded agency that can help the patient apply for benefits. Additionally, if the patient qualifies for SSI or SSDI, the income will significantly enhance the patient's ability to find housing.
4. Your hospital may provide the medications for free, or if there is a nearby pharmacy, check to see if they have generic equivalents in its \$4, \$5, or \$10 formulary. Your department may be able to provide gift cards for the pharmacy.
5. If the patient is insured, check to see if the managed care organization's case manager has engaged with the patient and/or is able to work with the patient, to help provide assistance and navigation to primary care.

Discussion Questions

1. What are the factors complicating this patient's care?
2. Why did this person, unknown to the ED, suddenly appear with a rapid spike in utilization? What would be some possible explanations why the patient has had frequent visits to the ED? How would these explanations affect the care he receives in the ED?
3. How can you, as an ED provider, better meet the health-related social needs of this patient? How can you engage their support?

References

1. Cass O. The cost-of-thriving index: reevaluating the prosperity of the American family New York. New York: Manhatten Institute; 2020.
2. Garon TD, Dunn A, Govala K, Wilson E. U.S. Financial Health Pulse: 2018 baseline survey results. https://s3.amazonaws.com/cfsi-innovation-files-2018/wp-content/uploads/2019/03/06213859/Pulse_Baseline_SurveyResults-jan2019-WEB-rev-1.pdf. 2019. Accessed.
3. Durkin E. US household back to pre-recession levels, census shows. 2018. https://www.theguardian.com/business/2018/sep/12/census-household-income-us-pre-recession-levels.

4. Mishel L, Gould E, Bivens J. Wage stagnation in nine charts. 2015. https://www.epi.org/publication/charting-wage-stagnation/.
5. McIntosh K, Moss E, Nunn R, Shambaugh J. Examining the black-white wealth gap. Washington, DC: The Brookings Institute; 2020.
6. CareerBuilder.com. Living paycheck to paycheck is a way of life for majority of U.S. workers, according to new careerbuilder survey. 2017. http://press.careerbuilder.com/2017-08-24-Living-Paycheck-to-Paycheck-is-a-Way-of-Life-for-Majority-of-U-S-Workers-According-to-New-CareerBuilder-Survey.
7. Larrimore JD, Durante A, Kreiss K, Park C, Sahm C. Report on the Economic Well-Being of U.S. Households in 2017. 2018. https://www.federalreserve.gov/publications/files/2017-report-economic-well-being-us-households-201805.pdf.
8. Kochhar R. Hispanic women, immigrants, young adults, those with less education hit hardest by COVID-19 job losses. Philadelphia: Pew Charitable Trust; 2020.
9. Lopez MH, Rainie L, Budiman A. Financial and health impacts of COVID-19 vary widely by race and ethnicity. Philadelphia: Pew Charitable Trust; 2020.
10. Adamczyk A. 32% of U.S. households missed their July housing payments. 2020. https://www.cnbc.com/2020/07/08/32-percent-of-us-households-missed-their-july-housing-payments.html.
11. Van Dam A. An indicator that presaged the housing crisis is flashing red again. 2020. https://www.washingtonpost.com/business/2020/07/14/new-mortgage-delinquencies-hit-record-high/.
12. Nova A. Looming evictions may soon make 28 million homeless in U.S., expert says. 2020. https://www.cnbc.com/2020/07/10/looming-evictions-may-soon-make-28-million-homeless-expert-says.html.
13. The Supplemental Nutrition Assistance Program (SNAP). Washington DC: Center for Budget and Policy Priorities; 2019.
14. Papanicolas I, Woskie LR, Jha AK. Health care spending in the United States and other high-income countries. JAMA. 2018;319(10):1024–39. https://doi.org/10.1001/jama.2018.1150.
15. Bradley EH, Elkins BR, Herrin J, Elbel B. Health and social services expenditures: associations with health outcomes. BMJ Qual Saf. 2011;20(10):826–31. https://doi.org/10.1136/bmjqs.2010.048363.
16. Promotion OoDPH. Social determinants of health. 2019. https://www.healthypeople.gov/2020/topics-objectives/topic/social-determinants-of-health.
17. U.S. Department of Housing & Urban Development. Housing Choice Vouchers Fact Sheet. https://www.hud.gov/topics/housing_choice_voucher_program_section_8 (2020). Accessed.
18. Amadeo K, Scott G. Federal poverty level guidelines and chart. 2020. https://www.thebalance.com/federal-poverty-level-definition-guidelines-chart-3305843.
19. Schott L, Finch I. TANF benefits are low and have not kept pace with inflation: benefits are not enough to meet families' basic needs. Washington, DC: Center on budget and policy priorities; 2010.
20. Administration USSS. Social security disability benefits. https://www.ssa.gov/benefits/disability/.
21. Lawrence BK. Is there a social security disability asset limit? https://www.disabilitysecrets.com/page7-5.html#:~:text=To%20be%20eligible%20for%20SSDI,you%20or%20your%20spouse%20makes.
22. Malecha PW, Williams JH, Kunzler NM, Goldfrank LR, Alter HJ, Doran KM. Material needs of emergency department patients: a systematic review. Acad Emerg Med. 2018;25(3):330–59. https://doi.org/10.1111/acem.13370.
23. Deaton A. Policy implications of the gradient of health and wealth. Health Aff (Millwood). 2002;21(2):13–30. https://doi.org/10.1377/hlthaff.21.2.13.
24. Galobardes B, Davey Smith G, Jeffreys M, McCarron P. Childhood socioeconomic circumstances predict specific causes of death in adulthood: the Glasgow student cohort study. J Epidemiol Community Health. 2006;60(6):527–9. https://doi.org/10.1136/jech.2005.044727.
25. Wang J, Geng L. Effects of socioeconomic status on physical and psychological health: lifestyle as a mediator. Int J Environ Res Public Health. 2019;16(2). https://doi.org/10.3390/ijerph16020281.

26. Alley DE, Asomugha CN, Conway PH, Darshak MS. Accountable health communities - addressing social needs through medicare and medicaid. N Engl J Med. 2016;274(8):8–11.
27. Kangovi S, Barg FK, Carter T, Long JA, Shannon R, Grande D. Understanding why patients of low socioeconomic status prefer hospitals over ambulatory care. Health Aff (Millwood). 2013;32(7):1196–203. https://doi.org/10.1377/hlthaff.2012.0825.
28. Beck AF, Huang B, Simmons JM, Moncrief T, Sauers HS, Chen C, et al. Role of financial and social hardships in asthma racial disparities. Pediatrics. 2014;133(3):431–9. https://doi.org/10.1542/peds.2013-2437.
29. Georgiades A, Janszky I, Blom M, Laszlo KD, Ahnve S. Financial strain predicts recurrent events among women with coronary artery disease. Int J Cardiol. 2009;135(2):175–83. https://doi.org/10.1016/j.ijcard.2008.03.093.
30. Szanton SL, Thorpe RJ, Whitfield K. Life-course financial strain and health in African-Americans. Soc Sci Med. 2010;71(2):259–65. https://doi.org/10.1016/j.socscimed.2010.04.001.
31. Tucker-Seeley RD, Li Y, Subramanian SV, Sorensen G. Financial hardship and mortality among older adults using the 1996–2004 Health and Retirement Study. Ann Epidemiol. 2009;19(12):850–7. https://doi.org/10.1016/j.annepidem.2009.08.003.
32. Barnow SLM, Lucht M, Freyberger H. The importance of psychosocial factors, gender, and severity of depression in distinguishing between adjustment and depressive disorders. J Affect Disord. 2002;72:71–8.
33. Centers for Disease Control. National hospital ambulatory medical care survey: 2011 emergency department summary tabs. 2011. https://www.cdc.gov/nchs/data/ahcd/nhamcs_emergency/2011_ed_web_tables.pdf.
34. Demiryoguran NS, Karcioglu O, Topacoglu H, Kiyan S, Ozbay D, Onur E, et al. Anxiety disorder in patients with non-specific chest pain in the emergency setting. Emerg Med J. 2006;23(2):99–102. https://doi.org/10.1136/emj.2005.025163.
35. Chou EY, Parmar BL, Galinsky AD. Economic insecurity increases physical pain. Psychol Sci. 2016;27(4):443–54. https://doi.org/10.1177/0956797615625640.
36. Elfassy T, Swift SL, Glymour MM, Calonico S, Jacobs DR Jr, Mayeda ER, et al. Associations of income volatility with incident cardiovascular disease and all-cause mortality in a US cohort. Circulation. 2019;139(7):850–9. https://doi.org/10.1161/CIRCULATIONAHA.118.035521.
37. Hedegaard HC, Curtin SC, Warner M. Suicide mortality in the United States, NCHS Data Brief. 2018;330:1–8.
38. Stack S. Why is suicide on the rise in the US - but falling in most of Europe? The conversation. Washington DC: The Conversation U.S.; 2018.
39. Dow WH, Godøy A, Lowenstein CA, Reich M. In: Research NBoE, editor. Can economic policies reduce deaths of despair? Cambridge, MA; 2019.
40. Pavao J, Alvarez J, Baumrind N, Induni M, Kimerling R. Intimate partner violence and housing instability. Am J Prev Med. 2007;32(2):143–6. https://doi.org/10.1016/j.amepre.2006.10.008.
41. Nguyen OK, Higashi RT, Makam AN, Mijares JC, Lee SC. The influence of financial strain on health decision-making. J Gen Intern Med. 2018;33(4):406–8. https://doi.org/10.1007/s11606-017-4296-3.
42. Griffin P. Waiting to feel better: survey reveals cost delays timely care. 2018. https://www.earnin.com/data/waiting-feel-better.
43. International O. Analysis on how healthcare costs impact patient behavior. 2018. https://www.accessonemedcard.com/wp-content/uploads/2018/10/AccessOne_Survey2018_final.pdf.
44. Cutshaw CA, Woolhandler S, Himmelstein DU, Robertson C. Medical causes and consequences of home foreclosures. Int J Health Serv. 2016;46(1):36–47. https://doi.org/10.1177/0020731415614249.
45. Hamel L, Norton M, Pollitz K, Levitt L, Claxton G, Brodie M. The burden of medical debt: results from the Kaiser Family Foundation/New York Times Medical Bills Study. 2016. https://www.kff.org/report-section/the-burden-of-medical-debt-introduction/.
46. Berger LM, Collins JM, Font SA, Gjertson L, Slack KS, Smeeding T. Home foreclosure and child protective services involvement. Pediatrics. 2015;136(2):299–307. https://doi.org/10.1542/peds.2014-2832.

47. Briesacher BA, Gurwitz JH, Soumerai SB. Patients at-risk for cost-related medication non-adherence: a review of the literature. J Gen Intern Med. 2007;22(6):864–71. https://doi.org/10.1007/s11606-007-0180-x.
48. Foundation KF. An overview of the Medicare part D prescription drug benefit. 2019. https://www.kff.org/medicare/fact-sheet/an-overview-of-the-medicare-part-d-prescription-drug-benefit/.
49. Cheung PT, Wiler JL, Lowe RA, Ginde AA. National study of barriers to timely primary care and emergency department utilization among Medicaid beneficiaries. Ann Emerg Med. 2012;60(1):4–10 e2. https://doi.org/10.1016/j.annemergmed.2012.01.035.
50. Azar A. The root of the problem: America's social determinants of health. 2018. https://www.hhs.gov/about/leadership/secretary/speeches/2018-speeches/the-root-of-the-problem-americas-social-determinants-of-health.html.
51. Johnson T. New rules to stop utilities from shutting off power to residents who use vital medical equipment. 2019.
52. Target.com. Target pharmacy: $4 and $10 generic medication list. 2019. https://tgtfiles.target.com/pharmacy/WCMP02-032536_RxGenericsList_NM7.pdf.
53. Walmart: $4 prescriptions. 2019. https://www.walmart.com/cp/$4-prescriptions/1078664.
54. NeedyMeds.org. NeedyMeds: find help with the cost of medicine. 2019. https://www.needymeds.org/.
55. CMS.gov. ICD-10. 2019 https://www.cms.gov/Medicare/Coding/ICD10/index.html?redirect=/icd10/.
56. Miner JR, Westgard B, Olives TD, Patel R, Biros M. Hunger and food insecurity among patients in an urban emergency department. West J Emerg Med. 2013;14(3):253–62. https://doi.org/10.5811/westjem.2012.5.6890.
57. Lee J, Korba C, Cohen AB, Sharma D. Social determinants of health: how are hospitals and health systems investing in and addressing social needs? 2017. www.deloitte.com/us/social-determinants-of-health.
58. Johnson SR. Kaiser to launch social care network. 2019. https://www.modernhealthcare.com/care-delivery/kaiser-launch-social-care-network.
59. CMS.gov. The Accountable Health Communities health-related social needs screening tool. 2019. https://innovation.cms.gov/Files/worksheets/ahcm-screeningtool.pdf.
60. CMS.gov: The Accountable Health Communities Health-Related Social Needs Screening Tool. https://innovation.cms.gov/Files/worksheets/ahcm-screeningtool.pdf (2019). Accessed.
61. NACHC.org: PRAPARE: Protocol for Responding to and Assessing Patients' Assets, Risks and Experiences. http://www.nachc.org/research-and-data/prapare/ (2019). Accessed.
62. Cartier Y, Fichtenberg C, Gottlieb L. Community resource referral platforms: a guide for health care organizations. San Francisco: Social Interventions Research & Evaluation Network (SIREN); 2019.
63. RxResource.org. Prescription assistance programs. 2019. https://www.rxresource.org/prescription-assistance/default.html.
64. Burley MH, Daratha KB, Tuttle K, White JR, Wilson M, Armstrong K, et al. Connecting patients to prescription assistance programs: effects on emergency department and hospital utilization. J Manag Care Spec Pharm. 2016;22(4):381–7. https://doi.org/10.18553/jmcp.2016.22.4.381.
65. Great Lakes Medicaid. https://greatlakesmedicaid.com.
66. Chazin S, Guerra V. Impact of the affordable care act on charity care programs. 2020. https://www.chcs.org/resource/impact-of-the-affordable-care-act-on-charity-care-programs-2/.
67. Bell ON, Hole MK, Johnson K, Marcil LE, Solomon BS, Schickedanz A. Medical-financial partnerships: cross-sector collaborations between medical and financial services to improve health. Acad Pediatr. 2020;20(2):166–74. https://doi.org/10.1016/j.acap.2019.10.001.
68. What is SOAR? 2020. https://soarworks.prainc.com/article/what-soar.
69. DemocracyCollaborative.org. Building community wealth: the anchor mission. 2019. https://democracycollaborative.org.
70. HealthcareAnchorNetwork: 40 leading healthcare systems building more inclusive and sustainable local economies. 2019. https://www.healthcareanchor.network.

71. WestsideUnited.org. Westside United: building blocks to better health. 2019. https://westsideunited.org.
72. Livingston S. UnitedHealthcare, AMA unveil more medical codes for social determinants. 2019. https://www.modernhealthcare.com/technology/unitedhealthcare-ama-unveil-more-medical-codes-social-determinants.
73. Axelson DJ, Stull MJ, Coates WC. Social determinants of health: a missing link in emergency medicine training. AEM Educ Train. 2018;2(1):66–8. https://doi.org/10.1002/aet2.10056.
74. Doran KM, Vashi AA, Platis S, Curry LA, Rowe M, Gang M, et al. Navigating the boundaries of emergency department care: addressing the medical and social needs of patients who are homeless. Am J Public Health. 2013;103(Suppl 2):S355–60. https://doi.org/10.2105/AJPH.2013.301540.
75. Goldberg R, Boss RW, Chan L, Goldberg J. Burnout and its correlates in emergency physicians: four years' experience with a wellness booth. Acad Emerg Med. 1996;3(12):1156–64.
76. Shanafelt TD, Boone S, Tan L, Dyrbye LN, Sotile W, Satele D, et al. Burnout and satisfaction with work-life balance among US physicians relative to the general US population. Arch Intern Med. 2012;172(18):1377–85. https://doi.org/10.1001/archinternmed.2012.3199.
77. Lu DW, Dresden S, McCloskey C, Branzetti J, Gisondi MA. Impact of burnout on self-reported patient care among emergency physicians. West J Emerg Med. 2015;16(7):996–1001. https://doi.org/10.5811/westjem.2015.9.27945.
78. Braveman P, Egerter S, Williams DR. The social determinants of health: coming of age. Annu Rev Public Health. 2011;32(1):381–98. https://doi.org/10.1146/annurev-publhealth-031210-101218.
79. Metzl JM. Structural competency. Am Q. 2012;64(2):213–8.
80. Neff J, Knight KR, Satterwhite S, Nelson N, Matthews J, Holmes SM. Teaching structure: a qualitative evaluation of a structural competency training for resident physicians. J Gen Intern Med. 2017;32(4):430–3. https://doi.org/10.1007/s11606-016-3924-7.
81. Neff J, Holmes SM, Knight KR, Strong S, Thompson-Lastad A, McGuinness C, et al. Structural competency: curriculum for medical students, residents, and interprofessional teams on the structural factors that produce health disparities. MedEdPORTAL. 2020;16:10888. https://doi.org/10.15766/mep_2374-8265.10888.
82. Development USDoHU: Continuum of Care (CoC) program. https://www.hudexchange.info/programs/coc/.
83. Spugnardi I. Addressing the social determinants of health: the role of health care organizations. Health Prog. 2016;97(6):80–3.
84. Community Benefit. Community health needs assessment. 2020. https://www.trinity-health.org/community-health-and-well-being/community-benefit/.
85. Kennedy M. Injuries at work may increase risk of losing one's job. 2016. https://www.reuters.com/article/us-health-workplace-injury/injuries-at-work-may-increase-risk-of-losing-ones-job-idUSKCN0VE2MW.
86. Pham BH. Healthcare Anchor Network's (HAN) anchor mission communications toolkit. 2020. https://democracycollaborative.org/learn/publication/healthcare-anchor-networks-han-anchor-mission-communications-toolkit.
87. Corporation for Supportive Housing. 2020. https://www.csh.org.
88. National Alliance on Mental Illness. 2020. https://www.nami.org.
89. National Alliance to End Homelessness. 2020. https://endhomelessness.org.

Food Insecurity: Hidden Problems, Real Remedies

13

Eric W. Fleegler, Deborah A. Frank, and Marisa B. Brett-Fleegler

Key Points

- Food insecurity is one of the most prevalent social problems in the US. It is common among ED patients and is a risk factor for increased ED utilization.
- Food insecurity is invisible unless actively inquired about; brief questionnaires that evaluate food insecurity are readily available.
- Multiple interventions exist for ameliorating the effects of food insecurity. The most basic include providing a meal within the ED. Referrals to local food pantries and soup kitchens can help meet additional needs. Food pantries within the hospital can make an immediate difference for patients who present with food insecurity.
- Federal programs such as Supplemental Nutrition Assistance Program (SNAP) and Special Supplemental Nutrition Program for Women, Infants, and Children (WIC) are important for long-term management of food insecurity. Creating systems to connect patients and patients' families to these programs either within the ED or through referrals to community resources are important system-level interventions.
- Emergency providers should advocate for improvements in the safety net to address the social and economic factors associated with food insecurity.

E. W. Fleegler (✉) · M. B. Brett-Fleegler
Department of Pediatrics, Division of Emergency Medicine, Boston Children's Hospital, Boston, MA, USA
e-mail: Eric.fleegler@childrens.harvard.edu; Marisa.Brett@childrens.harvard.edu

D. A. Frank
Department of Pediatrics, Boston Medical Center, Boston, MA, USA
e-mail: dafrank@bu.edu

H. J. Alter et al. (eds.), *Social Emergency Medicine*,
https://doi.org/10.1007/978-3-030-65672-0_13

Foundations

Background

Food insecurity is one of the most prevalent social problems in the US. In 2018, an estimated 1 in 9 people, over 37 million Americans, including 11 million children, were food insecure [2].

Defining and Measuring Food Insecurity

Since 1995, the US Department of Agriculture (USDA) has annually measured food security in the US. Currently, they define two levels of food insecurity [3]:

- Low food security: reports of reduced quality, variety, or desirability of diet; little or no indication of reduced food intake
- Very low food security: reports of multiple indications of disrupted eating patterns and reduced food intake, including skipping meals and going to bed hungry

The prevalence of food insecurity, describing households with difficulty meeting basic food needs, has fluctuated noticeably over the past 26 years [4]. In 1999, 10% of American households were food insecure. During the Great Recession, a significant spike occurred, with a peak of 14.9% of American households food insecure in 2011; by 2018 food insecurity had decreased to 11.1% of households. Households with very low food security, whose members may regularly skip meals or go to bed hungry, has ranged from 3% to a peak of 5.7% of American households in 2011. In 2018, 5.3 million households (4.3%) experienced very low food security [2].

Challenging financial circumstances frequently lead to food insecurity, and thus food insecurity has the highest prevalence among low-income households. Among families with incomes below 185% of the federal poverty level (which translates to an income less than or equal to $46,435 for a family of four in 2018), 29.1% were food insecure in 2018. The COVID-19 pandemic has the potential to cause enormous economic instability and, consequently, food insecurity across the US could increase to record high levels and remain there for years. Estimates suggest food insecurity in 2020 during the COVID-19 epidemic rose to 15.6% (50.4 million people) including 23.1% of all children (17.0 million children).[1] The prevalence of food insecurity also varies by race/ethnicity. In 2018, among non-Hispanic Black households, 21.2% were food insecure, among Hispanic households 16.2% were food insecure, while among non-Hispanic White households 8.1% were food insecure [2].

Households with children present a mixed picture of food insecurity. In 2018, 13.9% of families with children, representing 12.5 million children, were food

[1] Feeding America. The Impact of the Coronavirus on Food Insecurity in 2020 [Internet]. October 2020. Available from: https://www.feedingamerica.org/sites/default/files/2020-10/Brief_Local%20Impact_10.2020_0.pdf.

insecure. In nearly half of these households only the adults were food insecure, implying there were challenges procuring food but the children had enough to eat. However, in 51% of these households, representing 6 million children, the children also experienced food insecurity. Over 540,000 US children experienced very low food security, characterized by reduced food intake and disruptions in their eating patterns. In households with children led by a single mother, the food insecurity rate was 27.8% and the very low food security rate was 9.4% [2].

Programs That Address Food Insecurity

Programs that address food insecurity include federal programs, statewide and national nonprofit programming, as well as smaller local social programs. Supplemental Nutrition Assistance Program (SNAP) and Special Supplemental Nutrition Program for Women, Infant, and Children (WIC) are two of the better-known federal programs to help with food insecurity in low-income individuals and families.

SNAP, formerly known as "Food Stamps," is the largest federal program focused on food insecurity among poor individuals and families. Eligibility for SNAP at the national level is income-based and in some states asset-based as well. Overall, gross monthly income must be below 130% of the federal poverty line, $2183 per month for a family of four in 2020 ($26,200 annually). Forty states use "categorical eligibility," allowing the state to provide more households with benefits. Categorical eligibility gives states the option to automatically align gross income and asset requirements with Temporary Assistance for Needy Families (TANF) and other assistance programs (i.e., if you qualify for TANF, you are enrolled in SNAP) [5]. Though SNAP participation reduces the risk of food insecurity [47], more than half of households receiving SNAP are still food insecure due to the relatively limited amount of support SNAP provides. On average, families receive $1.40 per person per meal. The average SNAP household receives $256 per month [6].

WIC provides supplemental foods including formula for infants and healthy foods for pregnant and breastfeeding women and children up to age 5 years old. WIC also provides nutrition education, breastfeeding support, and referrals to healthcare and other services, free of charge, to families who qualify [7]. Income eligibility includes families with gross incomes below 185% of the Federal Poverty Level [8]. Some states have automatic income eligibility based on enrollment in TANF, Medicaid, or SNAP.

Food banks are nonprofit organizations that distribute food to hunger-relief charities. In 2018, Feeding America, which distributes food to 60,000 food pantries across the US, fed 40 million people, including 12 million children [9]. Meal programs, which are sometimes referred to as soup kitchens, offer prepared food and hot meals to the hungry for free or at reduced prices. They frequently have limited hours and days of the week of service and may serve only select geographic or demographic groups; they can reach only a fraction of people living with food insecurity on a constant basis.

Additional programs that provide meals to some low-income people include Meals on Wheels (for elderly and disabled) and school breakfast and lunch programs. The Child and Adult Care Food Program (CACFP) is a federal program that provides reimbursements for nutritious meals and snacks to eligible children and adults who are enrolled for care at participating child care centers, day care homes, and adult day care centers. Over 4.2 million children and 130,000 adults receive meals and snacks through this program on a daily basis [10].

Understanding these services can provide a framework to help address food insecurity, especially with the recognition that a meaningful proportion of those eligible for these programs are not enrolled. In 2017, 45 million people were eligible for SNAP and 84% of eligible participants used SNAP, up from 69% utilization in 2007 [6]. In 2014, 15 million people were eligible to receive WIC but only 55% enrolled; participation rates vary by state [11]. A little over two million of those eligible were infants, which is 62% of all infants born in the US. Participation rates were 80% for eligible infants compared to 50% of eligible pregnant women [11]. Participation lags behind eligibility for these programs for multiple reasons including (1) lack of understanding about eligibility, (2) social undesirability and stigma, (3) concerns about disqualification for immigration, (4) language and social barriers, (5) difficult forms, and (6) complicated asset tests and high burden of proof. Further, many of these programs are not entitlements but rather are contingent on funding legislation, and availability may vary by allocation and by state policy. The federal programs have strict eligibility based on income, assets, and immigration status while many food pantries and soup kitchens do not, which is especially important as changing federal regulations may cause people to lose their eligibility for programs they used to receive.

Evidence Basis

Food Insecurity and Health

The harmful health consequences of food insecurity have been well documented in the literature. The suspected mechanisms behind the negative relationship between food insecurity and health are numerous but not fully understood. Food insecure households are more likely to purchase inexpensive, energy-dense, and nutritionally low-quality food [12, 13]. This diet is associated with increases in body mass index, hypertension, elevated cholesterol, elevated HgbA1C, and other risk factors for poor health [14]. The strain associated with food insecurity likely has negative effects through non-dietary mechanisms which include (1) toxic stress that activates the hypothalamic-pituitary-adrenal axis, (2) decreased ability to manage other health-related social needs thus leading to accumulated social problems, (3) unhealthful coping behaviors such as smoking and excessive alcohol use, and (4) decreased ability to manage chronic diseases [15–18].

Among food insecure patients, there is a high rate of obesity, advanced hepatic fibrosis, and nonalcoholic steatohepatitis cirrhosis (NASH), which is a leading cause of liver transplantation in the US [19]. In a study of 13,518 adults, those living in very low food security households had more than a twofold increased risk of 10-year cardiovascular disease [20]. Children with food insecurity have nearly twofold higher risks of fair/poor health and 50% increased rates of parental concern for developmental delay compared to children without food insecurity [21, 22]. In very low food security households, families may not consume daily minimal requirements of calories and appropriate nutrients, in some cases leading to malnutrition. Food insecurity has also been associated with the presence of mental health disorders. Adults who are food insecure are about 3 times more likely to report frequent mental distress compared to those who are food secure [23]. Caregivers of children in food insecure households have over threefold higher rates of depressive symptoms [24]. In the US, children from food insecure households have 28% higher rates of depression than children from food secure households [25].

Food Insecurity and ED Utilization

People who are food insecure have higher rates of ED visits than those who are food secure. Simultaneously, surveys of patients in the ED have demonstrated higher prevalence of food insecurity among ED patients than in the general population [26, 27]. Adults with food insecurity have nearly a 50% increased rate of ED visits and hospitalizations, as well as longer hospitalization lengths, compared to those who are food secure [28, 29]. Among adults with type II diabetes, food insecure patients have a twofold increase in ED visits [30]. Children in food insecure households use the ED at up to 37% higher rates compared to food secure households [28, 35], and their risk of hospitalization is a third greater than children in food secure households [22]. Among vulnerable populations, people who are homeless and who have food insecurity have nearly threefold higher rates of ED utilization compared to homeless people who are not food insecure [31]. Among HIV+ homeless patients, those who were food insecure had 50% higher rates of ED utilization [32].

Food Insecurity and Health-Related Social Problems

As one of the most common health-related social problems, food insecurity is frequently an indicator of other health-related social needs. Food insecure patients are over four times more likely to have cost-related medication underuse compared to food secure patients, a major risk for poorly controlled health and worse disease outcomes [33]. A study of young adults found that patients with low and very low food security reported two- to fourfold higher rates of problems with healthcare access, education, housing, income security, and substance use compared to patients with high food security [34]. Among children with special healthcare needs, food insecurity is associated with a nearly two-fold increase in material hardships including unmet needs such as well-child checks, dental care, prescription medications, physical, occupational, or speech therapy, mental health counseling, or access to a range of medical equipment [35, 36]. In these contexts, food insecurity may be both a marker for other health-related social problems and a contributing factor.

Food Insecurity and Medical Cost

The medical costs in the US associated with food insecurity are likely far greater than realized by policy makers. A Massachusetts study in 2016 estimated the direct and indirect hospital costs of food insecurity at $1.9 billion [37]. Medicare patients with food insecurity have mean annual Medicare costs $5527 higher than food secure patients [28]. A large national study demonstrated higher average annual healthcare expenditures among food insecure individuals compared to food secure individuals ($6072 vs. $4208), equating to an estimated additional $77.5 billion in annual healthcare expenditures across the US associated with food insecurity [38].

Federal, state, and local programs are buffers for people with food insecurity. WIC has been shown to improve pregnancy and birth outcomes such as reduction in low birthweights [39, 40]. Children receiving WIC benefits have lower rates of anemia, and longer duration of WIC utilization is associated with enhancements in IQ scores [41, 42]. Similarly, the use of SNAP has been linked with many positive outcomes including improved diet, lower ED utilization [43], and better asthma control [44]. At a population level, SNAP may reduce all-cause mortality by 1–2% [45]. SNAP participants incur $1400 less in medical costs per year compared to other low-income adults [46].

The following sections aim to familiarize emergency providers with tools to identify and then alleviate food insecurity among the patients they serve. ED providers should recognize both the prevalence of food insecurity and the significant impact food insecurity has on the health of our patients. Ideally, the hospitals will embrace the importance of these interventions and recognize the essential safety net role of EDs for connecting patients to appropriate food resources.

Emergency Department and Beyond

Bedside

Many EDs serve as the entry point to healthcare for underprivileged individuals and families. Unless actively queried, food insecurity will remain an invisible, though prevalent, problem among ED patients. Standardized universal screening of ED patients has expanded in recent years to include concerns such as alcohol and drug abuse, depression and suicidal ideation, and intimate partner violence, among others. Such screening has become relatively common, albeit cumbersome, to ED staff. The use of screening tools such as the Hunger Vital Sign™ (below) would enable the identification of food insecure patients and families [1]. The integration of such screening represents an opportunity for culture change in our approach to food insecurity.

Children's HealthWatch established the two-question "Hunger Vital Sign™" which has a sensitivity of 97% and specificity of 82% for food insecurity compared to the gold standard USDA Household Food Security Scale [49]. The two Hunger Vital Sign™ questions are:

"Within the past 12 months we worried whether our food would run out before we got money to buy more." ("Often true" or "Sometimes true" vs. "Never true")
"Within the past 12 months the food we bought just didn't last and we didn't have money to get more." ("Often true" or "Sometimes true" vs. "Never true")

A response of "often true" or "sometimes true" to either of the two questions is considered a positive screen.

Unlike screeners for many other health-related social needs, the Hunger Vital Sign™ has been well-validated and is the clear first choice for food insecurity screening in healthcare settings. This questionnaire, originally designed for families with young children but also validated for adults [50], has become the standard in healthcare settings and is used in the Center for Medicare and Medicaid Innovation's Accountable Health Communities (CMS AHC) three million person ongoing innovation model [51]. Preliminary qualitative and quantitative studies of the CMS AHC questionnaire—which includes the two Hunger Vital Sign™ questions along with eight other social need questions—among primary care and ED patients show strong support from patients for asking these social needs questions [52, 53]. In a randomized trial of screening for food insecurity in the ED, 86% of families endorsed the concept of routine screening for food insecurity within the ED [54]. Within the ED, screening could be performed at triage, by the nurse during intake, by the primary clinician caring for the patient, or by social workers or other trained staff during the course of a patient's ED visit. Some methods of screening can even be done while patients are in the waiting room. A new study in North Carolina will evaluate the role of screening and referral for food resources from the ED [55], and an ongoing study in the ED at Boston Children's Hospital is evaluating the role of social screening and referral using patients' smart phones vs. tablets [56].

Studies have shown that both paper and electronic questionnaires are feasible in the healthcare setting [57–61]. However, the method by which patients are screened may influence the responses they provide. Administering sensitive questions one-on-one may be the easiest way to universally screen, but analyses comparing one-on-one, paper, and electronic formats suggest that screening via personnel asking patients questions directly may decrease positive responses (lowers sensitivity) for sensitive issues [62]. In one study, specific to food security screening, the change from an oral to a paper screening of patients using the Hunger Vital Sign™ increased positive response rates from 10.4% to 16.3% [63]. A randomized trial of screening for food insecurity in the ED showed that via tablets 23.6% of patients screened positive compared to 17.7% screening positive via verbal screening [54].

While understanding the prevalence of food insecurity within one's ED patient population is important, having a systematic approach to offer assistance is crucial. Information about local resources (e.g., food pantries, soup kitchens) that are geographically relevant can be readily provided via online tools (e.g., United Way 211 system, HelpSteps, Aunt Bertha, NowPow) [59, 64–66]. Limited literature from outpatient clinics has shown that providing pre-printed forms that list local food pantries and contact information is effective in connecting patients to community resources [67].

If the electronic medical record (EMR) is used to record the answers to the food security questionnaire, ideally positive responses can trigger a social work consult or the inclusion of a food resource referral sheet with the discharge paperwork. At a minimum, making referral sheets available within the EMR that include local food resources may ease the process of clinicians providing patients with this important information [68]. Food insecurity can be documented using ICD-10 codes (Z59.4) to help quantify the extent of this problem within a patient population that the ED serves.

When it comes to patients' desire for help with food, it is important to realize that screening positive for food security problems is not the same as food referral needs. Not all people who are food insecure will want referrals, and likewise, families that are "food secure" via screening questionnaires still may desire referrals. In a study of low-income families in a primary care clinic, 46% of food insecure families did not request referrals; among food secure families, 15% still requested food assistance [60]. Thus, screening processes should offer patients the ability to identify their referral needs (i.e., I want help with SNAP or WIC or finding a food pantry) even if such patients do not meet the standard definition of food insecurity.

If universal screening of food insecurity with validated questions has not been instituted, simply asking a patient in the ED (when clinically feasible) if they would like something to eat is an immediate and kind way to provide a meal. The price of food in many hospital cafeterias and the chain food stores located in hospitals are often too expensive for low-income families, and their ED visits may exacerbate hunger. Having food immediately available and/or hospital cafeteria vouchers that cover the cost of a meal can make a significant difference in families' lives during a stressful time and may also improve the therapeutic relationship. If clinicians ask about food needs at the bedside, it is critical to have access to food to respond to them.

Hospital/Healthcare System

At the hospital level, food insecurity screening procedures and interventions should be developed based on resources and partnerships with state, federal, and community organizations. Several large hospital networks and individual clinics have recognized the importance of food security on the health of their patients and perform universal screening for food insecurity [69, 70]. In a program supporting food insecure families with infants, parents were provided supplemental formula, educational materials and referrals to social workers, medical-legal partnerships, or food pantries directly from the clinic [71]. In other programs, hospitals provided vouchers for on-site or local farmers markets [72, 73]. Many medical centers such as Boston Medical Center, Massachusetts General Hospital, and St. Christopher's Hospital for Children in Philadelphia have on-site food pantries or partner with local organizations to bring mobile food pantries to their clinics on a weekly basis. Studies have shown that caregivers and patients find these programs both acceptable and desirable [74, 75].

Multi-stakeholder partnerships are especially important to link at-risk children to needed food service programs [61, 62]. An intervention at the Children's Hospital of Philadelphia in collaboration with federal and community partners provided free lunch to ED pediatric patients and their siblings during a summer food service program. In the 7-week pilot, 367 meals were distributed to children, and their families were referred to the US Department of Agriculture (USDA) Summer Food Service Program developed to bridge the summer food gap for those who receive free or reduced-price lunch during the school year [76]. Arkansas Children's Hospital provides access to free, nutritious meals for all children seen at the hospital as part of USDA's Child and Adult Care Food Program (CACFP) At-Risk Afterschool Meals Component [77].

Benefit Assistance and Referrals

Applications for SNAP and WIC can be completed and submitted from the hospital and/or by working with local community partners [66, 78, 79]. Boston Medical Center has an on-site WIC office and on-site SNAP application assistance [48]. Depending on the state and the cross-eligibility with other federal programs, SNAP may require documentation regarding income. The approach of filling out applications while at the hospital has been used in EDs with great success for enrollment in the State Children's Health Insurance Program [80] and could be similarly applied to SNAP and WIC applications. Local student organizations and other volunteers can help patients under the guidance of social workers in the ED.

Societal Level

As a society, it is important to recognize food insecurity not simply as an unfortunate problem of individuals and families but rather as a result of systematic efforts to keep wages low, leading to widespread poverty, while at the same time denying access to food resources for vulnerable populations. Increased expenditures on social services have been shown to reduce food insecurity in multiple countries. Unfortunately, the US ranks second to last among developed countries in public expenditure on families [81]. In the US, rather than supporting families' success in gaining small steps toward economic stability, small increases in income can lead to a loss of SNAP eligibility, thus increasing the risk of food insecurity and poor health [82].

Despite the vast unmet need for food resources and the success of many local and federal programs, multiple ongoing efforts to reduce access to SNAP and other programs will likely exacerbate food insecurity in the US. In December 2019, the Department of Agriculture gave its final approval to the first set of measures to cut more than 700,000 people from SNAP [83]. The loss of SNAP eligibility includes immigrants threatened with the loss of eligibility for permanent legal status ("green cards"), decreased eligibility for Able-Bodied Adults Without Dependents ("ABAWDs"), changes in broad-based categorical eligibility (i.e., TANF enrollment no longer qualifies as an automatic eligibility for SNAP), and changes in the calculation of the standard utility allowance (the household's heating and utility expenses) used to calculate SNAP benefits. A New York Times headline from

March 20, 2020, clearly captures the impending worsening food insecurity crisis in the US: "Coronavirus and Poverty: A Mother Skips Meals So Her Children Can Eat" [84].

At a societal level, it benefits our nation when the population has food, whether the benefit is measured through healthcare savings, educational gains, socioeconomic improvements, or the knowledge that children are not going to bed hungry. One practical needed reform is to eliminate disqualifications for those who otherwise meet SNAP eligibility requirements, such as low-income college students [85]. The move from paper forms to electronic applications for SNAP and WIC benefits could improve the ease and speed with which people receive benefits. Automating enrollment in SNAP and WIC for those receiving TANF, Medicaid, and other programs focused on low-income people would also increase utilization rates.

Emergency clinicians can advocate for these and other ways to mitigate poverty such as expanding access to the earned income tax credit, TANF, child and dependent care credit, Section 8, public housing, and Medicaid. The use of these programs are all mechanisms that would free up families' limited financial resources for the purchase of food [81]. Heating fuel subsidies are especially important in areas of the country that have significant fluctuations in the weather. Children's HealthWatch has noted the "Heat or Eat" phenomenon that leads to measurable stunting of children's growth in winter months [86].

Recommendations for Emergency Medicine Practice

To emphasize: the ED could play a powerful role as portal of entry to existing social service programs for a large population in need. While the limited availability of food resources such as food pantries may be an obstacle in some circumstances, for many patients what is lacking is simply the connection to already existing programs. To facilitate this connection in an effective and successful fashion, a comprehensive approach is needed, which ranges from education at the level of the individual clinician to systemic approaches to screening and to having institutional resources available to address identified food insecurity both immediately and longitudinally. The American Academy of Pediatrics and the Food Research & Action Center has a toolkit for pediatricians and others to address patients' food insecurity (https://frac.org/aaptoolkit) [87].

Basic

- Train clinicians. Despite the fact that clinicians routinely ask about personal and sensitive medical topics, they may feel uncomfortable asking about sensitive food security issues that may have stigma associated with them. A script can be helpful. For example, simply prefacing a question about food insecurity with

"Do you mind if I ask…" may allow clinicians to overcome the hurdle of initiating the conversation.
- Ask patients about their immediate food needs (i.e., are they hungry right now?). Provide food directly and/or have mechanisms that help families cover costs of in-hospital food if necessary.
- Extend the conversation. Understand whether patients have food insecurity beyond the ED visit, ask if they need help identifying food resources, and be prepared to connect them to experts such as social workers or provide referral sheets with food resources.

Intermediate

- Institute universal screening for and documentation of food insecurity in the ED. Consider use of the two-question Hunger Vital Sign™, which is well-validated and widely used.
- Provide ready access to information about food resources via the electronic medical record (EMR) or pre-printed sheets.
- Develop connections to local food pantries, soup kitchens, and WIC offices that serve people who need food resources.
- Connect patients to federal programs. Have application forms available and social workers, other experts, and even trained volunteers that can assist families.

Advanced

- Provide food resources directly to patients such as via in-hospital food pantries.
- Set up programs that help food security during high-risk times of need such as summer food programs for children and additional food resources in winter months when families might need to choose between paying heating bills and paying for food.
- Advocate at the state level to ease restrictions on SNAP applications.
- Advocate at the national level to increase eligibility for SNAP and WIC. Advocate for larger benefits for individuals and families in need. Advocate for living wages to help lower the number of patients living in poverty.

Teaching Case

Clinical Case

An 8-year-old girl with asthma presents to the ED at a tertiary care academic pediatric hospital. She has significantly increased work of breathing and is speaking in one-word sentences. She is tachypneic and hypoxic and requires immediate intervention.

While initiating albuterol treatments, steroids, and placing an IV for further management, the resident turns to the mother to obtain additional history. When asked how long the patient has been having trouble breathing, the mother responds this has been going on for a few days. She ran out of her albuterol inhaler 3 days ago. The resident does not say anything immediately, but has a questioning look on her face. The mother goes on to explain that the family had to choose between buying food or asthma medications for the patient, as they didn't have enough money for both so they chose to buy food.

The patient responds to the initial treatments but is persistently tachypneic with marked wheezing and poor aeration. She requires admission to the ICU for a higher level of care.

Prior to being transferred to the ICU, the mother asks to speak with the resident. She hesitates, the resident believes out of pride, but then finally asks if the ED has any meal vouchers for food from the hospital cafeteria. Even sacrificing their meager funds for food over albuterol, they are still hungry.

Teaching Points

Patients with circumstances similar to this present every day in the ED, but the contribution of food insecurity to a variety of clinical presentations is often invisible. Proactive approaches to making food insecurity a more visible health-related social problem are the first steps to addressing it.

1. In this situation, the patient's parent was able to articulate her needs. In other circumstances, the clinician must be able to:
   - Consider the possibility of food insecurity
   - Be able and willing to inquire about food insecurity
   - Have immediate (e.g., hospital food vouchers) and ideally longitudinal resources (WIC, SNAP, or local food pantries) to share
2. The institution should create programs to screen for food insecurity. Systematic screening has the potential to identify patients that would otherwise not self-identify as food insecure.
3. The department should create educational programs for all clinicians about the need to screen for food insecurity. Provider education about the way to approach these sensitive issues is important.
4. Clinicians should codify a spectrum of easily accessible resources to address food insecurity. Screening does not help a patient if there are no resources available to assist them, ideally in both the immediate- and long-term timeframes.

Discussion Questions

1. The resident was uncomfortable with exploring the reason why the patient was not using her medication. She did not know how to respond when the family disclosed severe food insecurity. How does this compare to your own experiences with following through with these types of questions? How do you determine if you should ask about food insecurity?

2. In what way does the patient's food insecurity affect the emergency physician's decision-making? What are the responsibilities of the ED for helping not just the patient but also the family supporting them?
3. What resources are available in your institution to help families with food insecurity? Is there free, nutritious, and satisfying food available? Many EDs limit their food immediately available for distribution to saltines, pudding, juice, cereal, and maybe a turkey sandwich. What message are we sending our patients? Is the food in the cafeteria financially accessible for families in need?
4. What opportunities are available for partnerships outside of the ED? Do you know the names and locations of food pantries and soup kitchens that are in your patient populations' neighborhood? Do you have social workers or other organizations that can help patients fill out SNAP or WIC forms?
5. What are the challenges to identifying and helping patients with food insecurity? How do they differ or feel the same compared to other social problems?

References

1. Goldman N, Sheward R, Ettinger de Cuba S, Black MM, Sandel M, Cook J, et al. The hunger vital sign: A new standard of care for preventitive health. Children's HealthWatch. 2014:1–4. Available at https://www.childrenshealthwatch.org/wp-content/uploads/FINAL-Hunger-Vital-Sign-4-pager.pdf.
2. Coleman-Jensen A, Rabbitt MP, Gregory C, Singh A. Household food security in the United States in 2018. United States Department of Agriculture: Economic Research Service; 2019. https://www.ers.usda.gov/publications/pub-details/?pubid=94848.
3. United States Department of Agriculture Economic Research Service. Definitions of food security. 2019. https://www.ers.usda.gov/topics/food-nutrition-assistance/food-security-in-the-us/definitions-of-food-security/.
4. United States Department of Agriculture Economic Research Service. Commemorating 20 years of U.S. Food Security Measurement. 2015. https://www.ers.usda.gov/amber-waves/2015/october/commemorating-20-years-of-us-food-security-measurement/. Accessed on 12 June 2019.
5. Falk G, Aussenberg RA. The supplemental nutrition assistance program (SNAP): categorical eligibility. Congressional Research Service. 2019. https://fas.org/sgp/crs/misc/R42054.pdf.
6. Center on Budget and Policy Priorities. Chart Book: SNAP helps struggling families put food on the table. Washington, DC. 2013. https://www.cbpp.org/sites/default/files/atoms/files/3-13-12fa-chartbook.pdf.
7. Special supplemental nutrition program for women, infants and children (WIC). Benefits.gov. 2019. https://www.benefits.gov/benefit/368.
8. Food and Nutrition Services. WIC frequently asked questions (FAQs). U.S. Department of Agriculture. 2019. https://www.fns.usda.gov/wic/frequently-asked-questions-about-wic.
9. Feeding America. What is a food bank? 2019. https://www.feedingamerica.org/our-work/food-bank-network.
10. Food and Nutrition Services. Child and adult care food program. U.S. Department of Agriculture. 2019. https://www.fns.usda.gov/cacfp/child-and-adult-care-food-program.
11. WIC eligibility and coverage rates. U.S. Department of Agriculture. 2017. https://www.fns.usda.gov/wic/wic-eligibility-and-coverage-rates.
12. Hanson KL, Connor LM. Food insecurity and dietary quality in US adults and children: a systematic review. Am J Clin Nutr. 2014;100(2):684–92.

13. Seligman HK, Schillinger D. Hunger and socioeconomic disparities in chronic disease. N Engl J Med. 2010;363(1):6–9.
14. Pan L, Sherry B, Njai R, Blanck HM. Food insecurity is associated with obesity among US adults in 12 states. J Acad Nutr Diet. 2012;112(9):1403–9.
15. Berkowitz SA, Baggett TP, Wexler DJ, Huskey KW, Wee CC. Food insecurity and metabolic control among U.S. adults with diabetes. Diabetes Care. 2013;36(10):3093–9.
16. Gundersen C, Ziliak JP. Food insecurity and health outcomes. Health Aff. 2015;34(11):1830–9.
17. Crews DC, Kuczmarski MF, Grubbs V, Hedgeman E, Shahinian VB, Evans MK, et al. Effect of food insecurity on chronic kidney disease in lower-income Americans. Am J Nephrol. 2014;39(1):27–35.
18. Castillo DC, Ramsey NLM, Yu SSK, Ricks M, Courville AB, Sumner AE. Inconsistent access to food and cardiometabolic disease: the effect of food insecurity. Curr Cardiovasc Risk Rep. 2012;6(3):245–50.
19. Golovaty I, Tien PC, Price JC, Sheira L, Seligman H, Weiser SD. Food insecurity may be an independent risk factor associated with nonalcoholic fatty liver disease among lowincome adults in the United States. J Nutr. 2020;150(1):91–8.
20. Vercammen KA, Moran AJ, McClain AC, Thorndike AN, Fulay AP, Rimm EB. Food security and 10-year cardiovascular disease risk among U.S. adults. Am J Prev Med. 2019;56(5):689–97.
21. Drennen CR, Coleman SM, De Cuba SE, Frank DA, Chilton M, Cook JT, et al. Food insecurity, health, and development in children under age four years. Pediatrics. 2019;144(4):e20190824.
22. Cook JT, Frank DA, Berkowitz C, Black MM, Casey PH, Cutts DB, et al. Food insecurity is associated with adverse health outcomes among human infants and toddlers. J Nutr. 2004;134(6):1432–8.
23. Liu Y, Njai RS, Greenlund KJ, Chapman DP, Croft JB. Relationships between housing and food insecurity, frequent mental distress, and insufficient sleep among adults in 12 US states, 2009. Prev Chronic Dis. 2014;11(3):1–9.
24. Cook JT, Black M, Chilton M, Cutts D, Ettinger de Cuba S, Heeren TC, et al. Are food insecurity's health impacts underestimated in the U.S. population? Marginal food security also predicts adverse health outcomes in young U.S. children and mothers. Adv Nutr. 2013;4(1):51–61.
25. Thomas MMC, Miller DP, Morrissey TW. Food insecurity and child health. Pediatrics. 2019;144(4):e20190397.
26. Kersey MA, Beran MS, McGovern PG, Biros MH, Lurie N. The prevalence and effects of hunger in an emergency department patient population. Acad Emerg Med. 1999;6(11):1109–14.
27. Mazer M, Bisgaier J, Dailey E, Srivastava K, McDermoth M, Datner E, et al. Risk for cost-related medication nonadherence among emergency department patients. Acad Emerg Med. 2011;18(3):267–72.
28. Berkowitz SA, Seligman HK, Meigs JB, Basu S. Food insecurity, healthcare utilization, and high cost: a longitudinal cohort study. Am J Manag Care. 2018;24(9):399–404.
29. Kushel MB, Gupta R, Gee L, Haas JS. Housing instability and food insecurity as barriers to health care among low-income Americans. J Gen Intern Med. 2006;21(1):71–7.
30. Becerra MB, Allen NL, Becerra BJ. Food insecurity and low self-efficacy are associated with increased healthcare utilization among adults with type II diabetes mellitus. J Diabetes Complicat. 2016;30(8):1488–93.
31. Baggett TP, Singer DE, Rao SR, O'Connell JJ, Bharel M, Rigotti NA. Food insufficiency and health services utilization in a national sample of homeless adults. J Gen Intern Med. 2011;26(6):627–34.
32. Weiser SD, Hatcher A, Frongillo EA, Guzman D, Riley ED, Bangsberg DR, et al. Food insecurity is associated with greater acute care utilization among HIV-infected homeless and marginally housed individuals in San Francisco. J Gen Intern Med. 2013;28(1):91–8.
33. Berkowitz SA, Seligman HK, Choudhry NK. Treat or eat: food insecurity, cost-related medication underuse, and unmet needs. Am J Med. 2014;127(4):303–10.e3. https://doi.org/10.1016/j.amjmed.2014.01.002.

34. Baer TE, Scherer EA, Fleegler EW, Hassan A. Food insecurity and the burden of health-related social problems in an urban youth population. J Adolesc Health. 2015;57(6):601–07. https://doi.org/10.1016/j.jadohealth.2015.08.013.
35. Fuller AE, Brown NM, Grado L, Oyeku SO, Gross RS. Material hardships and health care utilization among low-income children with special health care needs. Acad Pediatr. 2019;19(7):733–9. https://doi.org/10.1016/j.acap.2019.01.009.
36. Rose-Jacobs R, Fiore JG, De Cuba SE, Black M, Cutts DB, Coleman SM, et al. Children with special health care needs, supplemental security income, and food insecurity. J Dev Behav Pediatr. 2016;37(2):140–7.
37. Cook JT, Poblacion A. An Avoidable $2.4 Billion Cost: The estimated health-related costs of food insecurity and hunger in Massachusetts. Children's HealthWatch; 2018;1–20. http://macostofhunger.org/wp-content/uploads/2018/02/full-report.pdf.
38. Berkowitz SA, Basu S, Meigs JB, Seligman HK. Food insecurity and health care expenditures in the United States, 2011–2013. Health Serv Res. 2018;53(3):1600–20.
39. Bitler M, Currie J. Does WIC work? The effects of WIC on pregnancy and birth outcomes. J Policy Anal Manage. 2005;24(1):73–91.
40. Kowaleski-Jones L, Duncan GJ. Effects of participation in the WIC program on birthweight: evidence from the National Longitudinal Survey of Youth. Am J Public Health. 2002;92(5):799–804.
41. Sherry B, Mei Z, Yip R. Continuation of the decline in prevalence of anemia in low-income infants and children in five states. Pediatrics. 2001;107(4):677–82.
42. Hicks LE, Langham RA. Cognitive measure stability in siblings following early nutritional supplementation. Public Health Rep. 1985;100(6):656–62.
43. Arteaga I, Heflin C, Hodges L. SNAP benefits and pregnancy-related emergency room visits. Popul Res Policy Rev. 2018;37:1031–52.
44. Heflin C, Arteaga I, Hodges L, Ndashiyme JF, Rabbitt MP. SNAP benefits and childhood asthma. Soc Sci Med. 2019;220:203–11.
45. Heflin CM, Ingram SJ, Ziliak JP. The effect of the supplemental nutrition assistance program on mortality. Health Aff (Millwood). 2019;38(11):1807–15.
46. Berkowitz SA, Seligman HK, Rigdon J, Meigs JB, Basu S. Supplemental Nutrition Assistance Program (SNAP) participation and health care expenditures among low-income adults. JAMA Intern Med. 2017;177(11):1642–9.
47. Nord M. How much does the supplemental nutrition assistance program alleviate food insecurity? Evidence from recent programme leavers. Public Health Nutr. 2012;15(5):811–7.
48. Mabli J, Worthington J. Supplemental nutrition assistance program participation and child food security. Pediatrics. 2014;133(4):610–9.
49. Hager ER, Quigg AM, Black MM, Coleman SM, Heeren T, Rose-Jacobs R, et al. Development and validity of a 2-item screen to identify families at risk for food insecurity. Pediatrics. 2010;126(1):e26–32.
50. Gundersen C, Engelhard EE, Crumbaugh AS, Seligman HK. Brief assessment of food insecurity accurately identifies high-risk US adults. Public Health Nutr. 2017;20(8):1367–71.
51. Center for Medicare and Medicaid Services. The accountable health communities health-related social needs screening tool. Natl Acad Med Perspect. 2017:1–9. Available from: https://innovation.cms.gov/Files/worksheets/ahcm-screeningtool.pdf.
52. Byhoff E, De Marchis EH, Hessler D, Fichtenberg C, Adler N, Cohen AJ, et al. Part II: a qualitative study of social risk screening acceptability in patients and caregivers. Am J Prev Med. 2019;57(6):S38–46. https://doi.org/10.1016/j.amepre.2019.07.016.
53. De Marchis EH, Hessler D, Fichtenberg C, Adler N, Byhoff E, Cohen AJ, et al. Part I: a quantitative study of social risk screening acceptability in patients and caregivers. Am J Prev Med. 2019;57(6):S25–37. https://doi.org/10.1016/j.amepre.2019.07.010.
54. Cullen D, Woodford A, Fein J. Food for thought: a randomized trial of food insecurity screening in the emergency department. Acad Pediatr. 2019;19(6):646–51.

55. Morris AM, Anderson JKE, Schmitthenner B, Aylward AF, Shams RB, Hurka-richardson K, et al. Leveraging emergency department visits to connect older adults at risk for malnutrition and food insecurity to community resources: design and protocol development for the BRIDGE study. BMC Pilot Feasiblity Stud. 2020;6(36):1–7.
56. Kanak M, Fleegler E, Chang L, Curt A, Burdick K, Monuteaux MC, et al. Feasibility of personal phone use for pediatric emergency department social screening and referrals. Philadelphia: Pediatric Academic Societies; 2020 (Meeting canceled due to Covid-19).
57. Fleegler EW, Lieu TA, Wise PH, Muret-Wagstaff S. Families' health-related social problems and missed referral opportunities. Pediatrics. 2007;119(6):e1332–41.
58. Hassan A, Blood EA, Pikcilingis A, Krull EG, McNickles L, Marmon G, et al. Youths' health-related social problems: concerns often overlooked during the medical visit. J Adolesc Health. 2013;53(2):265–71. https://doi.org/10.1016/j.jadohealth.2013.02.024.
59. Hassan A, Scherer EA, Pikcilingis A, Krull E, McNickles L, Marmon G, et al. Improving social determinants of health: effectiveness of a web-based intervention. Am J Prev Med. 2015;49(6):822–31. https://doi.org/10.1016/j.amepre.2015.04.023.
60. Bottino CJ, Rhodes ET, Kreatsoulas C, Cox JE, Fleegler EW. Food insecurity screening in pediatric primary care: can offering referrals help identify families in need? Acad Pediatr. 2017;17(5):497–503. https://doi.org/10.1016/j.acap.2016.10.006.
61. Garg A, Butz A, Dworkin P, Lewis R, Thompson R, Serwint J. Improving the management of family psychosocial problems at low-income children's well-child care visits: the WE CARE project. Pediatrics. 2007;120(3):547–58.
62. Turner C, Ku L, Rogers S, Lindberg L, Pleck J, Sonenstein F. Adolescent sexual behavior, drug use, and violence: increased reporting with computer survey technology. Science. 1998;280:867–73.
63. Palakshappa D, Goodpasture M, Albertini L, Brown CL, Montez K, Skelton JA. Written versus verbal food insecurity screening in one primary care clinic. Acad Pediatr. 2020;20(2):203–7. https://doi.org/10.1016/j.acap.2019.10.011.
64. Tung EL, Abramsohn EM, Boyd K, Makelarski JA, Beiser DG, Chou C, et al. Impact of a low-intensity resource referral intervention on patients' knowledge, beliefs, and use of community resources: results from the CommunityRx trial. J Gen Intern Med. 2020;35(3):815–23. https://doi.org/10.1007/s11606-019-05530-5.
65. Goldstein Z, Krausman R, Howard-Karp M, Paulsen R, Nahm S, LeBeouf A, et al. Social determinants of health: overcoming the greatest barriers to patient care. HealthLeads, United Way, 2-1-1. 2017;1-7. https://healthleadsusa.org/wp-content/uploads/2018/10/SDOH-Overcoming-the-Greatest-Barriers-to-Patient-Care.pdf.
66. Rodgers JT, Purnell JQ. Healthcare navigation service in 2-1-1 San Diego: guiding individuals to the care they need. Am J Prev Med. 2012;43(6 S5):S450–6.
67. Garg A, Toy S, Tripodis Y, Silverstein M, Freeman E. Addressing social determinants of health at well child care visits: a cluster RCT. Pediatrics. 2015;135(2):e296–304.
68. Schickedanz A, Sharp A, Hu YR, Shah NR, Adams JL, Francis D, et al. Impact of social needs navigation on utilization among high utilizers in a large integrated health system: a quasi-experimental study. J Gen Intern Med. 2019;34(11):2382–9. https://doi.org/10.1007/s11606-019-05123-2.
69. Rottapel R, Sheward R. Children's HealthWatch Hunger Vital Sign™ case study # 2: why did Boston Medical Center begin screening for food insecurity? Boston. 2016. https://childrenshealthwatch.org/wp-content/uploads/CHW_bmc_FINAL.pdf.
70. Bash H. Food as medicine: food prescriptions coming to Cleveland community. News 5 Cleveland; 2018. https://www.news5cleveland.com/news/localnews/cleveland-metro/food-as-medicine-prescriptions-for-food-hope-to-fuel-cleveland-community.
71. Beck AF, Henize AW, Kahn RS, Reiber KL, Young JJ, Klein MD. Forging a pediatric primary care–community partnership to support food-insecure families. Pediatrics. 2014;134(2):e564–71.

72. Cavanagh M, Jurkowski J, Bozlak C, Hastings J, Klein A. Veggie Rx: an outcome evaluation of a healthy food incentive programme. Public Health Nutr. 2017;20(14):2636–41.
73. Bryce R, Guajardo C, Ilarraza D, Milgrom N, Pike D, Savoie K, et al. Participation in a farmers' market fruit and vegetable prescription program at a federally qualified health center improves hemoglobin A1C in low income uncontrolled diabetics. Prev Med Rep. 2017;7:176–9. https://doi.org/10.1016/j.pmedr.2017.06.006.
74. Saxe-Custack A, Lofton HC, Hanna-Attisha M, Victor C, Reyes G, Ceja T, et al. Caregiver perceptions of a fruit and vegetable prescription programme for low-income paediatric patients. Public Health Nutr. 2018;21(13):2497–506.
75. Health Research & Educational Trust. Social determinants of health series: Food insecurity and the role of hospitals. Chicago, IL: Health Research & Educational Trust. 2017. Accessed at: https://www.aha.org/foodinsecurity.
76. Cullen D, Blauch A, Mirth M, Fein J. Complete EATS: summer meals offered by the emergency department for food insecurity. Pediatrics. 2019;144(4):e20190201.
77. Arkansas Children's Hospital participating in USDA child and adult care food program at-risk afterschool meals component. Arkansas Children's. 2019. https://www.archildrens.org/news/releases/2019/ach-participating-in-usda-child-and-adult-care-food-program-at-risk-afterschool-meals-component.
78. Smith S, Malinak D, Chang J, Perez M, Perez S, Settlecowski E, et al. Implementation of a food insecurity screening and referral program in student-run free clinics in San Diego, California. Prev Med Rep. 2017;5:134–9.
79. Finkelstein A, Notowidigdo MJ. The effects of information and application assistance on take-up, targeting, and welfare: experimental evidence from SNAP. Proc Ann Conf Tax Minutes Ann Meet Natl Tax Assoc. 2017;110:1–50.
80. Gordon JA, Emond JA, Camargo CA. The state children's health insurance program: a multicenter trial of outreach through the emergency department. Am J Public Health. 2005;95(2):250–3.
81. Fernald LCH, Gosliner W. Alternatives to SNAP: global approaches to addressing childhood poverty and food insecurity. Am J Public Health. 2019;109(12):1668–77.
82. Gaines-Turner T, Simmons JC, Chilton M. Recommendations from SNAP participants to improve wages and end stigma. Am J Public Health. 2019;109(12):1664–7.
83. Fadulu L. Hundreds of thousands are losing access to food stamps. The New York Times. 2019. https://www.nytimes.com/2019/12/04/us/politics/food-stamps.html.
84. Fernandez M. Coronavirus and poverty: a mother skips meals so her children can eat. New York Times. 2020 Mar 20. https://www.nytimes.com/2020/03/20/us/coronavirus-poverty-school-lunch.html?referringSource=articleShare.
85. Freudenberg N, Goldrick-Rab S, Poppendieck J. College students and SNAP: the new face of food insecurity in the United States. Am J Public Health. 2019;109(12):1652–8.
86. Frank DA, Roos N, Meyers A, Napoleone M, Peterson K, Cather A, et al. Seasonal variation in weight-for-age in a pediatric emergency room. Public Health Rep. 1996;111(4):366–71.
87. American Academy Pediatrics; Food Research & Action Center. Addressing Food insecurity: a toolkit for pediatricians. 2019. https://frac.org/aaptoolkit.

Homelessness

14

Bisan A. Salhi and Kelly M. Doran

Key Points

- Homelessness is prevalent in the US and is commonly encountered in the ED.
- People who are homeless often do not fit the stereotypical picture of homelessness. Therefore, EPs should routinely ask patients about their housing situation.
- While EPs will usually be unable to immediately "fix" their patients' homelessness in the ED, they should still assess homelessness status since this is an important consideration in diagnosis, treatment, and disposition plans.
- Some hospitals have developed successful collaborations with local housing and other community organizations to better address patients' homelessness, with the goal of improving patients' health and reducing their future acute care use.
- EPs can have a powerful voice in local and national advocacy efforts to address the social and economic factors underlying homelessness.

Foundations

Background

Homelessness is a problem worldwide. Within the US alone, over 1.4 million people used homeless shelters in 2017 [1]. Counting both unsheltered (e.g., on the streets) and sheltered people, 567,715 people were homeless at the 2019 US "point-in-time" count [2]. Homelessness cuts across the US, with statistics for each state

B. A. Salhi (✉)
Department of Emergency Medicine, Emory University, Atlanta, GA, USA
e-mail: bsalhi@emory.edu

K. M. Doran
Departments of Emergency Medicine and Population Health, NYU School of Medicine, New York, NY, USA
e-mail: kelly.doran@nyulangone.org

H. J. Alter et al. (eds.), *Social Emergency Medicine*,
https://doi.org/10.1007/978-3-030-65672-0_14

detailed in Annual Homeless Assessment Reports to Congress (available online) [3]. The majority of people who experience homelessness do so for limited periods of time and are sometimes called transitionally homeless. In contrast, chronically homeless persons are defined by the federal government as individuals with a disabling condition (defined as chronic physical illness, substance use disorder, mental illness, or developmental disability) who have been continually homeless for at least 1 year or have had four or more episodes of homelessness in the past 3 years (totaling at least 12 months total time homeless) [1]. Importantly, homelessness is a condition to which people are exposed, rather than a marker of identity or a personal or moral failing. To convey this, we suggest using terms such as "people experiencing homelessness" or "people who are homeless" instead of "the homeless," which may reproduce stigma and blame.

Overall, approximately two-thirds of people who are homeless in the US use homeless shelters and the other third are unsheltered, though this varies by locality [1]. Around two-thirds of people who are homeless are individuals and one-third are members of families with children [1]. While emergency providers (EPs) may be most likely to recognize homelessness when patients are unsheltered and chronically homeless, this is a small subset of people experiencing homelessness.

The main causes of homelessness are structural and economic, with lack of affordable housing playing a primary role alongside growing income inequality, structural racism, and an inadequate and receding social safety net. Since 2001, the gap between rent and household income in the US has risen steadily, with only 25% of low-income renters receiving assistance for their housing needs [4]. The risk of experiencing homelessness is not distributed equally across society. People who are Black or African American, Hispanic/Latinx, Native American, and Pacific Islander are disproportionately represented among the US homeless population. Racial and ethnic disparities in homelessness reflect a long-standing history of structural racism in housing policy, economic opportunities, and access to social services [5–8]. Moreover, housing is a key mechanism by which structural racism is maintained and perpetuated in the US [9–11]. Youth leaving foster care and people with disabilities are also overrepresented among people who are homeless.

The COVID-19 pandemic will most likely further exacerbate these inequities, though the full scope of these impacts remains unclear at the time of this writing. Nevertheless, it is evident that the pandemic has exposed the vulnerabilities of people experiencing homelessness to infectious disease. The COVID-19 pandemic has dramatically illustrated the need for more robust health and social safety nets and the incompatibility of homelessness with individual and population health [12].

Other important and interrelated contributors to homelessness include unemployment and underemployment, social discord, incarceration, domestic violence, pregnancy, natural disasters, birth cohort effects, and many others [13–16]. Health issues—including substance use, mental health, and physical health—are also important contributors to homelessness for a subset of people [17–21]. Additionally, poor health leading to prolonged hospital stays or high medical expenses can result in loss of savings and income, with resultant inability to pay for housing and eviction or foreclosure.

Though homelessness is a complex issue, experts have a good understanding of its solutions. The US Interagency Council on Homelessness has previously outlined strategies for preventing and ending homelessness [22, 23]. The linchpin for these strategies is affordable, stable housing. One type of housing, called permanent supportive housing, provides non-time-limited affordable housing paired with supportive services such as case management. Housing First, developed in New York City in the early 1990s, is a term now used more generally to describe provision of permanent supportive housing without prerequisites around "housing readiness" (e.g., sobriety) or requirements that tenants accept or are compliant with services to remain in housing (though case management and other supportive services are offered as part of the program model). Permanent supportive housing overall, and Housing First specifically, have demonstrated success in ending homelessness, including among people with serious mental illness, and some studies have found resultant reductions in healthcare costs [24–28]. Unfortunately, the demand for permanent supportive housing and other forms of affordable housing exceeds supply [29].

People who are homeless have high rates of chronic medical conditions. A national study of publicly funded health center clients found that 41.9% of homeless patients had two or more chronic medical conditions, compared to 32.6% of low-income housed patients [30]. In addition, roughly one-third of people who are homeless have significant substance use and roughly one-third have a mental health condition, though estimates vary substantially depending on the samples studied and definitions used [31–35]. Further, people experiencing homelessness face many health risks unique to their situation, including threats to physical safety (e.g., physical and sexual assault), inadequate nutrition, exposure to the cold and other elements, poor sleep, and high stress. Multiple studies have found that homelessness is associated with significantly elevated age-adjusted mortality rates [36–38]. Currently, the most common cause of death is drug overdose, followed by cancer and heart disease [36]. A growing body of research has also found premature aging among people experiencing homelessness, with homeless people in their 50s having rates of dementia, incontinence, mobility limitations, and other geriatric conditions akin to those of non-homeless populations in their 70s [39]. Not surprisingly given their disproportionate burden of poor health, people experiencing homelessness are higher than average users of the healthcare system and particularly of acute, hospital-based care including emergency department (ED) care [40–44]. Indeed, in many US EDs, EPs are unlikely to work a single shift without encountering a patient who is homeless.

Evidence Basis

A large body of research has demonstrated that people experiencing homelessness use EDs more frequently than those who are not homeless. Homelessness becomes ubiquitous when considering the group of most frequent "high users" of EDs [40]. For these individuals, provision of permanent supportive housing is currently the best evidence-based intervention [24, 26, 27, 45]. However, it should be recognized that—while homelessness overall is associated with higher than average rates of ED use—the most frequent users of ED services represent only a small subset of the homeless population [46, 47].

Multiple single- and multi-site studies—spanning the country and including suburban as well as urban EDs—have examined prevalence of homelessness among ED patients, finding rates ranging from 2.5% to 13.8% [48]. National studies on homelessness among ED patients are limited by a lack of uniform data on homelessness outside of the Veterans Affairs (VA) system. Some authors have used the National Hospital Ambulatory Medical Care Survey (NHAMCS) to estimate homelessness among ED patients, but such efforts are hindered by significant limitations in NHAMCS' recording of homelessness [48]. In general, healthcare administrative data (such as NHAMCS, Healthcare Cost and Utilization Project [HCUP], and data from electronic health records) have unknown sensitivity and specificity for identifying homelessness and likely significantly underestimate homelessness.

Salhi et al. published a systematic review of the literature on homelessness and emergency medicine (EM) spanning 1990–2016, finding only 28 studies that met their inclusion criteria [42]. Most studies focused on ED use by people experiencing homelessness, as well as prevalence and basic characteristics of homeless ED patients [42]. National studies suggest that ED patients who are homeless have similar triage acuities and hospital admission rates compared to other patients but are more likely to arrive by ambulance and to have had a recent past ED visit [49–51]. Patients who are homeless also have higher 30-day hospital readmission rates [52]. Little past research has examined ED care from the perspective of homeless patients themselves; one exception is a study with patients who were homeless and had alcohol use disorders by McCormack et al. [53]. Another study found that EM residents struggled in providing care to their patients experiencing homelessness; residents noted that they lacked formal education on caring for patients who were homeless and felt frustrated by limitations in their ability to "make a true difference" for these patients [54, 55]. Significant gaps in the literature on homelessness and EM include research on how to best educate and support EM trainees and providers in caring for patients experiencing homelessness; characteristics and quality of emergency care received by homeless vs. non-homeless patients; best practices in ED care for patients experiencing homelessness; and how to most effectively partner with organizations outside the ED to optimize the ED's role and improve overall health for people who are homeless, including the concept of "in-reach" into EDs by homeless services organizations.

Emergency Department and Beyond

Bedside

Health risks, conditions, and treatment considerations for people experiencing homelessness have been well summarized, with clinical practice guidelines available in sources such as UpToDate [56]. While not specific to emergency care, these clinical practice resources may still be useful to EPs. Individuals and families may become homeless as a consequence of unexpected injury and illness. The psychosocial and structural stressors of homelessness also contribute to the poor health of this population. People who are homeless suffer from mortality rates three to six times those of the general population, and homelessness is an independent risk factor for mortality [38, 57–59]. People who are homeless also experience higher rates of chronic illness, injury, infectious disease (e.g., tuberculosis, HIV, hepatitis A, and hepatitis C), substance use, and mental illness than their low-income housed counterparts [60]. Moreover, patients experiencing homelessness may have difficulty storing and taking medications as prescribed due to their housing circumstances, and many report having medications stolen in shelters or on the streets [61, 62].

The US homeless population has aged considerably since the 1980s, which has important implications for EPs. The median age of single homeless adults is 50 years today, compared with 37 years in 1990, and is predicted to continue to rise [63]. More than one-third of ED visits by patients experiencing homelessness are by people over 50 years old [64, 65]. People experiencing homelessness are considered "elderly" at age 50, since homeless adults over age 50 suffer from "geriatric syndromes" commonly associated with aging (e.g., frequent falls, memory loss, vision or hearing impairment) at rates similar to or higher than housed adults 15–20 years older [66]. EPs should consider such geriatric syndromes among their patients who are homeless and aged 50 and older.

Substance use disorders and psychiatric illness contribute to a significant proportion of ED visits among people experiencing homelessness [47, 49, 65, 67]. Overdose is now the leading cause of death among people who are homeless throughout the US [36, 68–70]. EPs should ask all patients who present to the ED after a nonfatal overdose about their housing situation and recognize that homelessness may present barriers to initiation of and retention in substance use treatment. State and local ordinances create structural vulnerability to incarceration and assault. For example, the criminalization of public intoxication and drug use makes people experiencing homelessness more likely to be arrested (due to frequent encounters with police officers on the street) and assaulted (due to lack of shelter and pervasive stigma) [71–73]. A full exploration of the complex relationship between substance use and homelessness is beyond the scope of this chapter.

While substance use and mental illness represent an important subset of ED presentations by people who are homeless, EPs should recognize that people experiencing homelessness have a wide variety of physical health issues that require ED care. Evidence suggests that people experiencing homelessness suffer a disproportionate burden of injuries compared to housed persons, including unintentional, self-inflicted, and physical and sexual assault injuries [74, 75]. People experiencing homelessness are more likely to present to the ED with burns and injuries to the lower extremities and to have more severe injury patterns and longer injury-related hospitalizations than non-homeless persons [76, 77]. These injuries are sometimes related to the daily realities of homelessness, such as utilizing campfires for warmth. A high prevalence of violence against people who are homeless also drives injuries and ED use [78]. For example, studies have shown that 27–52% of homeless individuals were physically or sexually assaulted in the previous year [79, 80]. Although homeless men and women are equally likely to be victims of physical assault, homeless women have higher rates of sexual assault, with nearly 10% of homeless women reporting a sexual assault in the previous year [79].

While people experiencing homelessness do suffer from unique patterns of injury and illness, it is important to consider common diagnoses in the management and treatment of patients who are homeless. For example, one study found that one-fifth of ED visits among homeless women were related to pregnancy [81]. Another study found high levels of poor self-reported health status and chronic physical pain in homeless and unstably housed women [82]. Overall, after drug overdose, the most common causes of mortality among people who are homeless are cancer and heart disease [36]. In addition to having higher than average rates of many chronic diseases, homelessness makes it more difficult to carry out treatment plans, including regular attendance at primary care appointments, obtaining proper specialist care, arranging surgeries and postsurgical care, and adhering to medication regimens.

Finally, EPs should be especially cognizant of, and take steps to counter, the negative experiences that people experiencing homelessness may have when seeking care. People who are homeless often recount experiencing stigma in healthcare settings, which may have adverse effects on their health and likelihood of seeking healthcare in the future [83, 84].

Hospital/Healthcare System

Rates of hospital admission for ED patients who are homeless have varied across studies, but national studies show that admission rates are similar to those for patients who are not homeless [50, 85]. However, much attention has been given to the idea of "inappropriate" or "unnecessary" ED visits by homeless adults [86], which is sometimes attributed to lack of health insurance or access to primary care. People experiencing homelessness are more likely to lack health insurance or primary care than the general population. However, in states that have expanded Medicaid to low-income single adults, many people experiencing homelessness are enrolled in Medicaid and uninsured rates are lower [87]. While some studies have indicated a role for health insurance in increasing access to ambulatory care [47], insurance status and access to primary care on their own have consistently been shown not to be associated with decreased ED use among homeless adults [47, 81, 86, 88–92]. For example, studies of the VA health system—whose patients have robust access to outpatient health services—have shown that homelessness is still strongly associated with frequent ED use [40, 93, 94]. These and other studies strongly suggest that factors other than insurance and access to primary care are important drivers of ED visits among people experiencing homelessness. They further suggest that EDs provide the type of very low-barrier care that individuals experiencing homelessness can most easily access and that EDs address important unmet needs for this population.

Compared to non-homeless patients, patients who are homeless experience longer inpatient stays and are more likely to have repeat ED visits [49, 95, 96] and to visit the ED and/or be readmitted to the hospital within 30 days of hospital discharge [64, 97, 98]. In the process, their health deteriorates, and some may become embroiled in costly, sub-optimal cycles of care. This is especially pronounced among people discharged to shelters or unsheltered locations vs. other living situations (e.g., respite care, stable housing) and has been noted across age groups [52, 81, 99]. EPs may find themselves in the situation of readmitting homeless patients for whom a prior inpatient hospital discharge was inappropriate or otherwise unsuccessful (e.g., discharge to an unsheltered location or shelter without feasible plans for patients to recuperate and/or follow other discharge instructions). In some instances, individuals residing in shelters or transitional housing units are at risk of losing their bed or unit if a hospital stay is prolonged and demand for these services is high. Those with serious illness who are homeless may also have difficulty obtaining housing if building managers are concerned they may be "too sick" to be safely housed without intensive onsite medical and social services. Hospitals should work closely with community organizations to address these issues.

EDs provide not only healthcare for people experiencing homelessness but sometimes also vital resources (e.g., food, water, shelter, clothing) necessary for survival and subsistence [100–103]. Moreover, EDs are often a first-stop destination for people who have become newly homeless, with one study finding that nearly a quarter of homeless adults sought care in an ED upon first becoming homeless [104]. Uniquely accessible among healthcare settings due to their around-the-clock availability and the mandates of the Emergency Medical Treatment and Labor Act (EMTALA),[1] EDs represent a critical setting for intervening in the lives of people experiencing homelessness and linking them to available community resources [105, 106].

A growing number of hospitals and healthcare systems have successfully partnered with community organizations to address homelessness for a subset of their patients. These efforts have often focused on the most frequent or highest cost users of hospital systems. For example, Buchanan et al. found that discharging patients who were homeless to a medical respite program that provided 24-hour room and services was associated with decreased future hospitalizations [99]. Multiple studies have shown that permanent supportive housing (housing paired with case management) successfully ends homelessness and may also be associated with

[1]EMTALA does not specifically mandate the treatment of patients experiencing homelessness or other vulnerable populations beyond its general requirements for all people seeking emergency treatment. Nevertheless, this legislation has cemented the ED as the healthcare safety net and magnified the importance of the ED in providing healthcare for people experiencing homelessness. For further details of EMTALA, see https://www.cms.gov/regulations-and-guidance/legislation/emtala/.

reductions in future ED visits and hospitalizations [24–27, 107, 108]. For example, a randomized controlled trial in Chicago found that, after adjusting for baseline covariates, patients who were homeless and had chronic illnesses who were randomized to housing plus case management had reduced numbers of future hospitalizations, hospital length of stay, and ED visits [24]. While other studies have found no such reductions in hospital use, it is important to recognize that expecting housing to "pay for itself" via reductions in healthcare costs obscures the fundamental role that housing plays in creating an equitable society [27, 109, 110].

Citing mutual benefits to patients and hospitals, the American Hospital Association published a report providing case studies and guidance for hospitals seeking to collaborate with community and governmental organizations to address housing as a social determinant of health [111]. EPs can mobilize or join such efforts in conjunction with their hospitals and other local organizations [112]. In New York City, hospitals and other healthcare providers have joined together with homelessness and housing providers to form Health and Housing Consortia, which focus on training and resource development, cross sector communication, and research and advocacy to better serve their shared homeless populations [113].

Although these formal partnerships may not be replicable in all settings, EPs should maintain relationships with community organizations to better serve their patients who are experiencing homelessness and to seize opportunities for research, programmatic improvements, and advocacy when they arise.

Societal Level

The causes and circumstances of homelessness are mediated by larger social and economic conditions, including federal, state, and local funding priorities. There remains inadequate governmental funding for affordable housing, and in specific regions of the country rates of homelessness have increased sharply in recent years. EPs have a critical role to play in advocating at the federal, state, and local levels for increasing the supply of affordable, stable housing necessary to mitigate the health effects of homelessness. Given the in-depth experience of EPs with patients who are homeless and the outsized role that EDs play in caring for this population, EPs have a valuable voice to add to policy and advocacy efforts around homelessness. Advocacy efforts should take into account the local context, including the availability and accessibility of resources. This is especially important in rural communities, where homelessness is understudied and may differ from homelessness in urban areas [114–116].

In one recent effort to stem hospital discharges without adequate provisions for patients' health and safety, California instituted a law (SB 1152) outlining minimal standards in caring for and discharging patients who are homeless. The law requires, among other things, that hospitals screen all patients at every visit to determine if they are homeless; provide food and weather-appropriate clothing; supply patients with medications if the hospital has a retail pharmacy; coordinate care with community agencies and refer patients who are homeless to appropriate service providers; provide transportation to safe and appropriate locations within 30 miles (or 30 minutes) of the hospital; and document each of these processes in the electronic medical record [117]. This law may prevent hospitals from egregious discharge practices, improve the experience of homeless individuals within the healthcare system, and also improve the ability to document and track homelessness in healthcare settings. However, the law does not address the underlying drivers of homelessness.

The US Interagency Council on Homelessness' federal strategic plan to prevent and end homelessness, titled *Home, Together* [118], is a good starting point for EPs who may be interested in further understanding drivers of and solutions to homelessness in the US and becoming part of larger advocacy efforts. Principal among recommendations for preventing and ending homelessness are ensuring the availability of adequate affordable housing. EPs should also consider learning from or collaborating with national experts and advocacy groups including the Corporation for Supportive Housing, the National Alliance to End Homelessness, the National Low Income Housing Coalition, and the National Health Care for the Homeless Council.

Recommendations for Emergency Medicine Practice

Among the main difficulties in providing ED care for patients experiencing homelessness is that their needs are complex, are interdependent, exceed available resources, and defy traditional conceptions of health and healthcare delivery in EM. For example, housing is not traditionally considered a form of "healthcare"—despite our knowledge that housing is a prerequisite for health [119, 120]. Patients experiencing homelessness push the boundaries of healthcare and of EM in particular, thereby challenging EPs to engage with complex issues including social needs, stigma, racism, and social determinants of health [121, 122]. Despite these difficulties, there are tangible steps that EPs can take in improving the care of their patients experiencing homelessness.

Basic

- Provide care that is respectful, free of stigma, and equitable. Patients who are homeless are often required to deal with multiple bureaucratic systems in their daily lives. Providing healthcare that is respectful and compassionate is therefore especially important. When patients report physical symptoms, they should not be assumed to be "malingering" or seeking "secondary gain"; overall, homeless individuals are at higher risk for morbidity and mortality than their non-homeless counterparts.
- As for all patients, perform a tailored or full physical exam based on patients' presenting complaints. For patients who are intoxicated or otherwise incapable of providing a reliable history (e.g., due to dementia or other cognitive

impairment), EPs should perform a full physical exam, which may reveal occult injury or illness. This is especially important given the disproportionate burden of comorbid health conditions and risks to health faced by patients who are homeless and the increasing number of older homeless adults, many of whom may have difficulty with ADLs that one may not otherwise think to evaluate. Patients should be provided with assistance if needed to undress, store their clothes securely, and don a hospital gown.

- Ask patients about their housing status as part of the social history and adjust treatment plans accordingly. Many patients experiencing homelessness do not appear stereotypically "homeless," and many who are "doubled-up" or "couch surfing" (seeking temporary housing with a friend or family) may resist the label and its associated stigma. The most reliable way for EPs to identify homelessness is to ask their patients. There have not been studies to identify a "best" question to incorporate into providers' social histories. One version used successfully by the authors is, "Where are you staying these days?" with follow-up questions as needed. Even though EPs can rarely "solve" patients' homelessness immediately, knowledge of homelessness status should be considered in treatment, disposition, and follow-up plans. For example, consider need for refrigeration of medications; patients' ability to obtain medications (in terms of affordability and/or accessibility of outpatient pharmacies); complexity of medication dosing schedules (e.g., consider single-dose dexamethasone injection rather than steroid dose pack prescriptions for asthma); whether activity restrictions are needed (most homeless shelters do not allow patients to rest inside during the day and extremity elevation can be difficult in a shelter and impossible on the street); exposure to infectious disease in shelters; and spread of infectious disease to other shelter residents.
- Prepare to meet immediate needs (such as for clothing and food) and have at least one referral source for shelter or housing assistance. While the availability of resources will vary by the specific ED and location, EDs should identify at least one community or governmental organization to which patients can be referred for shelter or other assistance with housing. Patients should be referred to social work services if available. A meal and clean clothing should be available for those who need it; in California, this is now a legal requirement for hospitals prior to discharging patients who are homeless [117]. Although meals and clothing are not typically thought of as "medical care," they are nevertheless critical to human survival, as well as basic elements of compassionate patient care.

Intermediate

- Institute routine screening for and documentation of homelessness status. EDs that serve patients experiencing homelessness with some frequency could consider implementing routine, standardized screening of all ED patients for homelessness similar to how some EDs routinely screen for domestic violence. Screening could be instituted at triage or in another manner compatible with the

ED's workflow. Several options for standard screening questions for homelessness exist; UCSF's Social Interventions Research & Evaluation Network (SIREN) summarizes multiple options on their website (https://sirenetwork.ucsf.edu) [123]. Screening results can be documented in the electronic medical record in a standardized way, such as via ICD-10 Z codes (Z59.0 = homelessness), ideally with patient consent and accompanied by other efforts to minimize any associated stigma (see below). In the absence of universal screening, it is important to note that ICD-10 Z codes and electronic medical record address fields likely underestimate the number of ED patients experiencing homelessness [124, 125]. Accurately recording homelessness status is important not only for direct patient care but also for research and advocacy efforts. However, a critical caveat is that any efforts to screen for homelessness must take explicit steps to avoid reproducing the stigma and suffering of homeless patients. For example, patients who report homelessness should not be labeled differently (e.g., with a certain symbol on the patient electronic track board) nor should "homelessness" be listed in place of a medical chief complaint in the medical records. Any screening initiative should be developed in consultation with experts including local community organizations and people experiencing homelessness themselves. Screening and data collection should be part of larger initiatives to address housing and other needs (such as by partnering with and funding local community organizations to assist patients who are identified as homeless). In the absence of such plans, the risks of universal screening (e.g., stigma) may outweigh the benefits. More research is needed on this topic.

- Develop connections with organizations that serve people experiencing homelessness, including arrangements for mutual information sharing. Such relationships can be especially beneficial to assist with the housing needs of complex patients who frequently visit the ED [45]. Many patients may already be working with various community or governmental agencies serving people who are homeless. Learning that their clients are in the ED can be helpful to these agencies (of course following all applicable privacy rules and with permission of patients themselves to share any information), who in an ideal scenario can assist EPs in real-time in determining appropriate treatment plans and later can help to resolve patients' homelessness. Conversely, the ED may serve people experiencing homelessness who are not yet known to existing service organizations. The ED visit may therefore be a critical touchpoint at which a connection or referral can be made. Of course, the number and types of local organizations serving people who are homeless and resources for housing vary significantly across localities. In light of these variations, EDs in rural areas, for example, may need to rely on innovative programs (e.g., telemedicine-like services) to connect patients to housing and other social programs. Some EDs with high volumes of patients experiencing homelessness have collaborated with community-based organizations to employ "housing navigators" in the ED. These employees are sometimes affiliated with local homeless services organizations, have expertise in housing and homelessness, and can assist in identifying and addressing patients' housing needs.

Advanced

- Assist local Continuums of Care (CoCs). Continuums of Care exist in localities throughout the US and are the groups—consisting of multiple nonprofit and governmental agencies—tasked with local homelessness strategic planning. Local CoCs are required by the US Department of Housing and Urban Development (HUD) to establish and operate a "coordinated entry process" by which people are connected to organizations and interventions with the goal of rapidly ending their homelessness [126]. Hospitals and EDs may be valuable stakeholders and contributors to local CoCs given their unique and frequent interactions with people experiencing homelessness.
- Advocate for housing and other policies. The crisis of homelessness is felt acutely in EDs across the US. Providers can speak to the negative health effects they observe as a result of homelessness and advocate for the housing that is needed to address this critical root contributor to patients' health and ED use. Physicians and other EPs have the ability to contribute to existing local and national advocacy efforts for policies related to homelessness and housing, as well as related efforts to fight the criminalization of poverty and other manifestations of structural racism.

Teaching Case

Clinical Case

JW is a 67-year-old man with no past medical history who presents to the ED with a complaint of hemoptysis, dyspnea, and weight loss for 3 weeks. He is seen by a resident, who orders a CBC, chemistry panel, lactic acid, blood cultures, and chest x-ray. When asked about the patient's social history, the resident reports that the patient "does not look homeless." When probed further, he explains, "I just have a sense from working with these people the past few years. You know what they look like living on the street. You size them up—based on their clothes, their hygiene—just the way they look. Also, JW is 67 years old, so he has an income."

JW's x-ray shows a right upper lobe infiltrate. Laboratory tests are normal. Upon further questioning, JW reveals that he lost his job 6 months ago. He was unable to afford rent on his social security income and moved in with his daughter and two grandchildren in their one-bedroom apartment. Because he was not on the lease, their landlord threatened to evict them if he did not leave their home. He has been staying in his car and in homeless shelters throughout the city for the past month.

JW was given antibiotics for pneumonia. He was placed in a negative pressure room and admitted to the hospital for treatment of his pneumonia and evaluation of possible tuberculosis. After several days in the hospital, his sputum smears were negative for tuberculosis and his symptoms did not improve with antibiotics. A CT of the chest revealed a cavitating lung mass, which subsequent biopsy revealed to be a squamous cell carcinoma of the lung. The inpatient team discharged him to a medical respite program that collaborates with the hospital and provides 24-hour recuperative shelter, transportation to specialist appointments, and case management. His case manager is working to find JW permanent housing.

Teaching Points

1. Homelessness encompasses more people than those living "on the street." Assessment of homelessness should take into account those who are "doubled-up," living in hotels or motels, living in homeless shelters, or living in places not meant for human habitation (e.g., cars, abandoned houses).
2. Housing status should be routinely assessed in the social history. Relying on stereotypes, such as disheveled appearance, can perpetuate the stigma associated with homelessness and lead to under-recognition of homelessness among ED patients.
3. Patients experiencing homelessness face risks from a wide range of infectious disease and chronic illness. When treating patients experiencing homelessness, EPs should consider a broad range of possible diagnoses and not "anchor" on a single diagnosis based on a patient's housing status.
4. Standard decision guides commonly used by EPs, such as the CURB-65 score, do not account for homelessness. EPs should take into account the material challenges associated with homelessness, including the local contexts of housing difficulties, when deciding on treatment plans.

Discussion Questions

1. The resident makes an assumption about the patient's homelessness status based on his appearance. How does this compare with your own experiences in the ED? How do you determine if a patient is homeless?
2. Should the patient's homelessness influence the ED physicians' decisions about his workup, treatment, and disposition? If so, how? What resources can you use or modifications can you make to treatment plans to assist patients experiencing homelessness?
3. What opportunities exist for partnerships outside the ED (e.g., across the hospital and the community) that might be beneficial for this patient or others like him?
4. What are some challenges to caring for patients who are homeless in your ED? How are these unique (or similar) to challenges in other ED settings?

Acknowledgments The authors would like to acknowledge Emmanuella Asabor, Marcella Maguire, and Kazi Sumon for their contributions to this chapter.

References

1. Henry M, Watt R, Rosenthal L, Shijvi A. The 2017 annual homeless assessment report to congress. Washington, DC: HUD Office of Community Planning and Development; 2017.
2. Henry M, Watt R, Mahathey A, Ouellette J, Sitler A. The 2019 annual homeless assessment report (AHAR) to congress. Part 1: point-in-time estimates of homelessness. HUD Exchange AHAR reports. Washington, DC: US Housing and Urban Development; 2019.
3. HUD exchange AHAR reports. https://www.hudexchange.info/homelessness-assistance/ahar/#2018-reports. Accessed 8/14/2020.
4. Fischer W, Sard B. Chart book: federal housing spending is poorly matched to need. Washington, DC: Center on Budget and Policy Priorities; 2017.
5. Quadagno JS. The color of welfare: how racism undermined the war on poverty. New York: Oxford University Press; 1994.

6. Massey DS, Denton NA. American apartheid: segregation and the making of the underclass. Cambridge: Harvard University Press; 1993.
7. Gilmore RW. Fatal couplings of power and difference: notes on racism and geography. The Prof Geogr. 2002;54(1):15–24.
8. Desmond M. Eviction and the reproduction of urban poverty. Am J Sociol. 2012;118(1):88–133.
9. Cell JW. The highest stage of white supremacy : the origins of segregation in South Africa and the American South. Cambridge: Cambridge University Press; 1982.
10. Roscigno VJ, Karafin DL, Tester G. The complexities and processes of racial housing discrimination. Soc Probl. 2009;56(1):49–69.
11. Williams DR, Collins C. Racial residential segregation: a fundamental cause of racial disparities in health. Public Health Rep. 2016;116:404–16.
12. Doran KM, Cha S, Cho R, DiPietro B, Gelberg L, Kushel M. Housing as health care during and after the COVID-19 crisis. Ann Fam Med. Online COVID-19 collection. 2020. Available at: https://deepblue.lib.umich.edu/handle/2027.42/154767.
13. Lee CT, Guzman D, Ponath C, Tieu L, Riley E, Kushel M. Residential patterns in older homeless adults: results of a cluster analysis. Soc Sci Med. 2016;153:131–40. https://doi.org/10.1016/j.socscimed.2016.02.004.
14. Shelton KH, Taylor PJ, Bonner A, van den Bree M. Risk factors for homelessness: evidence from a population-based study. Psychiatr Serv. 2009;60(4):465–72. https://doi.org/10.1176/appi.ps.60.4.465.
15. Shinn M, Weitzman BC, Stojanovic D, Knickman JR, Jimenez L, Duchon L, et al. Predictors of homelessness among families in New York City: from shelter request to housing stability. Am J Public Health. 1998;88(11):1651–7.
16. Susser E, Moore R, Link B. Risk factors for homelessness. Epidemiol Rev. 1993;15(2):546–56.
17. Culhane DP, Averyt JM, Hadley TR. The rate of public shelter admission among Medicaid-reimbursed users of behavioral health services. Psychiatr Serv. 1997;48(3):390–2.
18. Edens EL, Kasprow W, Tsai J, Rosenheck RA. Association of substance use and VA service-connected disability benefits with risk of homelessness among veterans. Am J Addict. 2011;20(5):412–9. https://doi.org/10.1111/j.1521-0391.2011.00166.x.
19. Greenberg GA, Rosenheck RA. Mental health correlates of past homelessness in the National Comorbidity Study Replication. J Health Care Poor Underserved. 2010;21(4):1234–49. https://doi.org/10.1353/hpu.2010.0926.
20. Greenberg GA, Rosenheck RA. Correlates of past homelessness in the national epidemiological survey on alcohol and related conditions. Adm Policy Ment Health. 2010;37(4):357–66. https://doi.org/10.1007/s10488-009-0243-x.
21. Thompson RG Jr, Wall MM, Greenstein E, Grant BF, Hasin DS. Substance-use disorders and poverty as prospective predictors of first-time homelessness in the United States. Am J Public Health. 2013;103(Suppl 2):S282–8. https://doi.org/10.2105/ajph.2013.301302.
22. United States Interagency Council on Homelessness. Opening doors: federal strategic plan to prevent and end homelessness 2010; 2010.
23. United States Interagency Council on Homelessness. Opening doors: federal strategic plan to prevent and end homelessness. As amended in 2015; 2015.
24. Sadowski LS, Kee RA, VanderWeele TJ, Buchanan D. Effect of a housing and case management program on emergency department visits and hospitalizations among chronically ill homeless adults: a randomized trial. JAMA. 2009;301(17):1771–8. https://doi.org/10.1001/jama.2009.561.
25. Basu A, Kee R, Buchanan D, Sadowski LS. Comparative cost analysis of housing and case management program for chronically ill homeless adults compared to usual care. Health Serv Res. 2012;47(1 Pt 2):523–43. https://doi.org/10.1111/j.1475-6773.2011.01350.x.
26. Larimer ME, Malone DK, Garner MD, Atkins DC, Burlingham B, Lonczak HS, et al. Health care and public service use and costs before and after provision of housing for chronically homeless persons with severe alcohol problems. JAMA. 2009;301(13):1349–57. https://doi.org/10.1001/jama.2009.414.

27. National Academies of Sciences E, Medicine. Permanent supportive housing: evaluating the evidence for improving health outcomes among people experiencing chronic homelessness. Washington, DC: The National Academies Press; 2018.
28. Hunter SB, Harvey M, Briscombe B, Cefalu M. Evaluation of Housing for Health permanent supportive housing program. Santa Monica: RAND Corporation; 2017.
29. Corporation for Supportive Housing. Supportive housing 101: data. https://www.csh.org/supportive-housing-101/data/. Accessed 13 May 2019.
30. Lebrun-Harris LA, Baggett TP, Jenkins DM, Sripipatana A, Sharma R, Hayashi AS, et al. Health status and health care experiences among homeless patients in federally supported health centers: findings from the 2009 patient survey. Health Serv Res. 2013;48(3):992–1017. https://doi.org/10.1111/1475-6773.12009.
31. US Department of Housing and Urban Development. The 2010 annual homeless assessment report to congress. Washington, DC; 2011.
32. Burt MR, Aron LY, Douglas T, Valente J, Lee E, Iwen B. Homelessness: programs and the people they serve. Summary report. Findings of the national survey of homeless assistance providers and clients. Urban Institute; 1999.
33. Zlotnick C, Zerger S. Survey findings on characteristics and health status of clients treated by the federally funded (US) health care for the homeless programs. Health Soc Care Community. 2009;17(1):18–26. https://doi.org/10.1111/j.1365-2524.2008.00793.x.
34. Childress S, Reitzel LR, Maria DS, Kendzor DE, Moisiuc A, Businelle MS. Mental illness and substance use problems in relation to homelessness onset. Am J Health Behav. 2015;39(4):549–55.
35. Levitt AJ, Culhane DP, DeGenova J, O'Quinn P, Bainbridge J. Health and social characteristics of homeless adults in Manhattan who were chronically or not chronically unsheltered. Psychiatr Serv. 2009;60(7):978–81. https://doi.org/10.1176/appi.ps.60.7.978.
36. Baggett TP, Hwang SW, O'Connell JJ, Porneala BC, Stringfellow EJ, Orav EJ, et al. Mortality among homeless adults in Boston: shifts in causes of death over a 15-year period. JAMA Intern Med. 2013;173(3):189–95. https://doi.org/10.1001/jamainternmed.2013.1604.
37. Hwang SW, Wilkins R, Tjepkema M, O'Campo PJ, Dunn JR. Mortality among residents of shelters, rooming houses, and hotels in Canada: 11 year follow-up study. BMJ. 2009;339:b4036.
38. Hibbs JR, Benner L, Klugman L, Spencer R, Macchia I, Mellinger A, et al. Mortality in a cohort of homeless adults in Philadelphia. N Engl J Med. 1994;331(5):304–9. https://doi.org/10.1056/NEJM199408043310506.
39. Brown RT, Hemati K, Riley ED, Lee CT, Ponath C, Tieu L, et al. Geriatric conditions in a population-based sample of older homeless adults. Gerontologist. 2017;57(4):757–66. https://doi.org/10.1093/geront/gnw011.
40. Doran KM, Raven MC, Rosenheck RA. What drives frequent emergency department use in an integrated health system? National data from the Veterans Health Administration. Ann Emerg Med. 2013;62(2):151–9. https://doi.org/10.1016/j.annemergmed.2013.02.016.
41. Kushel MB, Perry S, Bangsberg D, Clark R, Moss AR. Emergency department use among the homeless and marginally housed: results from a community-based study. Am J Public Health. 2002;92(5):778–84.
42. Salhi BA, White MH, Pitts SR, Wright DW. Homelessness and emergency medicine: a review of the literature. Acad Emerg Med. 2018;25(5):577–93. https://doi.org/10.1111/acem.13358.
43. Weinreb L, Goldberg R, Perloff J. Health characteristics and medical service use patterns of sheltered homeless and low-income housed mothers. J Gen Intern Med. 1998;13(6):389–97.
44. Hwang S, Henderson M. Health care utilization in homeless people: translating research into policy and practice. Agency for Healthcare Research and Quality Working Paper No. 10002; October 2010.
45. Raven MC, Doran KM, Kostrowski S, Gillespie CC, Elbel BD. An intervention to improve care and reduce costs for high-risk patients with frequent hospital admissions: a pilot study. BMC Health Serv Res. 2011;11:270. https://doi.org/10.1186/1472-6963-11-270.

46. Bharel M, Lin WC, Zhang J, O'Connell E, Taube R, Clark RE. Health care utilization patterns of homeless individuals in Boston: preparing for Medicaid expansion under the Affordable Care Act. Am J Public Health. 2013;103 Suppl 2:S311–7. https://doi.org/10.2105/ajph.2013.301421.
47. Kushel MB, Vittinghoff E, Haas JS. Factors associated with the health care utilization of homeless persons. JAMA. 2001;285(2):200–6. jcu00007 [pii].
48. Malecha PW, Williams JH, Kunzler NM, Goldfrank LR, Alter HJ, Doran KM. Material needs of emergency department patients: a systematic review. Acad Emerg Med. 2017. https://doi.org/10.1111/acem.13370.
49. Ku BS, Scott KC, Kertesz SG, Pitts SR. Factors associated with use of urban emergency departments by the U.S. homeless population. Public Health Rep. 2010;125(3):398–405.
50. Oates G, Tadros A, Davis SM. A comparison of national emergency department use by homeless versus non-homeless people in the United States. J Health Care Poor Underserved.2009;20(3):840–5. https://doi.org/10.1353/hpu.0.0192.
51. Tadros A, Layman SM, Brewer MP, Davis SM. A 5-year comparison of ED visits by homeless and nonhomeless patients. Am J Emerg Med. 2016. https://doi.org/10.1016/j.ajem.2016.01.012.
52. Doran KM, Ragins KT, Iacomacci AL, Cunningham A, Jubanyik KJ, Jenq GY. The revolving hospital door: hospital readmissions among patients who are homeless. Med Care. 2013;51(9):767–73. https://doi.org/10.1097/MLR.0b013e31829fafbb.
53. McCormack RP, Hoffman LF, Norman M, Goldfrank LR, Norman EM. Voices of homeless alcoholics who frequent Bellevue Hospital: a qualitative study. Ann Emerg Med. 2015;65(2):178–86.e6. https://doi.org/10.1016/j.annemergmed.2014.05.025.
54. Doran KM, Curry LA, Vashi AA, Platis S, Rowe M, Gang M, et al. "Rewarding and challenging at the same time": emergency medicine residents' experiences caring for patients who are homeless. Acad Emerg Med. 2015;21(6):673–9. https://doi.org/10.1111/acem.12388.
55. Doran KM, Vashi AA, Platis S, Curry LA, Rowe M, Gang M, et al. Navigating the boundaries of emergency department care: addressing the medical and social needs of patients who are homeless. Am J Public Health. 2013;103 Suppl 2:S355–60. https://doi.org/10.2105/AJPH.2013.301540.
56. Baggett TP. Health care of homeless persons in the United States. In: UpToDate, editor, UpToDate, Waltham, MA. Accessed 10 Aug 2020.
57. Morrison DS. Homelessness as an independent risk factor for mortality: results from a retrospective cohort study. Int J Epidemiol. 2009;38(3):877–83.
58. O'Connell JJ. Premature mortality in homeless populations: a review of the literature. National Health Care for the Homeless Council: Nashville; 2005.
59. Roncarati JS, Baggett TP, O'Connell JJ, Hwang SW, Cook EF, Krieger N, Sorensen G. Mortality among unsheltered homeless adults in Boston, Massachusetts, 2000-2009. JAMA Intern Med. 2018;178(9):1242–48. https://doi.org/10.1001/jamainternmed.2018.2924.
60. D'Amore J, Hung O, Chiang W, Goldfrank L. The epidemiology of the homeless population and its impact on an urban emergency department. Acad Emerg Med. 2001;8(11):1051–5.
61. Coe AB, Moczygemba LR, Gatewood SB, Osborn RD, Matzke GR, Goode J-VR. Medication adherence challenges among patients experiencing homelessness in a behavioral health clinic. Res Social Adm Pharm. 2015;11(3):e110–e20.
62. Nyamathi A, Shuler P. Factors affecting prescribed medication compliance of the urban homeless adult. Nurse Pract. 1989;14(8):47–54.
63. Culhane DP, Metraux S, Byrne T, Stino M, Bainbridge J. The aging of contemporary homelessness. Contexts; 2013.
64. Brown RT, Steinman MA. Characteristics of emergency department visits by older versus younger homeless adults in the United States. Am J Public Health. 2013;103(6):1046–51. https://doi.org/10.2105/AJPH.2012.301006.
65. Coe AB, Moczygemba LR, Harpe SE, Gatewood SB. Homeless patients' use of urban emergency departments in the United States. J Ambul Care Manage. 2015;38(1):48–58. https://doi.org/10.1097/JAC.0000000000000034.

66. Brown RT, Kiely DK, Bharel M, Mitchell SL. Geriatric syndromes in older homeless adults. J Gen Intern Med. 2012;27(1):16–22.
67. Treglia D, Johns EL, Schretzman M, Berman J, Culhane DP, Lee DC, et al. When crises converge: hospital visits before and after shelter use among homeless New Yorkers. Health Aff. 2019;38(9):1458–67. https://doi.org/10.1377/hlthaff.2018.05308.
68. Thirteenth annual report on homeless deaths (July 1, 2017 – June 30, 2018). New York City: New York City Department of Health and Mental Hygiene, Bureau of Vital Statistics, New York City Department of Homeless Services; 2018.
69. Zevin B, Cawley C. Homeless mortality in San Francisco: opportunities for prevention. San Francisco: San Francisco Whole Person Care; 2019.
70. Los Angeles County Department of Public Health Center for Health Impact Evaluation. Recent trends in mortality rates and causes of death among people experiencing homelessness in Los Angeles County. Los Angeles; 2019.
71. Stuart F. Down, out, and under arrest: policing and everyday life in skid row. Chicago: University of Chicago Press; 2016.
72. Bourgois PI, Schonberg J. Righteous dopefiend. Berkeley: University of California Press; 2009.
73. Amster R. Patterns of exclusion: sanitizing space, criminalizing homelessness. Soc Just. 2003;30(1 (91)):195–221.
74. Hammig B, Jozkowski K, Jones C. Injury-related visits and comorbid conditions among homeless persons presenting to emergency departments. Acad Emerg Med. 2014;21(4):449–55.
75. Kushel MB, Evans JL, Perry S, Robertson MJ, Moss AR. No door to lock: victimization among homeless and marginally housed persons. Arch Intern Med. 2003;163(20):2492–9.
76. Mackelprang JL, Graves JM, Rivara FP. Homeless in America: injuries treated in US emergency departments, 2007–2011. Int J Inj Contr Saf Promot. 2014;21(3):289–97.
77. Kramer CB, Gibran NS, Heimbach DM, Rivara FP, Klein MB. Assault and substance abuse characterize burn injuries in homeless patients. J Burn Care Res. 2008;29(3):461.
78. Padgett DK, Struening EL, Andrews H, Pittman J. Predictors of emergency room use by homeless adults in New York City: the influence of predisposing, enabling and need factors. Soc Sci Med. 1995;41(4):547–56.
79. Fazel S, Geddes JR, Kushel M. The health of homeless people in high-income countries: descriptive epidemiology, health consequences, and clinical and policy recommendations. Lancet. 2014;384(9953):1529–40. https://doi.org/10.1016/s0140-6736(14)61132-6.
80. Meinbresse M, Brinkley-Rubinstein L, Grassette A, Benson J, Hall C, Hamilton R, et al. Exploring the experiences of violence among individuals who are homeless using a consumer-led approach. Violence Vict;(1):122–36. https://doi.org/10.1891/0886-6708.VV-D-12-00069.
81. Mackelprang JL, Qiu Q, Rivara FP. Predictors of emergency department visits and inpatient admissions among homeless and unstably housed adolescents and young adults. Med Care. 2015;53(12):1010–7.
82. Doran KM, Shumway M, Hoff RA, Blackstock OJ, Dilworth SE, Riley ED. Correlates of hospital use in homeless and unstably housed women: the role of physical health and pain. Womens Health Issues. 2014;24(5):535–41. https://doi.org/10.1016/j.whi.2014.06.003.
83. Henderson S, Stacey CL, Dohan D. Social stigma and the dilemmas of providing care to substance users in a safety-net emergency department. J Health Care Poor Underserved. 2008;19(4):1336–49.
84. O'Toole TP, Johnson E, Redihan S, Borgia M, Rose J. Needing primary care but not getting it: the role of trust, stigma and organizational obstacles reported by homeless veterans. J Health Care Poor Underserved. 2015;26(3):1019–31. https://doi.org/10.1353/hpu.2015.0077.
85. Tadros A, Layman SM, Brewer MP, Davis SM. A 5-year comparison of emergency department visits by homeless and non-homeless patients. Am J Emerg Med. 2016;34(5):805–8.
86. Han B, Wells BL. Inappropriate emergency department visits and use of the Health Care for the Homeless Program services by homeless adults in the northeastern United States. J Public Health Manag Pract. 2003;9(6):530–7.

87. DiPietro B, Artiga S, Gates A. Early impacts of the Medicaid expansion for the homeless population. The Kaiser Commission on Medicaid and the Uninsured; 2014.
88. Niska R, Bhuiya F, Xu J. National hospital ambulatory medical care survey: 2007 emergency department summary. Natl Health Stat Rep. 2010;26(26):1–31.
89. Wang H, Nejtek VA, Zieger D, Robinson RD, Schrader CD, Phariss C, et al. The role of charity care and primary care physician assignment on ED use in homeless patients. Am J Emerg Med. 2015;33(8):1006–11.
90. Burt MR, Sharkey P. The role of Medicaid in improving access to care for homeless people. Washington, DC: Urban Institute; 2002.
91. Lin W-C, Bharel M, Zhang J, O'Connell E, Clark RE. Frequent emergency department visits and hospitalizations among homeless people with Medicaid: implications for Medicaid expansion. Am J Public Health. 2015;105(s5):s716–s22. https://doi.org/10.2105/AJPH.2015.
92. Hwang SW, Chambers C, Chiu S, Katic M, Kiss A, Redelmeier DA, et al. A comprehensive assessment of health care utilization among homeless adults under a system of universal health insurance. Am J Public Health. 2013;103(S2):S294–301.
93. Tsai J, Doran KM, Rosenheck RA. When health insurance is not a factor: national comparison of homeless and nonhomeless US veterans who use Veterans Affairs Emergency Departments. Am J Public Health. 2013;103(S2):S225–S31.
94. Tsai J, Rosenheck RA. Risk factors for ED use among homeless veterans. Am J Emerg Med. 2013;31(5):855–8. https://doi.org/10.1016/j.ajem.2013.02.046.
95. Ku BS, Fields JM, Santana A, Wasserman D, Borman L, Scott KC. The urban homeless: super-users of the emergency department. Popul Health Manag. 2014;17(6):366–71.
96. Thakarar K, Morgan JR, Gaeta JM, Hohl C, Drainoni ML. Predictors of frequent emergency room visits among a homeless population. PLoS One. 2015;10(4):e0124552. https://doi.org/10.1371/journal.pone.0124552.
97. Pearson DA, Bruggman AR, Haukoos JS. Out-of-hospital and emergency department utilization by adult homeless patients. Ann Emerg Med. 2007;50(6):646–52. https://doi.org/10.1016/j.annemergmed.2007.07.015.
98. Buck DS, Brown CA, Mortensen K, Riggs JW, Franzini L. Comparing homeless and domiciled patients' utilization of the Harris County, Texas public hospital system. J Health Care Poor Underserved. 2012;23(4):1660–70. https://doi.org/10.1353/hpu.2012.0171.
99. Buchanan D, Doblin B, Sai T, Garcia P. The effects of respite care for homeless populations. Am J Public Health. 2006;96(7):1278–81. https://doi.org/10.2105/AJPH.2005.067850.
100. Rodriguez RM, Fortman J, Chee C, Ng V, Poon D. Food, shelter and safety needs motivating homeless persons' visits to an urban emergency department. Ann Emerg Med. 2009;53(5):598–602. https://doi.org/10.1016/j.annemergmed.2008.07.046.
101. Gordon JA. The hospital emergency department as a social welfare institution. Ann Emerg Med. 1999;33(3):321–5.
102. Salhi BA. Who are Clive's friends? Latent sociality in the emergency department. Soc Sci Med. 2020;245:112668.
103. Jackson TS, Moran TP, Lin J, Ackerman J, Salhi BA. Homelessness among patients in a southeastern safety net emergency department. South Med J. 2019;112(9):476–82.
104. O'Toole TP, Conde-Martel A, Gibbon JL, Hanusa BH, Freyder PJ, Fine MJ. Where do people go when they first become homeless? A survey of homeless adults in the USA. Health Soc Care Community. 2007;15(5):446–53.
105. Doran KM, Raven MC. Homelessness and emergency medicine: where do we go from here? Acad Emerg Med. 2018;25(5):598–600. https://doi.org/10.1111/acem.13392.
106. Salhi BA, White MH, Pitts SR, Wright DW. Homelessness and emergency medicine: furthering the conversation. Acad Emerg Med. 2018;25(5):597.
107. Levanon Seligson A, Lim S, Singh T, Laganis E, Stazesky E, Donahue S, et al. New York/New York III supportive housing evaluation: interim utilization and cost analysis. A report from the New York City Department of Health and Mental Hygiene in collaboration with the New York City Human Resources Administration and the New York State Office of Mental Health; 2013.

108. Linkins K, Brya J, Chandler D. Frequent users of health services initiative: final evaluation report. Oakland: Corporation for Supportive Housing; 2008.
109. Kertesz SG, Baggett TP, O'Connell JJ, Buck DS, Kushel MB. Permanent supportive housing for homeless people—reframing the debate. N Engl J Med. 2016;375(22):2115.
110. Cutler DM, Glaeser EL. Are ghettos good or bad? Q J Econ. 1997;112(3):827–72.
111. Health Research & Educational Trust. Social determinants of health series: housing and the role of hospitals. Health Research & Educational Trust: Chicago; 2017.
112. Doran KM, Greysen SR, Cunningham A, Tynan-McKiernan K, Lucas GI, Rosenthal MS. Improving post-hospital care for people who are homeless: community-based participatory research to community-based action. Healthcare. 2015;3(4):238–44. https://doi.org/10.1016/j.hjdsi.2015.07.006.
113. Freeman AL, Mohan B, Lustgarten H, Sekulic D, Shepard L, Fogarty M, et al. The development of health and housing consortia in New York City. Health Aff. 2020;39(4):631–8.
114. Lawrence M. Rural homelessness: a geography without a geography. J Rural Stud. 1995;11(3):297–307.
115. Whitley R. Fear and loathing in New England: examining the health-care perspectives of homeless people in rural areas. Anthropol Med. 2013;20(3):232–43. https://doi.org/10.1080/13648470.2013.853597.
116. Davidson C, Murry VM, Meinbresse M, Jenkins DM, Mindtrup R. Using the Social Ecological Model to examine how homelessness is defined and managed in rural East Tennessee. Nashville: National Health Care for the Homeless Council; 2016.
117. Senate Bill No. 1152, Hospital patient discharge process: homeless patients. (2017–2018). Chapter 981. An act to amend, repeal, and add Section 1262.5 to the Health and Safety Code, relating to public health. 2018. Available at: https://leginfo.legislature.ca.gov/faces/billTextClient.xhtml?bill_id=201720180SB1152. Accessed 8/10/2020.
118. United States Interagency Council on Homelessness. Home, together: the federal strategic plan to prevent and end homelessness. July 19, 2018. Available at: https://www.usich.gov/resources/uploads/asset_library/Home-Together-Federal-Strategic-Plan-to-Prevent-and-End-Homelessness.pdf. Accessed 8/10/2020.
119. Fullilove MT. Housing is health care. Am J Prev Med. 2010;39(6):607-8. https://doi.org/10.1016/j.amepre.2010.09.017.
120. Doran KM, Misa EJ, Shah NR. Housing as health care—New York's boundary-crossing experiment. N Engl J Med. 2013;369(25):2374–7.
121. Anderson ES, Lippert S, Newberry J, Bernstein E, Alter HJ, Wang NE. Addressing social determinants of health from the emergency department through social emergency medicine. West J Emerg Med. 2016;17(4):487.
122. Anderson ES, Hsieh D, Alter HJ. Social emergency medicine: embracing the dual role of the emergency department in acute care and population health. Ann Emerg Med. 2016;68(1):21–5.
123. Social Interventions Research & Evaluation Network (SIREN). University of California San Francisco(UCSF). https://sirenetwork.ucsf.edu/SocialNeedsScreeningToolComparisonTable. Accessed 12/22/20.
124. Lee SJ, Thomas P, Newnham H, Freidin J, Smith C, Lowthian J, et al. Homeless status documentation at a metropolitan hospital emergency department. Emerg Med Australas. 2019;31:639–45. https://doi.org/10.1111/1742-6723.13256.
125. Vickery KD, Shippee ND, Bodurtha P, Guzman-Corrales LM, Reamer E, Soderlund D, et al. Identifying homeless Medicaid enrollees using enrollment addresses. Health Serv Res. 2018;53(3):1992–2004. https://doi.org/10.1111/1475-6773.12738.
126. HUD Exchange. Coordinated entry core elements. June 2017. Washington, DC: US Department of Housing and Urban Development. Available at: https://www.hudexchange.info/resource/5340/coordinated-entry-core-elements/. Accessed 8/14/2020.

Housing Instability and Quality

15

Amanda Stewart and Megan Sandel

Key Points

- Housing instability, distinct from homelessness, is prevalent in the US and is commonly encountered in the emergency department (ED). Patients may move in between homelessness and unstable housing frequently, or may live in substandard housing because it is all that they can afford.
- Unlike homelessness—which has federal, state, and local definitions—housing instability, housing insecurity, and substandard housing quality have no standard definitions, which can make defining screening questions and protocols challenging.
- Even with some ambiguity in definitions and difficulty "treating" housing instability and poor housing quality in the ED, providers should assess these housing-related social risks since this is an important consideration for their treatment plans. Providers will often be surprised by who "screens positive" for housing instability and substandard housing.
- Some hospitals have developed successful collaborations with community-based organizations to better address patients' housing needs by providing services such as on-site housing search and stabilization services and medical-legal partnerships to address substandard housing.
- Housing instability and poor housing quality are ultimately systemic issues related to lack of investment in building affordable housing and enforcing housing codes related to health; emergency providers can play a role in larger advocacy efforts to address these issues.

A. Stewart (✉)
Division of Emergency Medicine, Boston Children's Hospital, Boston, MA, USA
e-mail: Amanda.stewart@childrens.harvard.edu

M. Sandel
Department of Pediatrics, Boston Medical Center, Boston, MA, USA
e-mail: Megan.sandel@bmc.org

H. J. Alter et al. (eds.), *Social Emergency Medicine*,
https://doi.org/10.1007/978-3-030-65672-0_15

Foundations

Background

Much of the literature on housing and health focuses on homelessness, which is defined as lacking a fixed, adequate nighttime residence [1]. Housing instability is on a spectrum with, but distinct from, homelessness. While far more common than frank homelessness, housing instability is just as damaging to health. Poor housing quality has been recognized as a cause of ill health for centuries, relating back to unsanitary water and slum conditions, but remains a common problem in urban, suburban, and rural settings to this day.

Definitions of Housing Instability

There is no single, standard definition of housing instability. Housing instability typically encompasses multiple issues including moving frequently, living in overcrowded homes ("doubling up"), staying with friends or relatives ("couch surfing"), being behind on rent, or spending a large portion on one's income on housing [1–3]. The US Department of Housing and Urban Development (HUD) defines housing cost burden as spending greater than 30% of income on housing, while spending more than 50% of income is termed severe housing cost burden [4].

Definitions of Housing Quality

As with housing instability, there is no standard definition of housing quality. Housing quality consists not only of the internal and external physical environment of the home itself, but also includes the social and physical environment the home is located within [2, 5]. Elements encompassed in more narrow definitions of housing quality include indoor air quality, dampness, issues with heating or cooling, substandard plumbing, and the presence of asbestos, lead, mold, or other allergens and pests in the home [2, 5]. Broader definitions frequently include neighborhood factors such as neighborhood safety, access to walking paths or exercise spaces, and community facilities [5].

Epidemiology

Estimates of the number of households with housing instability and substandard housing quality depend on which dimensions of the problem are included in the definitions. Despite the wide-ranging definitions, it is safe to say tens of millions of people in the US suffer from housing instability and poor housing quality annually. According to the American Census Survey in 2017, one-third of US households (over 42 million households) were cost burdened, spending over 30% of their income on housing [6, 7]. Nearly 20 million (one out of every 6) households were severely cost burdened, spending over 50% of their incomes on rent [6, 7]. Among renters, 2.7 million households reported being unable to pay all or part of their rent, over 800,000 households reported receiving eviction notices in the past 3 months [6], and 900,000 households (approximately 2.3 million people) were evicted in 2016 alone [8]. Due to systemic racism in the form of historical and current discriminatory housing policies, housing cost burden and evictions are highest in communities of color [9].

Besides rising housing costs and evictions, millions in the US live in substandard conditions or with energy insecurity that negatively impacts their health. According to the 2017 American Housing Survey, 15.3 million households (12.6%) reported seeing cockroaches in the home, 15.6 million (12.8%) reported seeing signs of mice or rats in the home, and 3.8 million (3.1%) reported mold in their homes [6]. Nearly 6 million homes (4.9%) reported moderately or severely inadequate housing related to plumbing, heating, electricity, and upkeep [6]. Further, over 18 million households reported receiving notice of utilities shut off due to missed payments, and 1.2 million had their utilities shut off in the last 3 months [6]. Nearly 7 million homes (5.7%) reported at least 24 hours of inadequate heating in the last 12 months [6].

Evidence Basis

Many studies evaluating unstable housing among ED patients have primarily focused on homelessness, rather than housing instability as defined above. A meta-analysis of material needs of ED patients revealed that the small number of studies that have evaluated housing instability (excluding homelessness) in ED patients found rates of 18–44% [10]. The same meta-analysis looked at issues with housing quality specifically in ED patients and found that 18–36% of patients reported concerns about their housing quality [10].

Though few studies have examined housing instability and quality among ED patients specifically, the connections between these housing conditions and chronic diseases often seen in ED settings are well documented. For example, housing quality has been shown to be associated with respiratory diseases across the lifespan, including asthma, chronic obstructive pulmonary disease (COPD), and lung cancer. One study estimated that 4.6 million cases (21%) of asthma in the US are attributable to dampness and mold, costing over $3.5 billion annually [11]. Census tracts with more housing code violations have higher pediatric asthma-related morbidity (even when controlling for poverty), and the density of asthma-relevant housing code violations was shown to explain 22% of variation in ED visits and hospitalizations by neighborhood [12]. Children are more likely to have repeat ED visits and hospitalizations if they live in high violation areas [12].

Children are particularly vulnerable to the negative health effects of unstable housing. Children living in households that are behind on rent have higher lifetime hospitalizations and their caregivers are more likely to report fair or poor child health [13]. Medical conditions can themselves contribute to housing insecurity. In a study of families with children with cancer, the proportion of families with household material hardship, including housing insecurity, increased during the first 6 months of chemotherapy, from one in five to one in three families [14].

Thus, there is a bidirectional relationship between unstable housing and chronic disease: those with chronic diseases are more likely to be unstably housed, and those who are unstably housed have higher morbidity related to their chronic diseases. One study of non-homeless adults with diabetes who received care at federally funded safety-net health centers found that 37% were unstably housed, defined as not having enough money to afford rent or mortgage, moving more than two times in the past year, or staying somewhere one does not rent or own [15]. Unstably housed individuals with diabetes were five times more likely to have a diabetes-related ED visit or hospitalization than stably housed individuals [15]. In a large survey conducted across 11 states, 36% of individuals with chronic illness (cancer, stroke, cardiovascular disease, and chronic lung disease) reported having experienced housing insecurity in the past year [16]. Having any chronic condition (versus none) was associated with 42% increased likelihood of housing insecurity; cardiovascular disease and chronic lung disease were associated with 69% and 71% increased likelihood of housing insecurity, respectively [16]. In a systematic review of adults living with HIV/AIDS, unstable or unaffordable housing was associated with worse outcomes on a broad array of measures, including access to and utilization of HIV care, antiretroviral medication adherence, high-risk behaviors, and HIV clinical outcomes including ED use, hospitalizations, and mortality [17]. When compared to stably housed individuals, people living with HIV in unstable or temporary housing situations have 30% more urgent care visits, 75% more ED visits, and 61% more hospitalizations than those living in more stable housing [18].

Poor housing quality has been associated with psychological distress and mental health problems in children and adults, including depression, anxiety, isolation, and mood/conduct disorders [19]. Housing insecurity is also associated with higher prevalence of mental distress and insufficient sleep [20]. Moving due to cost reasons and being behind on mortgage payments have both been associated with increased risk of anxiety attacks, and being behind on rent increases the risk of depression [21]. Among people who have ever owned a home, recent foreclosure has been associated with higher likelihood of major or minor depression and anxiety attacks [21]. In a systematic review, foreclosure and living near foreclosure were associated with anxiety, violent behavior, and declining health care utilization [22].

One feature of substandard housing that contributes to excess morbidity and mortality in ED patients is heat and cold exposure through inadequate heating in the winter or inadequate cooling in the summer. Extremes of both heat and cold have been shown to increase cardiovascular morbidity and mortality, and these risks are even greater in the elderly or those with pre-existing cardiovascular disease [23]. These effects are present for cerebrovascular disease as well, with a more profound effect of cold exposure than heat exposure [24]. Both high and low temperatures have been shown to increase emergency hospital admissions in patients with sickle cell disease [25]. Similarly, extremes of temperature worsen outcomes in COPD. Heat has been shown to increase emergency hospitalizations for COPD; for every 10 °F increase in outdoor temperature, one study found a 4% increase in same-day emergency hospitalizations [26].

Another element of housing that affects health is crowding, defined as more than 2 persons-per-bedroom [27]. Crowded housing conditions are thought to lead to adverse health effects via multiple mechanisms: increased risk of transmitting infectious diseases (including COVID-19 [28]) given the close proximity to others, increased noise, and reductions in sleep quality and duration. Crowding is also associated with psychological distress in children and adults [29]. Crowding is more pronounced in certain regions of the country. For example, in California, 13.3% of renter-occupied units are crowded [30], and in New York City over a third of children live in homes that are crowded [31]. Children bear the brunt of crowding. In 2017, 14% of children (over 10 million children) lived in homes that were crowded [31]. In children, crowded housing has been shown to significantly worsen physical and behavioral health, school performance, and academic achievement [32].

An additional factor that contributes to poorer health in those with housing instability may be delayed or deferred medical care. Delays in health care are common among housing unstable individuals, which can contribute to ED use. In a large survey study of low-income adults, housing instability was associated with not having a usual source of care, postponing needed medical care, and postponing taking prescribed medications. In this general population sample, housing insecure individuals had 43% more ED use and 30% more hospitalizations than low-income, stably housed individuals [33]. In a large survey study of adults living with chronic illness, individuals with housing insecurity were over twice as likely to have health care access hardship than those with stable housing [16]. One study found that people who reported difficulty paying for their housing were three times more likely to report cost-related health care nonadherence and over twice as likely to report cost-related prescription nonadherence [34]. Among families, those that are behind on rent are more likely to forego medical care and to make trade-offs (such as cut back on food or paying utility bills) to pay medical bills, compared to stably housed families and even compared to those living in homeless shelters [35]. Another study found that housing insecure individuals were over twice as likely as housing secure individuals to delay doctor visits because of costs, even after adjusting for socioeconomic measures and demographics [36].

Emergency Department and Beyond

Bedside

Given the contributions of housing instability and poor housing quality to disease and to higher ED utilization, it is critical that emergency providers are familiar with and inquire about the clinical manifestations of housing-related issues. There is no clear "best" question to identify housing status or problems with housing quality, and providers should familiarize themselves with various possible questions that may work well in their setting. One question used by the Center for Medicare and Medicaid Services (CMS) Accountable Health Communities initiative [37] is, "What best describes your living situation?" (with responses of "I have stable housing," "I have stable housing now but am worried I will lose it," or "I do not have stable housing, i.e. shelter, living outside, staying temporarily with others, etc."). Children's HealthWatch describes "Housing Vital Sign" questions, which ask about the last year: "1. Was there a time when you were not able to pay the mortgage or rent on time? 2. How many places have you lived? 3. Was there a time when you did not have a steady place to sleep or slept in a shelter (including now)?" These questions have been used for screening in clinics and EDs. Alternatively, ED providers can ask patients more open-ended questions as part of the social history such as, "where have you been staying recently?"

Providers should also consider asking specific questions about housing stability and quality based on an individual patient's presenting symptoms. First, respiratory and cardiovascular illnesses contribute substantially to housing-related morbidity. The best studied of these is asthma, which can be triggered by numerous household exposures such as dust, pests, tobacco smoke, and mold [38]. Dampness and mold are also associated with increased morbidity from asthma [11]. Multiple features of substandard housing (including lack of hot water for washing, pests which may harbor diseases, and crowding) increase the risk of respiratory infections [39]. In individuals with sickle cell disease (particularly children), respiratory infections are associated with acute chest syndrome, which is an important cause of mortality [40]. Thus, emergency providers should ask patients about exposure to pests, dampness, and mold for illnesses such as asthma and sickle cell disease, especially in cases that are difficult to control despite adequate medical management.

Depending on the presenting complaint, ED providers should also consider asking patients about excessive heat and cold exposure. Inadequate or substandard housing increases the risk of both heat and cold exposure via mechanisms including poor insulation and/or ventilation, issues with gas and electricity, and lack of central heating or cooling. Ambient temperature has been linked to cardiovascular morbidity and mortality, respiratory illnesses, and sickle cell pain episodes, in addition to the more obvious potential ED presentations for hypo- or hyperthermia. Cold temperatures trigger vasoconstriction, which slows the transit of red blood cells through tissues, leading to increased deoxygenation of hemoglobin, and predisposing individuals with sickle cell disease to red blood cell sickling [41]. Heat may also increase the risk of vaso-occlusive episodes via dehydration, but this risk is less profound than that of cold [41]. Children and adults with sickle cell disease have also been shown to have changes in thermal sensory processing, leading to increased sensitivity to heat and cold [42]. Thus, ED physicians should inquire about heat and cold exposures due to housing instability or quality in patients with sickle cell disease presenting with vaso-occlusive episodes, particularly in extreme weather.

COPD is exacerbated by outdoor weather and indoor temperature. In winter months, COPD morbidity and mortality are worse, including increased respiratory symptoms, decreased lung function, and increased rescue inhaler use [43]. Maintaining an indoor temperature of 21 °C (69.8 °F) in living spaces for the recommended 9 hours a day has been associated with improved health status in patients with COPD, regardless of age, smoking status, baseline respiratory status and lung function, or outdoor temperature [44]. ED clinicians should be aware of the effects of indoor temperature on worsening COPD morbidity and should consider recommending strategies for maintaining a comfortable indoor temperature to their patients, including the use of indoor heating, fans or air conditioning, and humidifiers or dehumidifiers as needed, particularly during extreme weather.

Both housing quality and housing instability have been shown to affect mental health, in renters and homeowners. Interventions aimed at improving housing quality and stability have been associated with improvements in psychological distress, anxiety, and depression [19]. Thus, ED providers should include questions about housing quality and stability concerns in their social history for patients presenting with mental health complaints, particularly for those that have worsened acutely or recently.

Given that housing insecurity also affects health care utilization, emergency providers should inquire about the high cost of housing and other necessities in patients presenting with missed outpatient visits or medication nonadherence, and refer them to case managers, social workers, or other available resources as indicated.

In pediatric populations, sleep-related death, housing-related injuries, and exposure to lead are all important housing-related causes of morbidity and mortality. Sleep-related death is a leading cause of death in infants, and has been associated with numerous housing conditions [45]. Exposure to tobacco smoke has been associated with an increased risk of sudden infant death syndrome [46]. The largest contributor to sleep-related infant death is an unsafe sleep surface (such as co-sleeping or not sleeping in a crib or bassinet) [45]. Sleep practices are associated with housing quality issues including crowding, pest infestation, and room temperature (e.g., inability to adequately heat or cool all rooms), which increase the likelihood of an infant being placed on an unsafe sleep surface [45].

Housing quality also affects the risk of injuries including, but not limited to, burns and falls. Exposed heating sources and building materials contribute to risk of injury related to burns and fires [39]. Housing conditions related to falls vary across the lifespan. In young children, low window sills, unprotected upper-story windows, and breakable window glass are risk factors for injuries from falls [39]. ED providers should inquire about the circumstances of injuries to determine whether any potentially remediable housing quality issues might have contributed and should consider the involvement of social workers or case managers if these issues are identified.

Lastly, exposure to lead, even at low levels, is known to increase the risk of intellectual disability, ADHD, and other behavioral disorders [47, 48]. The most common sources of lead exposure in children include living in an older home with lead-based paint which may be chipping and lead dust which can be ingested or inhaled [49], as well as proximity to demolished or remodeled homes [50]. Despite reductions in lead poisoning over the past four decades, over 500,000 children still had elevated blood lead levels as of 2014 [48], and 23 million homes continue to have significant lead-based paint hazards, including 1 in 3 homes with children under six [51]. The US Department of Housing and Urban Development estimated in 2016 that over 62,000 public housing units were known to be in need of lead abatement, and more than 450,000 other federally assisted Section 8 housing units were built before 1978 and have children living in them, putting those children at risk for exposure to adverse health effects related to lead-based paint [51]. Emergency providers should screen for lead exposure (e.g., recent moves, chipping paint) in children presenting to the ED with behavioral or developmental issues, especially those with subacute or acute worsening.

It is important for emergency providers to recognize that beyond housing quality and insecurity, where housing is located can directly influence health. These impacts are covered in more detail in the "Neighborhood and the Built Environment" chapter.

Even though emergency providers can rarely "solve" patients' housing issues in real time, knowledge of housing instability and quality should be considered in treatment, disposition, and follow-up plans. For example, physicians should recognize when social needs (including high housing costs) are contributing to medication nonadherence and consider lower cost regimens, or when issues with medication storage and refrigeration exist due to energy insecurity. Additionally, when housing issues are identified, ED physicians should know the available case management and social work resources in their practice setting which may help begin to address housing instability and quality issues. Research demonstrates that even providing lists of community resources can be important interventions to address the needs of patients, and feasible from ED settings [52, 53].

Hospital/Healthcare System

In addition to individual ED clinicians screening for housing-related social risks in high-risk patients, hospitals and healthcare systems should develop and implement systematic screening protocols for housing insecurity and substandard housing quality. Several options for standard screening questions for housing status exist, including Accountable Health Communities and Children's HealthWatch "Housing Vital Sign" questions described above. These tools should be adapted to the various clinical settings, and should capture as many patients as possible. Universal screening, as clinically appropriate, would alleviate some of the burden on clinicians to be aware of the myriad conditions that may be worsened by poor housing.

Healthcare systems should develop resource lists for housing resources in the local community, such as case management organizations outside of the healthcare setting, legal organizations (including medical-legal partnerships), and community organizations offering homelessness prevention services for people at risk for eviction. Programs such as the Low Income Home Energy Assistance Program [54] can help provide energy assistance for low-income individuals, and healthcare systems should ensure these resources are included in those provided to patients, especially those with health problems related to inadequate heating or cooling.

More and more hospital systems are working to develop housing-based interventions to address underlying health conditions that may be sensitive to housing quality and stability. Interventions related to improving indoor housing temperature have shown broad health improvements [55]. One meta-analysis demonstrated that interventions aimed at improving thermal comfort led to improvements in physical and mental health, improvement in chronic respiratory diseases, reduced absences from school and work, and improved social relationships [55]. Interventions to address asthma in homes using community health workers following ED visits by children have been well studied and recently adapted for adults with COPD [38].

Healthcare systems can improve access to services by co-locating social programs such as food pantries, legal aid, or housing navigators within hospitals, clinics, or the patient-centered medical home. Use of medical-legal partnerships (MLP) in ED settings to address problems including housing instability and quality has also been shown to be effective in reducing ED utilization [56]. Medical-legal partnerships are currently available in over 300 health care institutions, and the National Center for Medical-Legal Partnership offers information on how to start an MLP if one is not available nearby [57].

Societal Level

That millions of patients are affected by housing instability and quality problems are a direct result of lack of affordable housing. One study found that for every 100 extremely low-income renters, there were only 35 available affordable units [7]. Additionally, there is no city in the US where a person making minimum wage can afford (defined as spending under 30% of income on rent) to rent a 2 bedroom apartment [58]. As a result of systemic racism in the form of discriminatory policies such as redlining and segregated housing developments, issues of housing availability, affordability, and eviction rates are worst among racial and ethnic minorities [9]. In some cities, such as Richmond, Virginia, approximately 1 in 9 renter households was evicted in 2016, although rates were much lower in majority White neighborhoods [59]. Issues of housing instability, including evictions, are expected to worsen due to the economic fallout from the COVID-19 pandemic, with experts projecting that over 1.5 million families may become homeless because of the crisis [60]. The COVID-19 pandemic has disproportionately affected communities of color, just as housing instability had long before the novel coronavirus, in part due to higher likelihood of living in overcrowded homes [28]. Emergency departments are likely to be on the front line as this housing crisis unfolds, leading to the myriad health effects described in this chapter, including difficulties with accessing and affording care and medications.

Emergency providers should familiarize themselves with the status of affordable housing, eviction rates, and the housing policy landscape in their practice area. They should consider joining advocacy organizations working on the most pressing local issues in their area. For example, ED providers could advocate for zoning regulations and laws that would encourage their city to build more affordable housing, provide rent assistance to those who are behind on rent or are facing eviction, provide the right to legal representation in housing court, or improve conditions in local public housing. All of these issues, and many more, could benefit from emergency providers' unique combination of exposure to patient experiences and ability to compellingly convey data.

Emergency providers could also join their voices in advocating for housing voucher programs that would allow their patients to move to affordable housing in higher opportunity neighborhoods. Studies have shown that moving to higher opportunity neighborhoods may improve health outcomes [61, 62]. One study compared families randomly assigned to one of three groups: housing subsidy vouchers requiring them to move to low poverty neighborhoods, housing subsidy vouchers that did not require moving to low poverty neighborhoods, and no vouchers [61]. Children whose families moved to low poverty neighborhoods had 35% increased earnings as adults, 32% higher college attendance, and 26% reduced single motherhood later in life [61], even when compared to children whose families received vouchers without the requirement of moving to a low poverty neighborhood. This suggests a benefit of living in low poverty neighborhoods, beyond that of housing subsidies alone [61]. For adults, those who received vouchers requiring moving to a low poverty neighborhood had lower rates of obesity, severe obesity, and elevated HbA1C, even when compared to those who received traditional Section 8 housing vouchers [62]. However, subsequent analyses of this program found no significant change in healthcare utilization measures such as the rate of ED visits, hospitalizations, or hospital spending for children or adults in the program compared to those who did not receive vouchers [63, 64].

Recommendations for Emergency Medicine Practice

Frequently, ED providers feel helpless to influence housing instability and quality problems given the prevalence of these problems and the perceived inadequacy of responses from an ED setting. However, there are ways to integrate the identification of these issues into clinical care that will assist in tailoring care for unstably housed patients and those with housing conditions that exacerbate chronic medical conditions.

Basic

- Include questions about housing stability and quality in the social history, particularly for patients with presenting complaints that may be due to, or exacerbated by, housing instability or poor housing quality. Recognize that there are various screening tools and questions available and familiarize yourself with the questions that work best in your practice setting.
- When asking about a patient's housing status or condition, be respectful and normalize that this is an important aspect of care. It is important to explain that all patients are asked these questions and there is no judgment in them. Many patients may feel stigmatized because of their circumstances and therefore may wonder why they are being asked these questions.
- Consider housing-related exposures in your differential diagnosis. For example, for sickle cell disease, COPD, or cardiovascular events known to be related to heat/cold exposure, consider asking about energy insecurity, such as "have you recently received a shut off notice from a utility company?" or "have you had difficulty heating or cooling your home to comfort?" Similarly, in patients presenting with acute worsening of their physical or mental health conditions despite

seemingly adequate medical management, consider inquiring about changes in their housing situation, including housing quality and affordability. In patients who present with evidence of other social needs such as food insecurity or difficulty accessing primary care or medications, it may be especially important to consider whether housing instability or quality is affecting their medical conditions.

- Consider housing instability and quality in treatment, disposition, and follow-up plans. ED physicians should familiarize themselves with resources available in their practice setting or within their community so that when housing issues are identified, they can refer to these resources to begin to address housing instability and quality issues.

Intermediate

- EDs should implement screening for housing status using one of the tools described above, and should establish best practices around documentation of housing stability and quality. This is especially important in EDs that serve a high population of low-income patients, but should be considered in all EDs given the prevalence of and stigma around housing issues. Consider using ICD-10 codes for housing instability (such as Z59.1) to document for patients who are not literally homeless but who experience other forms of housing insecurity. Additional codes exist for extreme forms of housing code violations (such as lack of running water).
- EDs should screen for other health-related social needs, such as food insecurity, transportation insecurity, and inability to afford medical care or prescription drugs. Given the high overlap of these needs with housing concerns, this screening may identify additional patients at high risk for housing instability or substandard quality. Furthermore, many ED patients' housing-related health issues cannot be resolved by addressing housing concerns alone, without considering these other needs.

Advanced

- Emergency providers should educate themselves on local and state housing policies that may adversely affect their patients, and should advocate for improved policies. They can provide education to legislators and other stakeholders about the relationship between housing and health. They can participate in local, state, and national coalitions, speaking to the negative health effects they observe as a result of housing instability and quality problems. Emergency providers can be powerful advocates for the safe and affordable housing that is needed to address this critical root contributor to patients' health and ED use. Emergency providers should consider establishing relationships with community organizations to serve as effective housing advocates.

- Emergency providers should encourage their healthcare institutions to prioritize and advocate on housing issues. They can provide education to hospital leadership around the relationship between housing and health, and programs that have been effective at improving health through housing interventions. They should also educate themselves about their practice setting's role as an "anchor institution," and encourage investments in affordable housing for the benefit of patients, employees, and the broader community.
- Emergency providers should work with medical-legal partnerships to integrate legal services into health care settings. Over 300 hospitals and health centers have integrated legal aid attorneys and resources within medical facilities, including EDs, which allow for identification of housing issues, assistance with legal recourse (e.g., letters to landlords to encourage remediation of housing defects), and referral to other services. More information for EDs interested in starting medical-legal partnership programs is available from the National Center for Medical-Legal Partnership (medical-legalpartnership.org).

Teaching Case

Clinical Case

A 57-year-old man with a history of diabetes mellitus presents to your ED with a 3-day history of polyuria and polydipsia. On arrival to the ED, his blood sugar is 379, HbA1C is 11.4%, but VBG and electrolytes are reassuring without evidence of diabetic ketoacidosis. This is his third visit in the last 2 months for similar presentations. You confirm that he understands his medication regimen, including the sliding scale for his insulin, and that he has up-to-date prescriptions for these medications. You are preparing to discharge him, but prior to discharge you obtain additional history about barriers to taking his medications.

Upon further questioning, the patient endorses that about a year ago the price of his insulin increased and he began having difficulty affording it. He cut down his other expenses as much as possible, but about 6 months ago he had used up his savings and was still unable to afford his medications along with his other needs. At first, he tried to spread out his medications, but his endocrinologist told him that his labs showed poor diabetes control, so he dedicated himself to sticking strictly to his medication and insulin regimen. Unfortunately, he was then no longer able to afford his rent and was evicted for being behind on rent about 3 months ago. He has been staying on various friends' couches for the last 3 months, and has difficulty refrigerating his insulin at times. He recently moved to a new friend's home and mistakenly left his medications behind. He has been trying to get them back but has been having difficulty securing transportation, so has not been taking his medications for the past 4 days.

Given that you identified housing-related issues contributing to this patient's poor control of his diabetes, you provide a list of local housing resources and refer the patient to case management to link with housing search services and your hospital's medical-legal partnership (MLP). Case management and the MLP assisted the patient with his housing needs, helping him secure a low cost, subsidized apartment. The next time you see the patient in the ED for an unrelated complaint, he reports that he has had an easier time sticking with his insulin regimen and his diabetes control has been much better since moving into his new apartment.

Teaching Points

1. This patient demonstrates multiple types of housing instability and poor quality. He has a high rent burden, and when another important expense increases he is unable to afford his rent. He then endorses frequently moving ("couch surfing") which adds further barriers to medication adherence. He may also be experiencing issues with electricity or other housing quality issues given his difficulty refrigerating his insulin, but the provider would need to explore this more to confirm.
2. By recognizing that the underlying etiology of his poor diabetes control is related to cost and housing instability, the provider can address his diabetes more effectively, and hopefully reduce the likelihood of future ED visits for medication nonadherence. The provider should also screen for food insecurity, as this may also be contributing to poor diabetes control in this high-risk patient.
3. Providers should inquire about whether patients are receiving benefits that they are eligible for (e.g., the Supplemental Nutrition Assistance Program [SNAP], Medicaid, housing and utility subsidies), which may reduce the need for patients to choose between paying for their housing and paying for medication or other necessities. Patients in need of benefits should be referred to social workers, case managers, or community organizations to assist with the application process.
4. Providers should know the available resources to assist patients with housing instability and related needs in their ED/hospital (e.g., case management, medical-legal partnership, food pantry) or local community, and refer to those for patients who need them.

Discussion Questions

1. Are there specific diagnoses that should warrant consideration of housing-related issues as part of an ED history? Have you seen housing instability or quality issues affect patients in your own clinical experience?
2. Which features of this patient's presentation should clue the provider in to issues with housing that are contributing to his poor disease control? How might this case have played out if the provider hadn't recognized how housing instability and quality contributed to the patient's condition?
3. What tools exist in your current practice setting to assist with issues around housing instability and quality? What resources may be missing that could improve patients' needs and how could you advocate for them within your institution or community?

References

1. US Department of Health and Human Services. Housing Instability | Assistant Secretary for Planning and Evaluation [Internet]. Available from: https://aspe.hhs.gov/report/ancillary-services-support-welfare-work/housing-instability.
2. Office of Disease Prevention and Health Promotion. Housing Instability | Healthy People 2020 [Internet]. Available from: https://www.healthypeople.gov/2020/topics-objectives/topic/social-determinants-health/interventions-resources/housing-instability.
3. American Hospital Association. Housing and the Role of Hospitals Social Determinants of Health Series | Housing and the Role of Hospitals. 2017. Available from: www.aha.org/housing.
4. US Department of Housing and Urban Development (HUD). Affordable Housing [Internet]. [cited 2019 Apr 2]. Available from: https://www.hud.gov/program_offices/comm_planning/affordablehousing/.
5. Stats New Zealand. Measuring housing quality: Potential ways to improve data collection on housing quality in New Zealand [Internet]. [cited 2019 Mar 21]. Available from: http://archive.stats.govt.nz/browse_for_stats/people_and_communities/housing/measuring-housing-quality/intro-to-housing-quality-measuring.aspx.
6. United States Census Bureau. American Housing Survey Table Creator. [cited 2019 Apr 14]; Available from: https://www.census.gov/programs-surveys/ahs/data/interactive/ahstablecreator.html#?s_areas=a00000&s_year=n2017&s_tableName=Table5&s_byGroup1=a1&s_byGroup2=a1&s_filterGroup1=t1&s_filterGroup2=g1&s_show=S.
7. Joint Center for Housing Studies of Harvard University. The State of the Nation's Housing 2018 [Internet]. 2018. Available from: https://www.jchs.harvard.edu/sites/default/files/Harvard_JCHS_State_of_the_Nations_Housing_2018.pdf.
8. Brancaccio D, Long K. Millions of Americans are evicted every year — and not just in big cities [Internet]. [cited 2019 Apr 2]. Available from: https://www.marketplace.org/2018/04/09/economy/eviction-desmond-princeton-housing-crisis-rent.
9. Rothstein R. The color of law: a forgotten history of how our government segregated America. New York: Liveright Publishing Corporation; 2017.
10. Malecha PW, Williams JH, Kunzler NM, Goldfrank LR, Alter HJ, Doran KM. Material needs of emergency department patients: a systematic review. Acad Emerg Med. 2018;25(3):330–59.
11. Mudarri D, Fisk WJ. Public health and economic impact of dampness and mold. Indoor Air. 2007;17:226–35.
12. Beck AF, Huang B, Chundur R, Kahn RS. Housing code violation density associated with emergency department and hospital use by children with asthma. Health Aff NIH Public Access. 2014;33:1993–2002.
13. Sandel M, Sheward R, Ettinger de Cuba S, Coleman SM, Frank DA, Chilton M, et al. Unstable housing and caregiver and child health in renter families. Pediatrics. American Academy of Pediatrics. 2018;141:e20172199.
14. Bona K, London WB, Guo D, Frank DA, Wolfe J. Trajectory of material hardship and income poverty in families of children undergoing chemotherapy: a prospective cohort study. Pediatr Blood Cancer. John Wiley & Sons, Ltd. 2016;63:105–11.
15. Berkowitz SA, Kalkhoran S, Edwards ST, Essien UR, Baggett TP. Unstable housing and diabetes-related emergency department visits and hospitalization: a nationally representative study of safety-net clinic patients. Diabetes Care. 2018;41:933–9.
16. Charkhchi P, Fazeli Dehkordy S, Carlos RC. Housing and food insecurity, care access, and health status among the chronically ill: an analysis of the behavioral risk factor surveillance system. J Gen Intern Med. 2018;33:644–50.
17. Aidala AA, Wilson MG, Shubert V, Gogolishvili D, Globerman J, Rueda S, et al. Housing status, medical care, and health outcomes among people living with HIV/AIDS: a systematic review. Am J Public Health. 2016;106:e1–23.

18. Clemenzi-Allen A, Geng E, Sachdev D, Buchbinder S, Havlir D, Gandhi M, et al. Housing instability increases rates of urgent care visits, emergency department visits and hospitalizations among PLHIV. 2018.
19. Evans GW, Wells NM, Moch A. Housing and mental health: a review of the evidence and a methodological and conceptual critique. J Soc Issues. 2003;59(3):475–500.
20. Liu Y, Njai RS, Greenlund KJ, Chapman DP, Croft JB. Relationships between housing and food insecurity, frequent mental distress, and insufficient sleep among adults in 12 US States, 2009. Prev Chronic Dis. 2014;11:E37.
21. Burgard SA, Seefeldt KS, Zelner S. Housing instability and health: findings from the Michigan recession and recovery study. Soc Sci Med. 2012;75:2215–24.
22. Downing J. The health effects of the foreclosure crisis and unaffordable housing: a systematic review and explanation of evidence. Soc Sci Med. 2016;162:88–96.
23. Moghadamnia MT, Ardalan A, Mesdaghinia A, Keshtkar A, Naddafi K, Yekaninejad MS. Ambient temperature and cardiovascular mortality: a systematic review and meta-analysis. PeerJ. 2017;5:e3574.
24. Lin Y-K, Chang C-K, Wang Y-C, Ho T-J. Acute and prolonged adverse effects of temperature on mortality from cardiovascular diseases. PLoS One. 2013;8:e82678.
25. Mekontso Dessap A, Contou D, Dandine-Roulland C, Hemery F, Habibi A, Charles-Nelson A, et al. Environmental influences on daily emergency admissions in sickle-cell disease patients. Medicine (Baltimore). 2014;93:e280.
26. Anderson GB, Dominici F, Wang Y, McCormack MC, Bell ML, Peng RD. Heat-related emergency hospitalizations for respiratory diseases in the medicare population. Am J Respir Crit Care Med. 2013;187:1098.
27. Blake KS, Kellerson RL, Simic A. Measuring Overcrowding in Housing | U.S. Department of Housing and Urban Development Office of Policy Development and Research. 2007.
28. Does COVID-19's toll reflect social inequality? Early evidence from NYC. [cited 2020 Jun 30]; Available from: https://medium.com/@jmfeldman/does-covid-19s-toll-reflect-socialinequality-early-evidence-from-nyc-209c3b0a0ff7.
29. Evans GW. The built environment and mental health. J Urban Health. 2003;80(4):536–55.
30. Solari CD. America's housing is getting more crowded. How will that affect children? | The Urban Institute [Internet]. Available from: https://www.urban.org/urban-wire/americas-housing-getting-more-crowded-how-will-affect-children.
31. The Annie E. Casey Foundation. Children living in crowded housing | Kids Count Data Center [Internet]. [cited 2019 Mar 7]. Available from: https://datacenter.kidscount.org/data/tables/67-children-living-in-crowded-housing#detailed/1/any/false/871,870,573,869,36,868,867,133,38,35/any/368,369.
32. Solari CD, Mare RD. Housing crowding effects on children's wellbeing. Soc Sci Res. 2012;41:464.
33. Kushel MB, Gupta R, Gee L, Haas JS. Housing instability and food insecurity as barriers to health care among low-income Americans. J Gen Intern Med. 2006;21:71–7.
34. Pollack CE, Griffin BA, Lynch J. Housing affordability and health among homeowners and renters. Am J Prev Med. 2010;39:515–21.
35. Children's HealthWatch. Behind closed doors: the hidden health impacts of being behind on rent [internet]. 2011. Available from: https://childrenshealthwatch.org/wp-content/uploads/behindcloseddoors_report_jan11-.pdf.
36. Stahre M, VanEenwyk J, Siegel P, Njai R. Housing insecurity and the association with health outcomes and unhealthy behaviors, Washington State, 2011. Prev Chronic Dis. 2015;12:E109.
37. Alley DE, Asomugha CN, Conway PH, Sanghavi DM. Accountable health communities – addressing social needs through Medicare and Medicaid. N Engl J Med. 2016;374(1):8–11.
38. Krieger J, Jacobs DE, Ashley PJ, Baeder A, Chew GL, Dearborn D, et al. Housing interventions and control of asthma-related indoor biologic agents: a review of the evidence. J Public Health Manag Pract. 2010;16:S11–20.

39. Krieger J, Higgins DL. Housing and health: time again for public health action. Am J Public Health. 2002;92:758–68.
40. Vichinsky EP, Neumayr LD, Earles AN, Williams R, Lennette ET, Dean D, et al. Causes and outcomes of the acute chest syndrome in sickle cell disease. N Engl J Med. 2000;342:1855–65.
41. Tewari S, Brousse V, Piel FB, Menzel S, Rees DC. Environmental determinants of severity in sickle cell disease. Haematologica. 2015;100:1108–16.
42. Brandow AM, Stucky CL, Hillery CA, Hoffmann RG, Panepinto JA. Patients with sickle cell disease have increased sensitivity to cold and heat. Am J Hematol. 2013;88:37–43.
43. McCormack MC, Paulin LM, Gummerson CE, Peng RD, Diette GB, Hansel NN. Colder temperature is associated with increased COPD morbidity. Eur Respir J. 2017;49.
44. Osman LM, Ayres JG, Garden C, Reglitz K, Lyon J, Douglas JG. Home warmth and health status of COPD patients. Eur J Public Health. 2008;18:399–405.
45. Chu T, Hackett M, Kaur N. Housing influences among sleep-related infant injury deaths in the USA. Health Promot Int. 2016;31:396–404.
46. Athanasakis E, Karavasiliadou S, Styliadis I. The factors contributing to the risk of sudden infant death syndrome. Hippokratia. Hippokratio General Hospital of Thessaloniki. 2011;15:127–31.
47. American Academy of Pediatrics Council on Environmental Health. Prevention of childhood lead toxicity. Pediatrics. 2016;138:e20161493.
48. Centers for Disease Control and Prevention. Lead – Childhood Lead Poisoning Prevention Program [Internet]. [cited 2019 Apr 16]. Available from: https://www.cdc.gov/nceh/lead/default.htm.
49. Norris D, Olinger J, McKay M. The weight of lead — part I: how contaminated houses are poisoning the poor. 2018.; Available from: https://medium.com/the-block-project/the-weight-of-lead-part-i-how-contaminated-houses-are-poisoning-the-poor-ce3bc5fb9dda.
50. Jacobs DE, Wilson J, Dixon SL, Smith J, Evens A. The relationship of housing and population health: a 30-year retrospective analysis. Environ Health Perspect. 2009;117:597–604.
51. Benfer EA. Contaminated childhood: the chronic lead poisoning of low-income children and communities of color in the United States. Health Aff Blog. 2017.
52. Bottino CJ, Rhodes ET, Kreatsoulas C, Cox JE, Fleegler EW. Food insecurity screening in pediatric primary care: can offering referrals help identify families in need? Acad Pediatr. 2017;17(5):497–503.
53. Gurewich D, Garg A, Kressin NR. Addressing social determinants of health within healthcare delivery systems: a framework to ground and inform health outcomes. J Gen Intern Med. 2020;35(5):1571–5.
54. Office of Community Services. Low Income Home Energy Assistance Program (LIHEAP) [Internet]. [cited 2020 Jul 3]. Available from: https://www.acf.hhs.gov/ocs/programs/liheap.
55. Thomson H, Thomas S, Sellstrom E, Petticrew M. Housing improvements for health and associated socio-economic outcomes. Cochrane Database Syst Rev. 2013:CD008657.
56. Martin J, Martin A, Schultz C, Sandel M. Embedding civil legal aid services in care for high-utilizing patients using Medical-Legal Partnership. Health Aff Blog. 2015.
57. Medical-Legal Partnership FAQ [Internet]. [cited 2020 Jun 29]. Available from: https://medical-legalpartnership.org/about-us/faq/.
58. National Low Income Housing Coalition. Out of reach 2018 [Internet]. [cited 2019 Mar 12]. Available from: https://reports.nlihc.org/oor.
59. Badger E, Bui Q. In 83 million eviction records, a sweeping and intimate new look at housing in America [Internet]. [cited 2019 Apr 21]. Available from: https://www.nytimes.com/interactive/2018/04/07/upshot/millions-of-eviction-records-a-sweeping-new-look-at-housing-in-america.html.
60. Wulfhorst E. Coronavirus could put 1.5 million U.S. families on cusp of homelessness. Reuters [Internet]. [cited 2020 Jun 29]; Available from: https://www.reuters.com/article/us-health-coronavirus-housing-trfn/coronavirus-could-put-15-million-us-families-on-cusp-of-homelessness-idUSKBN21Q27M.

61. Chetty R, Hendren N, Katz LF. The effects of exposure to better neighborhoods on children: new evidence from the moving to opportunity experiment. Am Econ Rev. 2016;106:855–902.
62. Ludwig J, Sanbonmatsu L, Gennetian L, Adam E, Duncan GJ, Katz LF, et al. Neighborhoods, obesity, and diabetes — a randomized social experiment. N Engl J Med. 2011;365:1509–19.
63. Pollack CE, Pollack CE, Blackford AL, Du S, Deluca S, Thornton RLJ, et al. Association of receipt of a housing voucher with subsequent hospital utilization and spending. JAMA. 2019;322:2115–24.
64. Pollack CE, Du S, Blackford AL, Herring B. Experiment to decrease neighborhood poverty had limited effects on emergency department use. Health Aff. 2019;38:1442–50.

Transportation 16

Margaret B. Greenwood-Ericksen

Key Points

- Transportation barriers disproportionately affect the most vulnerable and have important geographic patterns.
- Inadequate transportation to outpatient care can contribute to ED visits.
- Transportation influences ED treatment and care plans. Specifically, transportation barriers can prevent routine and timely outpatient care after an ED visit. Further, safe disposition from the ED requires transportation to a destination with adequate social and medical resources.
- Current healthcare transportation interventions include taxi/ride-share vouchers, shuttle services, and payer-provided transportation for both emergency appointments (Medicare/Medicaid) and nonemergency appointments (Medicaid only).
- Emerging interventions include Medicare support for nonemergent ambulance transport, telehealth evaluations with emergency physicians prior to ED or primary care transport, and ride-share services for transportation after ED discharge and to follow-up outpatient appointments.

Foundations

Background

Transportation barriers affect access to healthcare services and are associated with missed appointments, increased health expenditures, worse health outcomes, and potentially inadequate chronic disease management [1]. Transportation is a key social determinant of health, acting as a facilitator or barrier to health management, and is intertwined with other social determinants of health such as poverty and

M. B. Greenwood-Ericksen (✉)
Department of Emergency Medicine, University of New Mexico Hospital, Albuquerque, NM, USA

H. J. Alter et al. (eds.), *Social Emergency Medicine*,
https://doi.org/10.1007/978-3-030-65672-0_16

social isolation [2, 3]. As geographic mobility is linked to economic success [4], addressing transportation barriers will not only improve health outcomes but will also reduce societal inequity. Transportation access is influenced by socioeconomic status, ethnicity, geography (e.g., rural vs. urban), age, mode of travel, and distance. Barriers to transportation disproportionately impact low-income patients, racial/ethnic minorities, and those facing long geographic distances to care and unsafe or inadequate public transportation infrastructure. Additionally, while emergency medical services (EMS) are available in all communities across the US, inequity still exists, with rural systems experiencing slower response times and longer transportation times than urban systems [5]. Further, rural EMS agencies have lower staff skill level, less oversight, and less public funding when compared to urban agencies [6].

In the US, 3.6 million people forgo medical care due to transportation barriers annually and are disproportionately poorer, older, less educated, and more likely to be from racial/ethnic minority groups [7]. Lack of transportation is often cited as a major barrier for low-income populations, with approximately 20–50% reporting missing or rescheduling outpatient appointments because of unreliable transportation [1, 8]. The elderly are particularly vulnerable to transportation barriers to healthcare access including inability to drive due to age-related disability or mobility issues making public transit a challenge [9]. Children are also at risk, with almost 10% of low-income children missing a healthcare appointment due to transportation issues, 30% of whom subsequently seek care in an emergency department (ED) [10]. Rural communities are also disproportionately impacted, because they have few available public transit resources compared to more urban areas. Finally, structural inequality as reflected by neighborhood racial segregation may result in difficulties accessing healthcare due to longer travel times or lack of bus routes crossing highways [11, 12].

Addressing transportation needs of low-income patients as a social determinant of health is critical. This can be accomplished though external structures such as increased investment in public transportation infrastructure or by increasing healthcare-supplied transportation options for patients. While the Medicaid benefit has included nonemergency medical transportation [NEMT] since its inception, until recently other payers and health systems have not focused on transportation. There is now increased recognition that patient transportation assistance can improve health outcomes and reduce healthcare costs [13].

Evidence Basis

Transportation is inextricably linked with both individual and community socioeconomic status. First, it is critical for ED providers to be well versed in the role that transportation plays in healthcare access and outcomes, as barriers to transportation prevent access to outpatient care and can result in ED visits [14, 15]. Second, ED providers make disposition decisions on every patient they evaluate and know first-hand the role transportation plays in arranging safe discharge plans – a topic largely unexplored in the ED literature. Third, disparities in EMS delivery by geography, race and ethnicity, and income [5, 16, 17] drive inequity. Knowledge of these disparities can help ED providers optimize emergency care in their communities. We will review each of these issues in more detail below.

Ambulatory Care Access Barriers and Interventions

Transportation barriers faced by patients, particularly the poorest, can result in ED visits for low-acuity needs due to preference, convenience, or medical or financial necessity [15, 18]. A recent systematic review on transportation interventions identified taxi/transport vouchers, free shuttle services, and bus passes as potentially effective methods to improve chronic disease management, reduce hospital utilization, and increase follow-up care [13]. Further, federal Medicaid regulations require that states ensure transportation to and from visits with healthcare providers, which is the Medicaid nonemergency medical transportation (NEMT) benefit. The scope of this benefit varies by state but generally covers a broad range of transportation options such as taxis, buses, vans, and personal vehicles [19]. In general, Medicaid beneficiaries are eligible for NEMT as long as the transportation is necessary and they have no other means of transportation. Some states rely on public transportation, but this approach varies within and across states based on public transportation availability. Other states limit the benefit through prior authorization requirements or place limits on the number of trips covered. Under Medicaid 1115 waivers, two states, Indiana and Iowa, have received approval to eliminate the NEMT benefit for the new group of Medicaid-eligible low-income non-disabled adults established under the ACA; additional research is needed to evaluate the impact of reducing or restricting this benefit [19].

The NEMT benefit is usually administered by brokers that coordinate and dispatch private cars, taxis, or specialized vehicles. Services can be initiated by patients or by individuals such as social workers or care coordinators. Challenges with NEMT delivery in some localities include poor customer service, inadequate responsiveness, and fraud and abuse [20]. In the face of these challenges, payers and healthcare delivery organizations have been experimenting with new strategies for delivering NEMT, including use of ride-share companies like Uber and Lyft to improve beneficiary experience and reduce costs [21].

NEMT is used most frequently by Medicaid beneficiaries for accessing behavioral health, dialysis, preventive services, specialist care, and physical therapy [22]. Medicaid beneficiaries who use NEMT services are significantly more likely to make the recommended number of annual visits for the management of chronic conditions than those who do not use NEMT [3]. A 2005 study done for the Transportation Research Board of the National Academies of Sciences, Engineering, and Medicine found NEMT services to be cost-effective for patients with chronic conditions [23]. There is increasing interest in expanding a NEMT benefit to Medicare as well as in other innovations. For example, in 2019, the Center for Medicare and Medicaid Innovation (CMMI) released a new payment model to pay ambulance providers to transport Medicare patients to either an ED or an alternative destination [like an urgent care] or to provide treatment in place via telehealth [24].

However, transportation access is only one part of improving health outcomes. This is best illustrated by a recent randomized control trial demonstrating that increased access to free, convenient transportation failed to reduce rates of no-show visits for routine care among Medicaid enrollees [25]. While evidence suggests that transportation services offered in combination with other services may improve patient health outcomes, it is critical to address transportation in the social and ecologic context of patients' communities. Important ongoing and future research includes identifying best practices in screening ED patients for transportation needs, examining patient perspectives on transportation interventions (vouchers, NEMT), and evaluating the impact of state waiver restrictions to NEMT, including for ED use.

ED Disposition and Transportation

The role of transportation in determining follow-up and ED disposition is an important area of research that remains understudied. Follow-up and disposition decisions are made on each patient treated in an ED, and these decisions are critical in providing high-quality emergency care.

To date, isolated transportation interventions (e.g., giving patients transportation vouchers or bus passes) designed to facilitate outpatient follow-up after an ED visit or increase primary care visits have shown mixed success [25–27]. Data is more favorable for multidisciplinary, community-based programs with care navigators, who can identify transportation solutions for patients; such programs have shown promise in increasing primary care use while reducing ED use and overall costs [28, 29]. This may indicate that helping patients navigate the complex health system, rather than providing stand-alone vouchers to fix the immediate transportation problem, may be more effective in the long term. We are already seeing increased use of care navigators and community health workers by outpatient clinics such as Federally Qualified Health Centers (FQHCs) [30, 31] and, less commonly but increasingly, EDs [32].

Emergency Medical Services (EMS) Disparities

Transportation for emergent conditions also has important health implications. Longer response and transport times are associated with worse outcomes for time-sensitive conditions such as trauma and cardiac arrest and disproportionately impact rural, low-income, and minority communities [33–35]. Patients from the poorest neighborhoods have longer EMS times compared to those from the wealthiest, contributing to health disparities [17]. Rural patients face inherently longer response and transport times [5, 36]. Other research has found that minority patients experience delays in transport and diagnosis for time-sensitive conditions such as stroke and myocardial infarction [37]. State initiatives establishing systems of care for conditions such as trauma, stroke, and cardiac arrest have demonstrated some improvements in time to definitive care across states with large urban and rural populations [38, 39]. Such systems may reduce disparities related to geography and race in transportation for emergency conditions [38, 39].

Emergency Department and Beyond

Bedside

Despite the importance of exploring a patient's transportation situation, this is often not a standard part of history taking in the ED. While the social history may involve asking about health-related risk factors such as smoking or alcohol use, it is infrequent that ED providers ask about patient transportation access. These conversations may feel uncomfortable for patients, particularly as transportation type (car vs. public transportation) can be a reflection of social and economic status. Thus, these conversations should be framed in the context of medical care access or safe disposition, such as a ride home and ability to attend follow-up visits. This may be best accomplished during the initial evaluation of the patient, rather than at the end of the visit, to anticipate any challenges related to transportation and avoid potential disposition delays. For example, in Studor's Acknowledge, Introduce, Duration, Explanation, and Thank you (AIDET) patient communication framework [40], this conversation should occur during the "explanation" portion when describing the ED visit plan. For patients in whom you anticipate discharge, asking "How will you get home today?" and "Do you have reliable transportation to your follow-up appointment?" is important in developing your initial ED care plan. Sometimes even asking "How did you get here today?" in the initial history is another way to start the conversation. If transportation barriers are identified, an appropriate referral can be placed in the ED for a social worker, community health worker, or care navigator to visit with the patient to assess transportation needs and provide resources. However, smaller (and many rural) EDs do not have such resources. In these settings, consulting the inpatient care coordinator and/or social workers is often effective, though may require you to hold a patient in the ED until business hours. For less complicated matters, asking the ED clerk or nurse may be adequate.

For patients with Medicaid, it may be possible to arrange a NEMT option by having the patient or social worker contact the NEMT provider to arrange transport. Other options include dial-a-ride (a transportation service for seniors, people with disabilities, and others who can't use the standard fixed route transit systems) or a public transportation voucher, or a cab/taxi voucher. Hospitals may assist patients by providing public transportation and/or cab/taxi vouchers when no other transportation options exist, but hospital policies are widely variable. Obtaining such vouchers may require ED providers to discuss the case with a hospital or nursing administrator.

In thinking about community connections, it is important to note that federally qualified health centers (FQHCs) are designed as "one-stop-shops" and provide patient services focused on health system navigation and social support. Referring patients to these sites of care can be very effective in improving their ability to connect with needed resources, which can include transportation benefits and affordable transportation options.

For rural patients who have been transported to tertiary care centers by way of air ambulance without family, arranging for transportation home is critical to appropriate disposition. There are several perspectives to consider. If acuity allows, the sending facility can discuss options for ultimately returning home with patients and their families prior to the transfer. For the receiving facility, it is important to determine quickly if the patient will be admitted or potentially discharged from the ED, as the latter requires mobilization of care coordination resources to determine a safe and timely method of transportation home.

Often, ED treatment and disposition plans rely on access to outpatient testing. However, such tests generally require timely follow-up and scheduling. It is important to consider transportation factors (reliability, distance to travel, rural geography) in making diagnostic testing decisions. For example, one might have a lower threshold to place a moderate-risk chest pain patient in observation status for a stress test in the morning – rather than discharge with 72-h follow-up for stress testing – if transportation barriers are identified. Other examples include obtaining an MRI for new-onset seizures or a CT scan for a new diagnosis of certain cancers. While these tests can certainly occur in the outpatient setting, transportation barriers might result in unacceptable delays to needed imaging and confirmatory diagnosis, potentially delaying specialist care and follow-up. Of course, additional ED testing must be balanced against cost to the patient. When possible, patients can be engaged in shared decision-making around costs vs. benefits of various diagnostic and treatment options, particularly when insurance coverage or other financial considerations may be an issue.

Hospital/Healthcare System

Health systems are traditionally arranged to be provider-centric, rather than patient-centric. This includes appointment scheduling at times convenient to providers, generally without consideration of how transportation access defines a patient's ability to access healthcare. More recently, health systems have focused on reducing admissions and ED visits for primary care-sensitive conditions [41], with increased understanding that addressing transportation barriers is a key component in these efforts [1]. Hospitals are increasingly investing in transportation solutions to improve outpatient follow-up and routine care. However, while hospitals may consider routine screening for patient transportation barriers as an important social determinant of health, this is not yet a common practice.

One example of a hospital-driven transportation effort is a partnership between Denver Health, an urban safety net public hospital, and Lyft, a for-profit ride-share company, to develop a platform to allow the hospital to order rides for patients in need of transportation [42]. The service is offered to recently discharged patients and those who need transport for outpatient appointments. Initially piloted in the ED, it has expanded to hospital inpatients and now several outpatient clinics as well. Denver Health is linking this program to a larger initiative involving patient navigators and social workers, who help raise awareness of the service and provide ride coordination. The program has reduced the number of patient complaints regarding inadequate transportation. Additionally, the hospital can track the number of rides by location and thereby identify where community needs are the greatest. They have found that the program is of particular benefit to patients for whom English is a second language, who may find navigating public transportation to be a challenge.

One example of a health system-community partnership to overcome barriers to transportation in a rural community is the Grace Cottage Family Health & Hospital's volunteer driver program [43]. The rural Vermont hospital identified transportation as a common barrier to care access and started the volunteer driver program in 2016. Grace Cottage partnered with the Green Mountain RSVP, a nationwide program of volunteers over the age of 55. To improve access to medical appointments, the volunteer drivers from the RSVP program are stationed at the hospital or are at home "on-call" and use their personal cars to assist those struggling to attend primary care medical appointments or to return home after an inpatient admission. Similar programs could be replicated or expanded to include post-ED follow-up care.

Societal Level

Transportation is intertwined with environment, social isolation, and state and federal policy. Geography impacts transportation with respect to travel distances in rural settings and safety in some urban settings. Social structures are inherently linked with transportation, with private vehicle ownership being related directly to income, and access to a family member's car linked to familial support. Commonly used public transportation services are becoming increasingly expensive (e.g., NYC's subway and D.C.'s Metro), with others falling into disrepair due to lack of needed state funding [44, 45]. Rural areas have little public transportation and areas with poor infrastructure investment may experience unsafe and inadequate road maintenance [43].

One example of a community-level intervention is the ETHAN (Emergency Telehealth and Navigation) Project, a community-based paramedicine approach that uses telehealth assessment to determine the best location of healthcare services [46]. ETHAN is a community-wide collaboration led by the Houston Fire Department that uses mobile technology, community-based paramedicine (paramedics operating in expanded roles by assisting with public health, primary, and preventive services), and local partnerships to triage and connect low-acuity 9-1-1 callers with primary care. Once the crew arrives at the call location, they determine if the patient needs ED transport; if not, the ETHAN program is activated with a tablet connecting the patient with an emergency physician via HIPAA-compliant video conferencing software. After a telehealth assessment, the physician provides a recommendation for an alternate location of care if urgent or primary care is most appropriate – these include referral to community primary care or urgent care via taxi, referral to ED via taxi (for patients who refuse primary or urgent care), or referral to the patient's PCP or home care. The patient's information is forwarded to a Houston Health Department program that provides care navigation services such as assistance with insurance coverage, health literacy, or material needs like food and shelter. This collaborative model has been well received and has demonstrated reductions in ambulance transports to the ED [47]. Community paramedicine encompasses a broad range of programs in which paramedics provide elements of primary and preventive care [48]. These programs are varied, and more research is still needed as to their safety and efficacy. Community paramedicine is rapidly evolving [49–51], and continued research in this field will yield insight into ideal deployment of community paramedicine to improve healthcare delivery and outcomes, as well as safety and efficacy of such programs.

Recommendations for Emergency Medicine Practice

Basic

- Physicians and advanced practice providers can query patients on their transportation options (public vs. private; reliance on others for transportation) as a standard piece of the social history for the purpose of facilitating a safe discharge.
  - Identify how transportation may be a barrier or facilitator to disposition following today's ED visit. Examples of how to ask about transportation include: "How did you get here today?" and "How will you get home today?"
- Have dedicated social work staff who can support ED providers in using available local transportation resources to assist patients with safe discharge.
  - Options include public transportation, taxi vouchers, or ride-share services.

Intermediate

- Moving beyond assuring a safe disposition, EDs can identify and address patients' transportation access, which may improve patients' ability to access primary and preventive care. This may require additional investment as it falls beyond the traditional scope of ED care provision. Such efforts could take the form of:
  - Routine screening for patients' transportation needs at the time of triage. EDs should consider documenting screening results in the patient chart.
  - Investing in dedicated ED social work resources to assist in identifying resources for transportation.
  - Investing in ED care navigators to screen for transportation barriers to primary care and link patients to transportation resources.
- Assure ED providers are educated on their state's nonemergency medicine transportation (NEMT) benefits, including what benefits might cover transportation home from the ED.

Advanced

- Collaborate with state and local public health and/or transit departments to identify methods to expand transportation options for vulnerable populations. Such methods may include free/reduced-cost public transportation assistance, which some cities make available to low-income communities.

- Collaborate with EMS to research and elucidate the causes behind differential response times in low-income or minority neighborhoods in order to intervene.
- As appropriate for the setting, collaborate with EMS, local departments of health, and others to consider establishing a community paramedicine program or telehealth program like ETHAN.
- Conduct research on the role of transportation in driving disposition decisions from the ED, including the cost of "social" admissions related to transportation barriers, to provide data to support larger societal investments in improved transportation access and systems of care.

Teaching Case

Clinical Case

An elderly Native American female presents to a tertiary ED after being diagnosed with possible EKG changes at a small, rural ED in New Mexico. She has had two troponin levels drawn that are undetectable. She has a HEART score of 4 due to risk factors, age, and EKG changes. She is evaluated by the cardiologist in the ED who recommends either an observation stay for stress testing or a stress test within the next week as an outpatient. She had been flown to the tertiary ED by an air ambulance and has no family present currently. She declines observation overnight and wishes to return home.

The emergency physician requests that social work and case management contact the patient's family; they confirm that the family is driving to meet their grandmother at the tertiary care center. When the family arrives, they report their grandmother prefers to live independently. She has a small herd of sheep which she cares for and infrequently travels away from home due to her responsibilities. She does not drive, but family is often available to help if she needs transportation.

The physician at the tertiary ED calls the rural ED from where she was transferred to get an understanding of outpatient stress test capabilities at the rural hospital and in that community. It is found that stress testing is available once a week at the hospital. The family and patient are eager to return home. The tertiary ED physician makes a plan to discharge the patient with transportation with family and follow-up at their local clinic the following day to schedule the needed stress test. The patient and family are pleased with this plan and the patient is discharged into the care of the family.

Teaching Points

1. Transportation is critical to ED disposition. For patients who have arrived via EMS, it is important for the emergency provider to determine early on in the ED visit if transport home will be needed. Options may include contacting family or identifying benefits available to the patient, such as NEMT.
2. Transportation to follow-up testing is equally important. While many tests can be conducted in the outpatient setting, it is critical that the patient has reliable

transportation to these tests and other follow-up appointments. Some patients experience significant social isolation and have little family support in regard to transportation and assistance in navigating the health system.

3. Rural transfers are a unique situation, as many result in discharge from the tertiary ED. At times, it may be most prudent to place rural patients in observation while a safe method of transportation home is identified. Additionally, it is critical that the follow-up plan takes local resources into account. This may be a unique role for community paramedicine, which can conduct home visits to assure the patient is recovering as expected.

Discussion Questions

1. What are the local transportation challenges where you work? Is there good access to public transportation options? If not, what other transportation resources exist?
2. What types of transportation support does your ED provide? Cab or ride-sharing vouchers? Public transport vouchers?
3. Are there any local outreach efforts in your community to help vulnerable patients get to routine appointments? How can your ED collaborate with these efforts?
4. How does transportation influence your disposition plan? Do you ask about transportation early in the ED visit? If so, what resources can you deploy to help the patient obtain the transportation they need both immediately and in the future for follow-up?
5. When working in a rural setting, what transportation issues should be considered prior to transfer of a patient? How do you discuss transportation issues with patients and their family prior to transfer?

References

1. Syed ST, Gerber BS, Sharp LK. Traveling towards disease: transportation barriers to health care access. J Community Health. 2013;38:976–93.
2. Ruggiano N, Shtompel N, Whiteman K, Sias K. Influences of transportation on health decision-making and self-management behaviors among older adults with chronic conditions. Behav Med. 2017;43:61–70.
3. Thomas LV, Wedel KR. Nonemergency medical transportation and health care visits among chronically ill urban and rural medicaid beneficiaries. Soc Work Public Health. 2014;29:629–39.
4. Chetty R, Hendren N. The impacts of neighborhoods on intergenerational mobility I: childhood exposure effects. Q J Econ. 2018;133:1107–62.
5. Mell HK, Mumma SN, Hiestand B, Carr BG, Holland T, Stopyra J. Emergency medical services response times in rural, suburban, and urban areas. JAMA Surg. 2017;152:983–4.
6. Patterson D, Skillman S, Fordyce M. Prehospital emergency medical services personnel in rural areas: results from a survey in nine states. Final report #149. Seattle: WWAMI Rural Health Research Center, University of Washington; 2015.
7. Wallace R, Hughes-Cromwick P, Mull H, Khasnabis S. Access to health care and nonemergency medical transportation: two missing links. Transp Res Rec. 1924;2005:76–84.
8. Silver D, Blustein J, Weitzman BC. Transportation to clinic: findings from a pilot clinic-based survey of low-income suburbanites. J Immigr Minor Health. 2012;14:350–5.

9. Fitzpatrick AL, Powe NR, Cooper LS, Ives DG, Robbins JA. Barriers to health care access among the elderly and who perceives them. Am J Public Health. 2004;94:1788–94.
10. Grant R, Gracy D, Goldsmith G, Sobelson M, Johnson D. Transportation barriers to child health care access remain after health reform. JAMA Pediatr. 2014;168:385–6.
11. Vaughan Sarrazin MS, Campbell ME, Richardson KK, Rosenthal GE. Racial segregation and disparities in health care delivery: conceptual model and empirical assessment. Health Serv Res. 2009;44:1424–44.
12. Caldwell JT, Ford CL, Wallace SP, Wang MC, Takahashi LM. Racial and ethnic residential segregation and access to health care in rural areas. Health Place. 2017;43:104–12.
13. Starbird LE, DiMaina C, Sun C-A, Han H-R. A systematic review of interventions to minimize transportation barriers among people with chronic diseases. J Community Health. 2019;44:400–11.
14. Cheung PT, Wiler JL, Ginde AA. Changes in barriers to primary care and emergency department utilization. Arch Intern Med. 2011;171:1393–400.
15. Kangovi S, Barg FK, Carter T, Long JA, Shannon R, Grande D. Understanding why patients of low socioeconomic status prefer hospitals over ambulatory care. Health Aff. 2013;32:1196–203.
16. Govindarajan P, Friedman BT, Delgadillo JQ, Ghilarducci D, Cook LJ, Grimes B, et al. Race and sex disparities in prehospital recognition of acute stroke. Acad Emerg Med. 2015;22:264–72.
17. Hsia RY, Huang D, Mann NC, Colwell C, Mercer MP, Dai M, et al. A US national study of the association between income and ambulance response time in cardiac arrest. JAMA Netw Open. 2018;1:e185202.
18. Capp R, Kelley L, Ellis P, Carmona J, Lofton A, Cobbs-Lomax D, et al. Reasons for frequent emergency department use by Medicaid enrollees: a qualitative study. Acad Emerg Med. 2016;23:476–81.
19. Chaiyachati KH, Moore K, Adelberg M. Too early to cut transportation benefits from Medicaid enrollees. Health Serv Insights [Internet]. 2018 [cited 2019 Jun 4];12. Available from: https://www.ncbi.nlm.nih.gov/pmc/articles/PMC6194915/.
20. United States Government Accountability Office. Nonemergency medical transportation: updated Medicaid guidance could help states. 2016 [cited 2019 Sep 3]; Available from: https://www.gao.gov/products/GAO-16-238.
21. Shifting non-emergency medical transportation to Lyft improves patient experience and lowers costs | health affairs [Internet]. [cited 2019 Jul 24]. Available from: https://www.healthaffairs.org/do/10.1377/hblog20180907.685440/full/.
22. Budget proposal would allow states to drop Medicaid transportation benefits across the entire program | Health Affairs [Internet]. [cited 2019 Jun 6]. Available from: https://www.healthaffairs.org/do/10.1377/hblog20180608.971229/full/.
23. Transportation Research Board, National Academies of Sciences, Engineering, and Medicine. Cost-benefit analysis of providing non-emergency medical transportation [Internet]. Washington, D.C.: Transportation Research Board; 2005 [cited 2019 Jun 4]. Available from: http://www.nap.edu/catalog/22055.
24. Emergency Triage, Treat, and Transport (ET3) Model. Center for Medicare & Medicaid Innovation. [Internet]. [cited 2019 Jun 5]. Available from: https://innovation.cms.gov/initiatives/et3/.
25. Chaiyachati KH, Hubbard RA, Yeager A, Mugo B, Lopez S, Asch E, et al. Association of rideshare-based transportation services and missed primary care appointments: a clinical trial. JAMA Intern Med. 2018;178:383–9.
26. Baren JM, Boudreaux ED, Brenner BE, Cydulka RK, Rowe BH, Clark S, et al. Randomized controlled trial of emergency department interventions to improve primary care follow-up for patients with acute asthma. Chest. 2006;129:257–65.
27. Schwab RA, Radakovich R, Tate K. Does payment improve follow-up of pelvic inflammatory disease? Mo Med. 2002;99:286–8.
28. Capp R, Misky GJ, Lindrooth RC, Honigman B, Logan H, Hardy R, et al. Coordination program reduced acute care use and increased primary care visits among frequent emergency care users. Health Aff. 2017;36:1705–11.

29. Moffett ML, Kaufman A, Bazemore A. Community health workers bring cost savings to patient-centered medical homes. J Community Health. 2018;43:1–3.
30. Rogers EA, Manser ST, Cleary J, Joseph AM, Harwood EM, Call KT. Integrating community health workers into medical homes. Ann Fam Med. 2018;16:14–20.
31. Adashi EY, Geiger HJ, Fine MD. Health care reform and primary care—the growing importance of the community health center. N Engl J Med. 2010;362:2047–50.
32. Diffusion of community health workers within Medicaid managed care: a strategy to address social determinants of health | Health Affairs [Internet]. [cited 2019 Jul 26]. Available from: https://www.healthaffairs.org/do/10.1377/hblog20170725.061194/full/.
33. Wilde ET. Do emergency medical system response times matter for health outcomes? Health Econ. 2013;22:790–806.
34. Mercuri M, Velianou JL, Welsford M, Gauthier L, Natarajan MK. Improving the timeliness of care for patients with acute ST-elevation myocardial infarction: implications of "self-transport" versus use of EMS. Healthc Q. 2010;13:105–9.
35. Newgard CD, Fu R, Bulger E, Hedges JR, Mann NC, Wright DA, et al. Evaluation of rural vs urban trauma patients served by 9-1-1 emergency medical services. JAMA Surg. 2017;152:11–8.
36. King N, Pigman M, Huling S, Arrt BS, Hanson B. EMS services in rural America: challenges and opportunities. 14.
37. Moser DK, Kimble LP, Alberts MJ, Angelo A, Croft Janet B, Kathleen D, et al. Reducing delay in seeking treatment by patients with acute coronary syndrome and stroke. Circulation. 2006;114:168–82.
38. Durham R, Pracht E, Orban B, Lottenburg L, Tepas J, Flint L. Evaluation of a mature trauma system. Ann Surg. 2006;243:775–85.
39. Jollis JG, Mehta RH, Roettig ML, Berger PB, Babb JD, Granger CB. Reperfusion of acute myocardial infarction in North Carolina emergency departments (RACE): study design. Am Heart J. 2006;152:851.e1–11.
40. AIDET Patient Communication [Internet]. [cited 2019 Jul 15]. Available from: https://www.studergroup.com/aidet.
41. Green LA, Chang H, Markovitz AR, Paustian ML. The reduction in ED and hospital admissions in medical home practices is specific to primary care–sensitive chronic conditions. Health Serv Res. 2018;53:1163–79.
42. American Hospital Association. Case study: Denver Health Medical Center collaborates with Lyft to improve transportation for patients. 2018. [cited 2019 Jun 6]. Available from: https://www.aha.org/news/insights-and-analysis/2018-03-01-case-study-denver-health-medical-center-collaborates-lyft.
43. American Hospital Association. Social determinants of health series: transportation. 2017. [Internet]. [cited 2019 Jun 6]. Available from: http://www.hpoe.org/resources/ahahret-guides/3078.
44. Fitzsimmons EG. Subway fares are rising again. But that won't solve the MTA's crisis. The New York Times [Internet]. 2019 Feb 27 [cited 2019 Jul 28]; Available from: https://www.nytimes.com/2019/02/27/nyregion/mta-fares-hike.html.
45. The Wall Street Journal. Washington, D.C.'s five-letter problem: metro. [Internet]. [cited 2019 Jul 28]. Available from: https://www.wsj.com/articles/washington-d-c-s-five-letter-problem-metro-1519304400.
46. Houston EMS advances mobile integrated healthcare through the ETHAN program. 2018. [Internet]. [cited 2019 Jun 6]. Available from: https://www.jems.com/articles/print/volume-40/issue-11/features/houston-ems-advances-mobile-integrated-healthcare-through-the-ethan-program.html.
47. Langabeer JR, Gonzalez M, Alqusairi D, Champagne-Langabeer T, Jackson A, Mikhail J, et al. Telehealth-enabled emergency medical services program reduces ambulance transport to urban emergency departments. West J Emerg Med. 2016;17:713–20.
48. Glenn M, Zoph O, Weidenaar K, Barraza L, Greco W, Jenkins K, et al. State regulation of community paramedicine programs: a national analysis. Prehosp Emerg Care. 2018;22:244–51.

49. Agarwal G, Pirrie M, McLeod B, Angeles R, Tavares W, Marzanek F, et al. Rationale and methods of an evaluation of the effectiveness of the community paramedicine at home (CP@home) program for frequent users of emergency medical services in multiple Ontario regions: a study protocol for a randomized controlled trial. Trials. 2019;20:75.
50. Snooks HA, Anthony R, Chatters R, Dale J, Fothergill R, Gaze S, et al. Support and Assessment for Fall Emergency Referrals (SAFER) 2: a cluster randomised trial and systematic review of clinical effectiveness and cost-effectiveness of new protocols for emergency ambulance paramedics to assess older people following a fall with referral to community-based care when appropriate. Health Technol Assess. 2017;21:1–218.
51. Community Paramedicine in the Pueblo of Laguna, New Mexico. 2018. [Internet]. [cited 2019 Jul 28]. Available from: https://www.jems.com/articles/print/volume-43/issue-1/features/community-paramedicine-in-the-pueblo-of-laguna-new-mexico.html.

Legal Needs and Medical Legal Partnerships

17

Dennis Hsieh

Key Points

- Health-harming legal needs negatively affect the health of patients and arise from a violation of a patient's legal rights. These legal needs can be identified and addressed in healthcare settings including emergency departments (ED).
- Emergency department providers can refer patients with health-harming legal needs to lawyers via several mechanisms: (1) the local bar association attorney referral service; (2) local Legal Services Corporation-funded legal service organizations; and/or (3) other nonprofit law firms that provide services at low or no cost.
- Medical legal partnerships represent an established mechanism to integrate identification and response to health-harming legal needs into emergency department care.

Foundations

Background

Fundamentally, legal needs are a key subset of the 50% of nonclinical health factors (socioeconomic factors and physical environment) that affect health outcomes [1]. Laws, regulations, and policies create legal rights. When an individual's legal right is violated, resulting in conditions that harm their health, that individual is said to have a health-harming legal need (see Fig. 17.1). Examples include poor housing conditions, unexplained food stamp termination, and challenges with immigration. Lawyers bring a different approach to health-harming legal needs compared to health navigators, case managers, or social workers. For example, if an individual is

D. Hsieh (✉)
Contra Costa Health Plan, Harbor UCLA Medical Center, Martinez, CA, USA

H. J. Alter et al. (eds.), *Social Emergency Medicine*,
https://doi.org/10.1007/978-3-030-65672-0_17

I-HELP* Issue	Civil Legal Aid Interventions	Related Social Determinant of Health	Impact on Health / Health Care Talking Points
Income	**Benefits Units:** Appeal denials of food stamps, health insurance, cash benefits, and disability benefits	Increase the availability of resources to meet daily basic needs	1. Increasing someone's income means s/he makes fewer trade-offs between affording food and health care, including medications.[3] 2. Being able to afford enough healthy food helps people manage chronic diseases and helps children grow and develop.[4,5]
Housing & utilities	**Housing Units:** Secure housing subsidies; Improve substandard conditions; Prevent eviction; Protect against utility shut-off	Ensure healthy physical environments	1. A stable, decent, affordable home helps invididuals avoid costly emergency room visits related to homelessness.[6,7] 2. Consistent housing, heat and electricity helps people follow their medical treatment plans.[8]
Education & Employment	**Education & Employment Units:** Secure specialized education services; Prevent and remedy employment discrimination and enforce workplace rights	Create equal access to the opportunity to learn and work	1. A quality education is the single greatest predictor of a person's adult health.[9] 2. Consistent employment helps provide money for food and safe housing, which also helps avoid costly emergency health care services.[10] 3. Health insurance is often linked to employment.[11]
Legal Status	**Veterans & Immigration Units:** Resolve veteran discharge status; Clear criminal / credit histories; Assist with asylum applications	Create equal access to the opportunity to work	1. Clear a person's criminal history or helping a veteran's change their discharge status helps make consistent employment and access to public benefits possible.[12] 2. Consistent employment provides money for food and safe housing, which helps people avoid costly emergency health care services.[13]
Personal Stability	**Family Law Unit:** Secure restraining orders for domestic violence; Secure adoption, custody and guardianship for children	Reduce exposure to violence and create a safe and healthy community	1. Less violence at home means less need for costly emergency health care services.[14] 2. Stable family relationships significantly reduce stress and allow for better decision-making, including descions related to health care.[15]

Fig. 17.1 Framing legal care as healthcare: a guide to help legal civil aid practitioners message their work to healthcare audiences [2]

being evicted, a health navigator or case manager could help them look for alternative housing options. A social worker could provide a shelter list or help the individual think through alternative housing options. A legal team could examine the basis for the eviction and fight the eviction or negotiate a settlement that allows the individual to transition to a different housing option without becoming homeless.

Health-harming legal needs can arise from any of the different social determinant domains. Table 17.1 provides examples of common health-harming legal needs that healthcare providers may encounter. The list is not exhaustive, as laws, regulations, and policies vary by municipality and by state. However, when a patient has a conflict with another legal entity (such as a person, an organization, a business/corporation, or a unit of government) that affects their health, a health-harming legal need is likely present. Even within this framework, identifying needs as legal needs as opposed to general social needs can be challenging. Implementing specific screening tools and providing healthcare teams training can help increase identification and awareness of patients' health-harming legal needs [3].

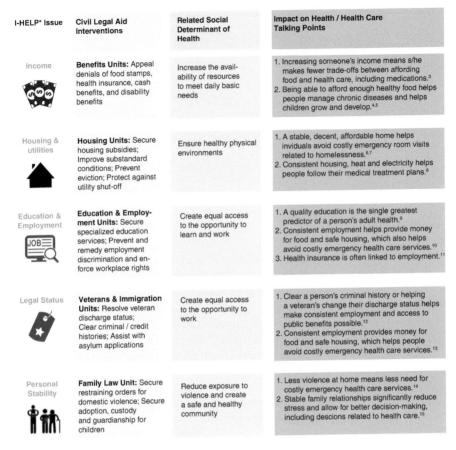

Table 17.1 Examples of health-harming legal needs by social determinants domain

Social determinants domain	Health-harming legal need
Housing	Eviction Poor housing conditions (no heat, mold, roaches/rodents/pests) Loss of Section 8 voucher Problems with public housing Discrimination Reasonable accommodation Foreclosure
Food	Loss of SNAP benefits (food stamps) Incorrect amount of SNAP benefits Denial for SNAP benefits
Healthcare	Loss of benefit (Medicaid, CHIP, Medicare, VA, others) Denial for benefit Coverage of medications or durable medical equipment Help with medical bills/debt/charity care
Employment	Improper termination Unpaid wages Discrimination FMLA/reasonable accommodation
Income support	Loss of benefit (TANF, SSI, unemployment benefit, state disability benefit, social security retirement, VA benefit) Incorrect amount of benefit Termination of benefit
Family	Divorce Custody Child support Guardianship Conservatorship
Education	Truancy Suspension Expulsion Individualized Educational Plan (IEP)/Accommodation
Immigration	Naturalization Deportation Adjustment of status: T-Visa, U-Visa, VAWA, SJIS, Refugee/Asylee
Transportation	Problems with transportation benefits
Law enforcement	Unpaid tickets Outstanding warrants Criminal prosecution Expungement
Consumer protection	Consumer debt Educational debt Bankruptcy Fraud/loss Identity theft

Addressing these issues requires referral to a lawyer or legal team who can evaluate the case and provide assistance as appropriate. In general, there are four types of legal assistance available for patients: (1) referral through the local bar association legal referral services for paid, low-cost, or free services; (2) federally funded Legal Services Corporation (LSC) legal aid organizations dedicated full-time to providing no cost services for low-income individuals; (3) nonprofit legal organizations or law school clinics without LSC funding that provide services at low or no cost for low-income individuals; and (4) private attorneys who have full-time paid practices who partner with LSC legal aid or nonprofit legal organizations to provide pro bono (free) services for specific cases.

In addition to referral out to attorneys, providers, hospitals, and healthcare systems can develop a more formal relationship with attorneys. This model is known as the medical legal partnership (MLP), which embeds a legal team into the healthcare setting and makes them a part of the healthcare team to address patients' health-harming legal needs. MLPs started at Boston Medical Center (BMC) in 1993, when Dr. Barry Zuckerman, then Chair of Pediatrics, realized that patients had problems that social workers were not trained to solve (e.g., getting denied for food stamps, fighting illegal evictions) [4]; instead, they needed a lawyer to address the social factors affecting their health and stability [5]. Thus, just as a physician consults a cardiologist for an acute myocardial infarction, the same can be done through an MLP for a pending eviction. Through legal advocacy, the MLP attempts to meet patients' needs with the goal of improving health. Advocacy can range from simple advice and counsel to written advocacy to full representation in a lawsuit. When truly integrated, the legal team functions as a specialist service working in tandem with social workers and case managers, allowing each team member to function at the top of their license with fluid interdisciplinary referrals as they identify different patient needs.

MLPs have various levels of integration with the healthcare system: full integration where the healthcare system hires a lawyer or legal team and has them on-site and on-call; a relationship with a legal partner that offers on-site services; and a looser agreement with a legal partner through a memorandum of understanding where lawyers are available to take referrals but are not physically on-site. In most cases, the attorneys are not employed by the healthcare system but instead are employed by nonprofit organizations and legal aid organizations or are law professors and students working in clinical programs. Although private-practice lawyers may partner with nonprofit and legal aid organizations to occasionally take on pro bono MLP cases, they do not provide a sustainable model for legal representation for patients of a clinic, hospital, or health system.

Since the first MLP at BMC, the concept has spread widely, and MLPs are recognized by the federal government as a mechanism to improve patient care [6]. There are over 300 MLPs across the country, serving inpatient, outpatient, and emergency departments (EDs) [7]. The number continues to grow. This is not surprising given that the LSC estimates over 70% of low-income families have at least one unmet legal need during the year [8]. Thus, as healthcare shifts to a population health framework with a focus on improving outcomes while lowering costs, health systems must address their patients' social needs. MLPs are a key part of the solution in addressing these needs when they qualify as health-harming legal needs.

Evidence Basis

Addressing health-harming legal needs has been shown to be effective in decreasing acute care utilization, increasing access and adherence to primary and preventive care, and improving physical and mental health outcomes. Three systematic reviews provide a comprehensive overview of the evidence behind the effect of addressing health-harming legal needs. One examined studies that show the effect MLPs have on health disparities [9]. A second looked at different social needs and the effect of MLPs on these various areas, including income and insurance, housing and neighborhood, education, employment, legal status, and personal safety [10]. A third summarized the level of evidence for MLPs addressing different types of health-harming legal needs [11].

Some key studies from those included in the abovementioned systematic reviews are discussed here. A randomized, controlled trial of 330 families at Boston Medical Center who were randomized to an MLP versus no MLP showed that MLPs have a significant impact on improving access to public benefits including income, utility, housing, and food assistance [12]. Furthermore, this study showed that families randomized to MLPs had improved access to healthcare as they were more likely to have had their 6-month immunizations, had 5 or more routine preventive visits by age 1, and were less likely to have visited the emergency department by age 6 months. Another prospective cohort study showed that with MLP intervention, there was a significant reduction in the proportion of children with recent hospitalizations [13]. Similarly, a retrospective chart review of patients with poorly controlled asthma showed a decrease in emergency department visits, hospital admissions, overall need for systemic steroids, medication dosage, and asthma severity after an MLP intervention to improve housing conditions [14].

In addition to improved access to resources that address social needs, improved adherence to preventive care, and decreased acute care utilization, MLPs have also been shown to improve patients' physical and mental health. Engagement with an MLP resulted in a significant decline in perceived stress and well-being [15, 16].

For health systems, compelling data also exists in terms of return on investment (ROI) for MLPs. One study showed a 419% ROI and 319% ROI annually, over the course of 2 years, in terms of Medicaid reimbursement for the healthcare partner from 2007–2009, not taking into account the benefits for the patient [17]. This ROI was based on establishing Medicaid eligibility for individuals who were initially denied Medicaid, thus improving reimbursement to the hospital. Similarly, other studies have shown a ROI of three dollars for each dollar invested in the MLP [18]. These figures make a compelling case for healthcare institutions to invest in MLPs in addition to their demonstrated benefits to patients.

Emergency Department and Beyond

Bedside

Traditionally, EDs do not systematically screen for social and legal needs. Instead, care team members might haphazardly identify these needs during a visit. Two specific questions regarding patients' social history can begin to help identify both basic social and legal needs: (1) Do you have somewhere safe to stay? (2) Do you have enough money to pay for your food and medications? [19]. To identify health-harming legal needs systematically, EDs can implement a uniform social needs screen that has additional questions about these legal needs, such as using the I-HELP [20], PREPARE [21], or the LA County Social and Behavioral Determinants of Health Screening Tool [22]. EDs have a number of options as to when and how to implement systematic screening. Formats for these screens can vary from patient self-administered on paper or via tablet/computer/kiosk to staff-administered on paper or via tablet/computer [23]. Screening can occur in the waiting room, during the medical screening exam (MSE), after the MSE while patients are waiting, or inpatient rooms. Given staffing constraints in many EDs, one feasible option may be a patient self-administered screen via paper or tablet/computer/kiosk in the waiting room or patient room. Providers can then review the results of these screens.

When a social or health-harming legal need is identified by a care team member, the ED should have a protocol for addressing these needs. Often, this entails a referral to a social worker who will evaluate the need and if there is a legal component, refer the patient to a lawyer. If there is no social worker available, the care team should make a direct referral to a lawyer when they believe there is a health-harming legal need. Referring to a lawyer can entail one of the four mechanisms discussed above (via the local bar attorney referral service, to an LSC-funded organization, to another nonprofit or law school-based clinic, or to an MLP). Regardless of the established local referral path, the key for emergency providers is to understand that health-harming legal needs exist and can be addressed by lawyers. This way, instead of simply expressing empathy when the patient brings up a challenging issue, the provider can refer the patient to a lawyer and help them to take action.

In terms of triaging cases, timeliness is a major factor. There are emergent cases that should see an attorney the same day or the next business day and those that are urgent and should see an attorney within a week or a few weeks. Emergent cases are often housing cases, such as when a patient is being threatened with eviction or termination of a federal Section 8 housing voucher because these cases move on very strict timelines and may result in homelessness. These cases are the legal equivalent of an ST-elevation myocardial infarction and must be addressed immediately. Other emergent cases include any case with a deadline in the next few days, as well as any case where physical safety is involved. Other cases may be judged to be urgent, but not emergent.

Examples of linkages to legal services providers include the emergency departments at Harbor-UCLA Medical Center, Los Angeles County+USC Medical Center, and Alameda Health System Highland Hospital where formal relationships with local legal services organizations exist and providers can refer patients via phone, fax, or email.

Hospital/Healthcare System

The main challenge for addressing legal needs within hospitals and healthcare systems are creating hospital/systems-level linkages to attorneys that allow for direct referral by providers. Three main barriers exist. First, hospitals and health systems have traditionally been suspicious of attorneys due to issues surrounding malpractice actions and HIPAA. Education about health-harming legal needs and about the types of cases that LSC and other nonprofit organizations provide assistance with (not malpractice) can help assuage these concerns. Obtaining the patient's permission to share information with a legal partner for the purposes of treatment addresses the HIPAA concern.

The second barrier is support for organizational and system-wide capacity to address legal needs. Although traditionally, legal aid and nonprofits have received federal, state, local, and/or charitable funds to provide free legal services, there is not sufficient legal capacity to meet all of the need: In 2017, more than 60 million Americans were low-income and eligible for legal assistance from legal aid [8]. The Legal Services Corporation estimates that 71% of low-income households (below 125% of the Federal Poverty Level) have a legal need each year and one quarter of this population has six or more legal needs in a year [24]. Nationally, there is one legal aid attorney (an attorney who provides free or low-cost legal services) per 6415 low-income individuals, while attorneys who charge market rates are much more available with one for every 429 people in the general population [25]. Given this dearth of legal aid attorneys, 86% of civil legal problems faced by low-income Americans receive inadequate or no legal help [24]. Given this "justice gap," much more capacity is needed for poor individuals. Relying on existing LSC-funded legal aid corporations, local nonprofits, and pro bono referrals from the local bar association will simply perpetuate this unmet need. To increase capacity and ensure that patients' needs are assessed and addressed, the hospital or health system will need to pay for attorney time by providing payment to legal aid organizations, local nonprofits, or private bar attorneys, creating a MLP, and/or hiring attorneys to address patients' health-harming legal needs.

The third barrier is information sharing. Both healthcare providers and lawyers have information sharing concerns: The healthcare partner is wary of HIPAA regulations, while the legal partner is thinking about the attorney-client relationship. These concerns can be resolved by asking the patient to sign a bilateral information release for the purposes of facilitating the patient's care and case. Once this is in place, the healthcare and legal teams can share information necessary to address the health-harming legal need. The legal team can be educated to understand that the medical team does not need nor want details of the case that could compromise attorney-client privilege. Instead, the medical team simply wants to be in communication to close the loop and know the outcome and recommendations from the legal referral that will impact the health and social services the healthcare team provides. The legal team can share this information in a way that preserves attorney-client privilege. The health system does need to be clear with the legal team that members of the medical team are mandated reporters and must report issues such as domestic violence, child abuse, and elder abuse if they learn about such issues from the attorneys, whereas attorneys are not mandated reporters even though they otherwise act as part of the medical team. From the medical perspective, the legal team is part of the care team and all information shared with the legal team is for the care of the patient and thus HIPAA should not be a barrier.

There are many examples around the country of successful, sustained healthcare and legal organization partnerships with hospitals and health systems to address patients' health-harming legal needs, as well as written guidance for health systems interested in developing MLP programs [26, 27]. The New York Legal Assistance Group (NYLAG) is an example that spans numerous hospitals in the New York City (www.nylag.org). Other examples can be found at the National Center for Medical-Legal Partnership's website (https://medical-legalpartnership.org/).

Societal Level

Although all people in the US are protected by the Constitution, many rights vary based on state and local laws and policies. Thus, the level and type of legal advocacy available to patients may be limited based on their jurisdiction. For example, different states have different eligibility requirements for Medicaid. Yet even if a remedy varies based on local policies and laws, patients across jurisdictions will face similar types of health-harming legal needs. Similarly, all states, whether rural or urban, have local bar association referral services, legal aid organizations, and nonprofit law firms that assist low-income individuals with legal issues, and 46 states have MLPs. However, distribution, ease of access, and availability are major considerations that differ for urban versus rural areas. Finally, depending on the gross domestic product (GDP) and wealth of the state, the available resources and number of individuals needing services will vary.

Community-wide solutions that are integrated with the healthcare delivery system to address health-harming legal needs are not common outside of the existing nonprofit structure. In Los Angeles, the Whole Person Care (WPC) medical legal partnership is a rare example of an MLP that is integrated into a large healthcare delivery system spanning over 4700 square miles [28]. This MLP came out of the need to serve high-risk individuals across Los Angeles County who are enrolled in the WPC program and is made possible by funding through California's Medicaid §1115 Waiver Program. This model integrates lawyers with community health workers, medical case workers, and social workers who are case managing high-risk individuals across *multiple* healthcare facilities in Los Angeles. It leverages technology and allows referrals to be made via an online portal supplemented by on-site case reviews by the legal services providers with the care team.

Currently, no model exists to allow health systems or legal providers to bill insurance for provision of legal services, and this can render such services difficult to sustain. Some programs are run through grants, while others come out of health system general funds. Making legal services that address health-harming legal needs reimbursable, similar to how other consultative services (e.g., cardiology, orthopedics) are paid for by insurance, would allow for widespread dissemination and implementation of ED-linked or ED-based legal services. This dissemination would be amplified by education of emergency medicine residents on the importance of addressing health-harming legal needs.

Recommendations for Emergency Medicine Practice

Basic

- Learn about at least one type of health-harming legal need and how to discuss this need with patients.
- Identify one location to refer patients with such needs in your community.
- Invite legal service provider(s) to educate ED staff about legal needs. Contacting providers from LSC-funded legal aid organizations and other nonprofits is a reasonable first step. An ED champion can work with these experts to craft a presentation that healthcare providers can easily understand and relate the presentation to their daily work.
- Check in with ED providers regularly on how referrals are working and make adjustments as needed.

Intermediate

- Work with ED case management staff and a physician ED champion to develop a list of local legal service provider(s) that includes contact information and what specific legal needs each provider can help with.
- Reach out to form a relationship with local legal services provider(s) to create a direct referral system, where ED providers can refer patients when legal needs are identified in the course of clinical practice. Participate in real-time dialogue with the legal team and assist the legal team by providing relevant information and documentation with patients' permission.
- Consider systematic screening for social and legal needs depending on the needs of your patient population and available resources. A self-administered paper or electronic screen for providers to review can eliminate the need for additional screening at triage. The chatbot screening system being piloted at Harbor-UCLA, the University of Washington, and Columbia University is one promising approach [23].

Advanced

- Develop and implement an MLP on-site to provide legal services by providing space and funding to support the legal team. Bringing legal providers on-site for regular intake and face-to-face consultation allows for further integration to identify and address legal needs into providers' daily practice. It also allows the legal team to interface with patients in real-time.
- Advocate for policy change. By addressing overarching policy issues that relate to health coverage rights, housing, and other social needs, we can prevent re-addressing similar problems (e.g., changing housing laws or streamlining the local Medicaid renewal process) for different individuals repeatedly. Working upstream will help prevent the problem from occurring in the future and open up capacity to address other health-harming legal needs.
- Advocate for increased resources outside of the healthcare system. For example, addressing homelessness requires more than only legal advocacy. In addition to preventing evictions, physicians can work with legal partners and advocacy organizations to demonstrate the importance of social services that affect health. In addition to more affordable housing, there are larger issues such as increasing the minimum wage, creating more jobs, and improving mental health and substance use disorder treatment to help different groups of individuals who are unstably housed access housing.

Teaching Case

Clinical Case

Paramedics bring in Anna from the local housing authority for threatening to overdose on her antipsychotic medication. She is tearful and combative. Through her sobbing, Anna tells you that she does not want to be homeless and that she would rather die than be homeless. She has been part of the federal low-income housing subsidy program, Section 8, for over 10 years. About a month ago, she received a notice in the mail that her Section 8 voucher was going to be terminated. Fearful of losing her housing, Anna had been going to the housing authority daily to try to get help. However, after a month of daily trips, Anna still could not get an answer as to why her voucher was being terminated. Desperate, frustrated, and emotionally drained, Anna just wanted to get some help with her voucher when she threatened to take a few extra quetiapine tablets.

The Housing Authority Police placed Anna on an involuntary psychiatric emergency hold, a "5150" in California, for being a danger to herself, and the paramedics brought her to your emergency department. When you consult social work, they are able to offer help placing her into a shelter if the 5150 is dropped, but the social worker does not have a good idea about how to resolve Anna's Section 8 challenges. Hearing this, Anna becomes more tearful, telling you that her son is graduating from high school next month and going to college. She cannot bear the thought of being homeless when her son graduates from high school. She knows that the waitlist to get a new Section 8 voucher is 5–10 years once one gets onto the waitlist, and without the voucher, she does not have the resources to pay her rent. What do you do?

Case Resolution

Anna was able to be medically cleared during her ED evaluation and she was transferred for evaluation by psychiatry. Psychiatry was able to drop her 5150 psychiatric hold as Anna was able to contract for safety and explained she did not want to die but felt a sense of desperation about her impending eviction. The providers connected Anna to a legal services attorney who is part of the hospital's medical legal partnership. The attorney provided free services in representing Anna at her Section 8 appeal hearing where Anna challenged the proposed termination of her Section 8 voucher. Anna prevailed at her hearing and was able to keep her Section 8 voucher, move into a new house, and see her son graduate from high school while remaining housed.

Teaching Points

1. Lawyers can provide a different type of expertise and level of service compared with other members of the healthcare team when it comes to challenges around social determinants of health. In this case, the social worker could offer to help Anna find a shelter but did not have the expertise to help fight the termination of her Section 8 voucher and prevent her from becoming homeless. The lawyer is able to assess the legality of the proposed Section 8 voucher termination and challenge it in court to preserve Anna's housing.
2. Housing is one of the most challenging social determinants for healthcare providers as it is relatively expensive, and low-income housing is in short supply despite a high need. Unlike other public benefits programs, such as Medicaid or food stamps, where if one is eligible, one can receive benefits, the federal Section 8 voucher is limited in supply and often the wait to get a voucher can be 5–10 years, if not longer. Preserving a Section 8 voucher is thus a legal emergency, as the voucher is often the difference between housing and homelessness as patients who are eligible for Section 8 are unlikely to be able to afford rent without the voucher.
3. Navigating public benefits systems, such as the housing authority, can be complicated, challenging, and time consuming. This is even more challenging for individuals who have lower literacy, limited English proficiency, and/or limited educational background. Although beneficiaries have due process available to them, the system does not always function as intended, and thus beneficiaries may need a lawyer to provide advocacy in order to preserve their benefits.

Discussion Questions

1. What health-harming legal needs does Anna have and what are the benefits of consulting a lawyer as opposed to social work in Anna's case?
2. What is the ideal workflow for a medical legal partnership in the emergency department? Should providers be expected to identify legal issues and decide what provider to refer to or is this another team member's role?
3. Given that public benefits systems are so complex and challenging to navigate that a lawyer may be required to assist patients, can you conceptualize how to integrate legal needs screening and referral for your hospital or ED?

References

1. Peppard PE, Kindig DA, Dranger E, Jovaag A, Remington PL. Ranking community health status to stimulate discussion of local public health issues: the Wisconsin health rankings. Am J Public Health. 2008;98(2):209–12. https://doi.org/10.2105/AJPH.2006.092981.
2. The National Center for Medical Legal Partnership. 2015. Available at: https://medical-legalpartnership.org/wp-content/uploads/2015/01/Framing-Legal-Care-as-Health-Care-Messaging-Guide.pdf. Used with permission from Kate Marple.
3. Trott J, Regenstein M. Screening for health-harming legal needs. The National Center for Medical Legal Partnership 2016. https://medical-legalpartnership.org/wp-content/uploads/2016/12/Screening-for-Health-Harming-Legal-Needs.pdf. Accessed 15 Feb 2010.
4. Zuckerman B, Sandel M, Smith L, Lawton E. Why pediatricians needs lawyers to keep children healthy. Pediatrics. 2004;114(1):224–8. https://doi.org/10.1542/peds.114.1.224.
5. American Bar Association: Medical-Legal Partnerships Pro Bono Project. https://www.americanbar.org/groups/probono_public_service/projects_awards/medical_legal_partnerships_pro_bono_project/. 2019. Accessed 01 May 2019.
6. Trott J, Peterson A and Regenstein M. Financing Medical-Legal Partnerships: View From the Field. Issue Brief Two; National Center for Medical Legal Parnership at the George Washington Univeristy; April 2019. https://medical-legalpartnership.org/wp-content/uploads/2019/04/Financing-MLPs-View-from-the-Field.pdf
7. Regenstein M, Trott J, Williamson A. Report: findings from the 2016 NCMLP national survey on MLP activities and trends. National Center for Medical Legal Partnerships 2017. https://medical-legalpartnership.org/wp-content/uploads/2017/07/2016-MLP-Survey-Report.pdf. Accessed 30 Apr 2019.
8. Creekmore L, Hughes R, Jennings L, John S, LaBella J, et al. The justice gap: measuring the unmet civil legal needs of low-income Americans. The Legal Services Corporation. 2017. https://www.lsc.gov/sites/default/files/images/TheJusticeGap-FullReport.pdf. Accessed 15 May 2019.
9. Martinez O, Boles J, Munoz-Laboy M, Levine EC, Chukwuemeka A, et al. Bridging health disparity gaps through the use of medical legal partnerships in patient care: a systematic review. J Law Med Ethics. 2017;45:260–75. https://doi.org/10.1177/1073110517720654.
10. Matthew DB. The law as healer: how paying for medical-legal partnerships saves lives and money. Center for Health Policy at Brookings 2017. https://www.brookings.edu/wp-content/uploads/2017/01/es_20170130_medicallegal.pdf. Accessed 15 May 2019.
11. County Health Rankings & Roadmaps. Medical-legal partnerships. https://www.countyhealthrankings.org/take-action-to-improve-health/what-works-for-health/policies/medical-legal-partnerships. 2019. Accessed 15 May 2019.
12. Sege R, Preer G, Morton SJ, Cabral H, Morakinyo O, et al. Medical-legal strategies to improve infant health care: a randomized trial. Pediatrics. 2015;136(1):97–106. https://doi.org/10.1542/peds.2014-2955.
13. Weintraub D, Rodgers MA, Botcheva L, Loeb A, Knight R, et al. Pilot study of medical-legal partnership to address social and legal needs of patients. J Health Care Poor Underserved. 2010;21(2 Suppl):157–68. https://doi.org/10.1353/hpu.0.0311.
14. O'Sullivan MM, Branfield J, Hoskote SS, Segal SN, Chug L, et al. Environmental improvements brought by the legal interventions in the homes of poorly controlled inner-city adult asthmatic patients: a proof of concept study. J Asthma. 2012;49(9):911–7. https://doi.org/10.3109/02770903.2012.724131.
15. Ryan AM, Kutob RM, Suther E, Hansen M, Sandel M. Pilot study of impact of medical-legal partnership services on patients' perceived stress and well being. J Health Care Poor Underserved. 2012;23(4):1536–46. https://doi.org/10.1353/hpu.2012.0179.
16. Rosen Valverde JN, Backstrand J, Hills L, Tanuos H. Medical-legal partnership impact on parents' perceived stress: a pilot study. Behavioral Med. 2019;45(1):70–7. https://doi.org/10.1080/08964289.2018.1481011.

17. Teufel JA, Werner D, Goffinet D, Thorne W, Brown SL, et al. Rural medical-legal partnership and advocacy: a three year follow-up study. J Health Care Poor Underserved. 2012;23(2):705–14. https://doi.org/10.1353/hpu.2012.0038.
18. Atkins D, Heller SM, DeBartolo E, Sandel M. Medical-legal partnership and health start: integrating civil legal aid services into public health advocacy. J of Legal Med. 2014;35:195–209. https://doi.org/10.1080/01947648.2014.885333.
19. Adapted from Dr. Mark Richman's suggestions during a presentation at Olive View Medical Center in 2009.
20. Kenyon C, Sandel M, Silverstein M, Shakir A, Zuckerman B. Revisiting the social history for child health. Pediatrics. 2007;120(3):e734–8. https://doi.org/10.1542/peds.2006-2495.
21. National Association of Community Health Centers. PRAPARE. http://www.nachc.org/research-and-data/prapare/about-the-prapare-assessment-tool/. 2019. Accessed 25 Apr 2019.
22. Johnson S, Liu P, Campa D, Dorsey C, Hong C, Hsieh D. on behalf of the Los Angeles County Health Agency, Social and Behavioral Determinants of Health Workgroup. Los Angeles County Health Agency Social and Behavioral Determinants of Health Screening Report. Los Angeles County Health Agency. 2019. (forthcoming).
23. Kocielnik R, Agapie E, Argyle A, Hsieh DT, Yadav K, Taira B, Hsieh G. HarborBot: A Chatbot for Social Needs Screening. AMIA Annu Symp Proc. 2020:4;2019:552–61. PMID: 32308849; PMCID: PMC7153089.
24. Creekmore L, Hughes R, Jennings L, John S, LaBella J, et al. The justice gap: measuring the Unmet civil legal needs of low-income Americans. The Legal Services Corporation. 2017. https://www.lsc.gov/media-center/publications/2017-justice-gap-report.
25. American Bar Association. The justice gap and pro bono legal. https://www.americanbar.org/groups/litigation/committees/commercial-business/spotlight/2017/justice-gap-pro-bono-legal/. 2019. Accessed 25 Apr 2019.
26. Brewer, J. LMH Health News: LMH health and KU school of law team up to provide healthcare and legal aid to patients, June 12, 2020. https://www.lmh.org/news/2020-news/lmhhealth-and-ku-school-of-law-team-up-to-provide-healthcare-an/.
27. Medical-legalpartnership.org. 2020. [online] Available at: https://medical-legalpartnership.org/wp-content/uploads/2019/02/Complex-Care-Fact-Sheet-FINAL.pdf.
28. Los Angeles County Department of Health Services. Whole Person Care Medical Legal Partnership. https://dhs.lacounty.gov/whole-person-care/our-services/whole-person-care-los-angeles/homeless-high-risk/#1604002603288-9025b0f0-9205. Accessed 16 Feb 2020.

Part V

Individual and Structural Violence

Community Violence and Its Implications for Emergency Providers

18

Theodore Corbin and Nathan Irvin

Key Points

- Violence is a leading cause of death for youth, particularly African American and Latino males, and is a recurrent problem with high rates of recidivism.
- Injuries associated with trauma extend beyond physical wounds and include psychological trauma, which is often overlooked yet is important to address for patient well-being and to break the cycle of recurrent injury.
- Hospital-based violence intervention programs show promise as a means of addressing the unique needs of victims of violent injury.
- Trauma-informed care is a framework that can be helpful in better caring for patients as well as health care workers that provide care for this population.

Foundations

Background

Community violence is a significant issue in many cities across the US and is particularly burdensome to poorer, disadvantaged minority—often African American and Latino-neighborhoods [1]. Cities such as Philadelphia, Baltimore, St. Louis, and Chicago are commonly sensationalized in the media as having high violent

T. Corbin
Department of Emergency Medicine, St. Christopher's Hospital for Children/Drexel University College of Medicine, Philadelphia, PA, USA
e-mail: tcorbin@drexelmed.edu

N. Irvin (✉)
Department of Emergency Medicine, Johns Hopkins University School of Medicine, Baltimore, MD, USA
e-mail: nirvin1@jhmi.edu

H. J. Alter et al. (eds.), *Social Emergency Medicine*,
https://doi.org/10.1007/978-3-030-65672-0_18

crime rates, but aggregate numbers only tell part of the story. Within each of these cities, there is significant variability at the neighborhood level. Such variability can tell a more illuminating narrative about parts of a city that endure a disproportionate share of these events, while others are relatively spared.

In 2014, for example, public health tracking in Philadelphia identified 246 homicide victims; these homicides, however, were not evenly distributed across the geographic landscape. Cartographic modeling estimated that the firearm assault rate was 5 times higher for African American residents and that the violence was located primarily in several low-income "hot spots" in Philadelphia [2]. Similar findings are documented in Chicago, where a significant proportion of its firearm violence occurred within socioeconomically depressed African American and Latino communities [3]. In contrast, lower rates of violence were noted in other areas of the city, including other African American and Latino communities. This observation has led some researchers to posit that socioeconomic status was a larger driver of the observed disparity than were race and ethnicity [3].

And yet, young men of color are affected by violent victimization at a higher rate than any other group [4]. Centers for Disease Control and Prevention (CDC) statistics show that homicide is the leading cause of death for African American males between the ages of 15 and 34. Violence is the leading cause of death for young African American males between the ages of 10 and 24, and the second leading cause of death for young Latino males. By contrast, violence ranks as the fifth leading cause of death among White males in the same age group [5]. Between 1999 and 2014, more than 47,000 African American males and more than 18,000 Latino males between the ages of 10 and 25 were victims of homicide [6]. While homicides for young men of any race or ethnicity are tragic, these statistics highlight the dramatic racial and ethnic disparity in homicide across the country.

Despite the tendency to focus on homicide as the prime indicator of violence, it is only the tip of the iceberg. The CDC estimates that for every homicide there are 94 nonfatal violent injuries, another disease burden shared disproportionately [6]. In 2013, African American males between the ages of 10 and 25 suffered nearly three times the rate of nonfatal assaults as similarly aged White males, while Latino males in this age range suffered 1.5 times their rate [6].

Violence is a cyclical problem. Being the victim of a non-fatal violent injury increases the risk of re-injury, retaliation, premature death, and incarceration. In urban settings, it is estimated that up to 45 percent of victims treated for violent injury are re-injured within 5 years [7]. One study of survivors of violence at 5 years follow-up found that 20% had died [8]. The risk of re-injury has been found to be greatest within 30 days of the initial incident [9]. These disparities are occurring on a global scale as well. A national study in New Zealand found that individuals hospitalized for violent injuries were readmitted for assault at a rate three-fold higher than non-assault injured patients within the ensuing 30 days [9].

The economic and societal costs of interpersonal violence have been estimated in rigorous analyses. Corso and her colleagues calculated a lifetime lost productivity cost per homicide of $1.3 million and a lost productivity cost of more than $57,000 for each nonfatal injury. These lost productivity costs combined with the medical costs of interpersonal injury totaled an estimated $33 billion in 2007 [10].

Another important consideration is the fact that the US has long struggled with accepting the fact that racism permeates the health, well-being, and safety of communities of color. The majority of American society has yet to accept race as a social construct that is perceived to be representative of one's culture and socioeconomic status and that it has no biological basis. Despite this fact, race and racism drastically influence one's functioning in society [11]. It is therefore imperative to develop an understanding of how racism affects individuals and fosters an environment where violence can flourish.

Dr. Camara Jones describes racism as working at three distinct levels: institutionalized, personally mediated, and internalized. Institutionalized racism is defined as "differential access to the goods, services, and opportunities of society by race" [12]. In her typology, institutional racism is often built into the fabric of our systems and can create differential access to resources like health care, quality education, safe housing, wealth, and power. Within the context of institutionalized racism, certain privileges are conferred to some at birth while denied to others. In contrast, personally mediated racism is the experience of prejudice and discriminatory behaviors by an individual as a result of their race and includes experiences like being treated suspiciously in stores, hate crimes, and devaluation of one's intellect and abilities. Internalized racism is the acceptance of these stigmatizing messages and can manifest in a myriad of ways including the abandonment of one's culture, feeling hopeless, and resigning to failure [12].

This framework is important to understand how racism affects community violence. Slavery and the mistreatment of other racial and ethnic groups within American society, including indigenous and Latino populations, were followed by other systemic and institutionalized practices and policies that resulted in a disproportionate share of people of color being socioeconomically marginalized and saddled with access to poor health care, under-performing schools, high rates of incarceration, and other barriers to thriving in society. This legacy helped set the stage for the circumstances in which violence thrives today.

Evidence Basis

While hospitals routinely treat the physical consequences of assault, many victims of violence experience psychological effects that go unaddressed [13]. Despite the psychological trauma of violent injury, many victims are not offered or do not seek mental health services [14]. Barriers to accessing services include, but are not limited to, perceived stigma of mental illness, distrust of mental health professionals, and lack of knowledge about and logistical barriers to accessing assistance services. This is troubling as the psychological trauma of violent injury and its biological correlates may lead violently injured youth to obtain weapons or engage in the use of illicit substances to restore feelings of safety. These coping mechanisms potentially increase the risk of re-injury and retaliation, and thus re-hospitalization and incarceration [15].

Population-based surveys in urban settings estimate that between 15% and 23% of victims of non-sexual assault meet the criteria for post-traumatic stress disorder (PTSD) at some point in their lives [16–19]. Studies in hospital settings have produced similar results. A study of males hospitalized for aggravated assault found that 27% had PTSD at 3 months follow-up and 18% had PTSD 1 year later [20]. Fifty-two percent of patients treated at an urban trauma center for gunshot wounds screened positively for possible PTSD [21]. A cross-sectional study of clients participating in Healing Hurt People (HHP), a hospital-based violence intervention program at Hahnemann Hospital in Philadelphia, found that 75% met the diagnostic criteria for PTSD 6 weeks after their injury [22]. Another survey of 541 victims of violent crime at an urban hospital found that 35% had major depressive disorder 1 month after the crime [23]. Despite this knowledge, few studies have demonstrated how to address psychological effects resulting from violent victimization, especially among young males of color.

Given the prevalence of violent injury, its recurrent nature, and its mental health sequelae, hospital-based violence intervention programs (HVIPs) have emerged as a strategy to prevent violent re-injury and retaliation, and improve the life course trajectories of survivors of violence [24].

Emergency Department and Beyond

HVIPs have demonstrated effectiveness in reducing rates of re-injury and violent crime, minimizing risk factors, and cultivating protective factors to prevent violent re-injury and retaliation [25]. HVIPs are staffed by a team of violence intervention specialists with varying backgrounds (social workers, peer specialists, community health workers) who conduct psychosocial needs assessments, provide therapeutic case management services, and offer psychoeducation to violently injured individuals shortly after they sustain their injuries. HVIPs operate out of trauma divisions or emergency departments (EDs), as well as community-based organizations that have affiliations with trauma centers. HVIPs have emerged as a promising locus for violence prevention and intervention. These programs have proliferated across the country in recent years-many coming together formerly under the National Network of Hospital-Based Violence Intervention Programs, now known as the Health Alliance for Violence Intervention (HAVI).

Although HVIPs have demonstrated varying degrees of success across a range of outcomes, best practices for program design, service delivery, and evaluation are yet to be elucidated [26]. The focus of most HVIPs has also been limited to outcomes such as re-injury, retaliation, and involvement with the criminal justice system. While critically important, HVIPs, as well as other community-based violence prevention interventions, have thus far largely neglected the psychological consequences of violent injury-outcomes which may mediate the relationship between violent injury and re-injury/retaliation and may have substantial impacts on the lives of violently injured individuals, their families, and communities.

Bedside

The Office for Victims of Crime of the US Department of Justice found that health care and criminal justice systems respond less sympathetically to violently injured youth, particularly African American male victims of gun violence, than to other crime victims. They noted that, "Whatever the reason for the disparate treatment of these victims, we must not ignore them. Assumptions about the blameworthiness of young African-Americans and Latinos' shortchange a large segment of the population and perpetuate racial stereotyping." [27] These experiences only reinforce their trauma and augment their mistrust of the systems that are designed to assist victims. The Vision 21: Transforming Victim Services Report concurs, noting that services for groups such as boys and men of color "may be unavailable, inadequate or difficult to access" despite the fact that this population suffers victimization at higher rates than the general US population [28].

Having an understanding of how patients being treated for violent injuries both perceive and experience health care is critical. As an emergency medical provider, it is imperative that we know our own biases so that we can address them in order to provide appropriate care for our patients. As soon as providers start to make assumptions about patients or this is perceived to be the case, patients may become guarded and mistrust the provider. Whether intentional or not, these interactions can negatively impact patient outcomes and the likelihood of adherence to treatment plans [29]. Given that emergency medical providers are at the forefront of violence in our communities, it is imperative that we be equipped to respectfully and competently engage in and advocate for the delivery of equitable and quality care to all patients. Yet several studies have found that many health care providers, including emergency medicine providers, have pro-White (i.e., preference for White patients) [30, 31]. Thus medical education facilities and hospitals can implement their own implicit bias training, beginning by having providers take an Implicit Association Test (IAT), which is freely available online (https://implicit.harvard.edu/implicit/takeatest.html). It is important for providers to first acknowledge that they have biases and then through an IAT determine what these biases are [29, 32, 33]. Once biases are identified, medical education facilities and hospitals should develop and incorporate trainings to establish a broad cultural competence appropriate to the populations that are being seen in their programs and institutions, work on building authentic relationships with trusted community leaders outside the institution, and, lastly, cultivate respect and empathy for all patients.

Additionally, it is important to understand trauma in the context of the patient's life story. As noted in the literature, it is not feasible in the ED to obtain a complete trauma history; though it is important to recognize that in many instances patients' injuries that bring them to the ED are often preceded by past traumatic experiences [34]. The word "trauma" refers to not just physical injuries, rather includes "an event, series of events, or set of circumstances that is experienced by an individual as physically or emotionally harmful or life threatening and that has lasting adverse effects on the individual's functioning and mental, physical, social, emotional, or spiritual well-being" [35]. Emergency providers can alter their clinical approach ensuring that they promote a culture of safety, empowerment, and healing. For instance, a provider should assume that every patient they see may have experienced trauma in their past: instead of asking patients to remove their clothes they may tell a patient why they are asking them to remove their clothes; instead of demanding to know a patient's past, explain how their past can illuminate their presenting problem. Providers can also ask patients if there is anything in their past that makes being seen by a provider or interacting with the health care system difficult for them.

Trauma-informed care, an approach already adopted by some emergency providers, incorporates how a patient's previous experiences influence how he or she may perceive and react to medical care [34]. As one can imagine, trauma is something that many individuals receiving care within the health system experience, particularly in emergency settings, and affects both patients and providers. Moderate symptoms of PTSD, such as hyper-arousal, irritability, and compassion fatigue, have been found in as high as 80% of victims of violence post injury. Furthermore, in a study of 118 providers, almost 40% developed clinical symptoms from secondary traumatic stress—from working with victims of violence and experiencing abuse and other negative exposures [36–38]. Such symptoms may degrade the care that all patients receive and can also drive burnout within health care systems. Instituting trauma-informed care systems is therefore an important part of providing appropriate and holistic care, and to fostering self-care and wellness in the providers caring for these patients. Emergency physicians can be powerful voices to ensure all patients receive equitable care.

Emergency providers can adopt a practice of trauma-informed care for violently injured patients through a set of principles, which can be remembered as four "R"s: Realizing the widespread influence of trauma and understanding the potential paths for recovery; Recognizing the signs and symptoms of trauma in patients, families, and staff involved in the system; Responding by fully integrating knowledge about trauma into policies, procedures and practices; and Resisting re-traumatization [34]. Training on trauma-informed care should be required by state licensing boards and the federal government to be adopted into all medical education programs, including medical school and residency curricula [39]. Lastly, ensuring that trauma-informed care is incorporated into the mission, quality assurance and quality improvement activities and metrics at an institutional level enhances the culturally appropriate environment in which patients can be seen.

On a related note, it is also important that providers recognize specific challenges that victims of violence may face with respect to interactions with health care and law enforcement while in the ED. A study exploring challenges in engaging African American male victims of community violence in health care revealed a fear of police involvement, an impression of "snitching" when disclosing personal information, mistrust of research motives, suspicion of the informed consent process, the emotional impact of the trauma itself, in addition to practical and logistical issues these victims faced [40]. Emphasizing the confidentiality of the patient-doctor interaction, talking patients through standard procedures, and being sensitive to situational realities are important considerations to help address these issues.

Hospital/Healthcare System

Victims of violent injury meet many struggles when interacting with the health care system. While violence and violence-related injuries are a frequent reason for visiting and being admitted to a health care facility, evidence suggests that only a minority of institutions have access to social services programs to assist these individuals [41]. This is in contrast to the current screening programs and resources for intimate partner violence that hospitals have [42, 43], which could be potentially useful as a model for other types of interpersonal violence. Physicians and nurses alike working with violent-injured individuals are often also deeply affected—often negatively-by their interactions, and these feelings can be amplified in the absence of appropriate resources to assist either patients or staff [41]. To our knowledge, there is a dearth of validated and reliable instruments to screen for interpersonal and/or community violence for adult victims. However, several questions have been developed by Harris et al. to capture exposure to community violence in children by expanding the Adverse Childhood Experiences (ACEs) survey [44]. Examples include: Has your child ever witnessed violence in your neighborhood, community, or school? (e.g., bullying, or organized violent crime, or police action, war, or terrorism); Has your child ever been a victim of violence in your neighborhood, community, or school? (e.g., bullying, or organized violent crime, or police action, war, or terrorism). Like other ACEs items, these questions may be adapted to try to engage adult patients about their exposure and/or risk to community violence exposure.

While many intervention strategies exist to break cycles of violence, hospitals present a unique opportunity to reach patients at the highest risk [45]. The presumption is that a violent injury is a teachable moment when patients may be particularly responsive to intervention. One thing hospitals can do is to implement hospital-based violence intervention programs. An HVIP one might use as an exemplar is Boston Medical Center's Violence Intervention Advocacy Program (VIAP), based out of its Department of Emergency Medicine. This program has a team that helps victims of violence recover from physical and emotional trauma through crisis intervention, support, and advocacy; ongoing case management and connections to community resources; and family support services. These services all play central roles in the healing process, and all have as an intentional emphasis the importance of the social determinants of health.

Young male victims, particularly African American and Latino victims, face barriers to health and human services that undermine their future life choices, health, and well-being. Consequences of violent injury often hurt victims long after initial treatment and hospital discharge, especially young people of color [24, 27]. The impact on these victims can be profound, affecting mental and physical health and altering their interactions with others. In addition, literature has found that health and human service systems that serve boys, young men, and their families are fragmented, do not share common knowledge or language, compete for limited resources, and are under stress. When victims interact with staff in stressed systems, trauma-related issues can be downplayed, negatively affecting service access and success [46].

In hospitals and health care systems, victims of violence, particularly young men of color, face persistent barriers to finding and accessing victim services that meet their needs, in part because of their own exposure to significant trauma and in part because institutions and programs designed to serve them often lack relevant, culturally appropriate services and, therefore, the capacity to fully engage such victims [45, 47]. Often, because of the combined influence of trauma, the culture of masculinity, and the persistent adversity and threats to basic safety that these male victims of violence face, they may not identify as victims and therefore may not seek out and engage with traditional victim-centered services [47, 48]. Because of poverty and lack of resources, they may be also disconnected from other types of resources, such as insurance [49] and adaptive modifications for homes, which hinders their ability to focus on their physical, emotional, and behavioral health and overall well-being. Despite this reality, few studies have examined ways to overcome barriers to accessing needed services for young male victims of violence, particularly in light of their complex social contexts and physical environments.

Trauma-informed care can also be practiced at a systems level [45]. An innovative program in Philadelphia, the Youth Nonfatal Injury Review Panel, brings together representatives from 23 agencies (e.g., police, schools, human services) quarterly to share information confidentially to discuss nonfatal violent injury cases that presented to EDs. Each meeting of the panel begins and ends with a session on trauma-informed care, and how this lens influences the outcome of the case. Such discussions provide insight on how to proactively, rather than reactively, assist patients and their families in their healing process, engaging the institution in trauma-informed care by providing spaces and a culture that facilitates healing [50].

Societal Level

Communities operate within the larger society, which is inherently unequally structured. Systemic biases in policy and practice create structural inequities, which manifest as advantages for one social group over another [51]. These structural inequities often parallel and reinforce many of the social determinants of health that drive community violence [52]. Two particularly salient examples include redlining, which has created the social circumstances for violence perpetration, and police violence, which shows distrust and hampers communities' capacity to heal.

An important history for medical professionals to acknowledge is the intentional implementation of federal, state, and local level policies that explicitly forced and continues to systematically force African Americans to live in separate neighborhoods from White Americans. Redlining began in the 1930s when President Franklin Roosevelt signed the Home Owners Loan Corporation Act (HOLC), which was intended to prevent homeowners from defaulting on loans [53]. As part of this program, HOLC drew maps to depict the risk of real estate investments. Such maps were color-coded, with red representing poor investments. Neighborhoods that housed "undesirable" populations like African Americans and Jews were thus outlined in red, or "redlined" [53–55]. During the mid-twentieth century, the federal government enacted two major policies: (1) demolition of integrated neighborhoods and creation of segregated public housing; and (2) subsidization for White Americans only to move into suburbs, adding text into the home deeds prohibiting resale to African Americans [56]. The Federal Housing Authority (FHA) thereby influenced where Americans lived, and had a stronghold in preventing African Americans from buying and owning homes throughout the US—not just in southern areas but also in the North, West, and Midwest. Very quickly, African American neighborhoods became undesirable due to unfavorable zoning laws. For example, the federal government systematically permitted industrial and waste disposal plants to operate in African American neighborhoods and explicitly disallowed them in White neighborhoods. Even with the Fair Housing Act, established segregation became more entrenched; within two generations, White Americans benefited greatly, as they could purchase homes in the suburbs, thus gaining intergenerational equity appreciation. At the same time, African Americans, forbidden from moving into the suburbs, were forced to live in rental properties in urban areas, thus gaining no equity appreciation over generations [56].

In an environment of systematic racial segregation, enforced by government officials, bankers, real estate agents, and landlords, African Americans in cities like Baltimore, Philadelphia, and Chicago were shunted to these redlined areas. Starved of capital investment, many redlined neighborhoods became blighted and toxic—a reality which persists today [57]. As a result of these policies, today, nationwide, we see that African American incomes on average are only 60% that of White incomes and that African American wealth is only 5–7% of White wealth—the majority of that difference attributable to racist federal, state, and local level policies [56]. Furthermore, the vestiges of these policies live on today, fueling inequity and community violence in our society.

Growing distrust of law enforcement is another driver of continued violence. Federal Bureau of Investigations data on police shootings from 2010 to 2012 show that young African American males were 21 times more likely to be shot and killed by the police than young White males [58, 59]. Between 2010 and 2012, 1217 people were fatally shot by police; African American males aged 15–19 were killed at a rate of 31.17 per million while among young White men, 1.47 per million suffered the same fate [58, 59].

Long recognized among some, traumatizing experiences of violence at the hands of the police are only more recently publicized to the broader public. Dramatic video of police killings of Michael Brown, Eric Garner, Freddie Gray, Tamir Rice, George Floyd, Ahmaud Arbery, Elijah McClain, Breonna Taylor, and too many others have sent the chilling message to young people of color that their lives are valued less than the lives of their White counterparts. Beyond the brutal, excessive, and unjustified force used in these incidents, the disregard for the dignity of the bodies of these young people after their fatal injuries carries a message of dehumanization. Despite the growing condemnation of police violence, these incidents reinforce the perception that the police likely regard young males of color as perpetrators first and are, therefore, more likely to do harm to them.

Mistrust of the police and lack of faith in their efforts to protect young males of color sometimes leads victims of violence to engage in retaliation against those responsible for their injury [15]. A review of needs-assessment data collected for 78 violently injured youth participating in Healing Hurt People found that 65% of clients reported that the index incident was still unresolved and 60% had thoughts of retaliation [60]. This deep mistrust of the police potentiates the psychological trauma of violent victimization which, if unaddressed, may lead individuals to obtain weapons—thereby potentially increasing the risk of violent re-victimization and violent offending [46].

To respond, US policymakers have looked for solutions around the world. One such response is the adoption of the Cardiff Model from Wales, advocated by the World Health Organization, which has demonstrated substantial and sustained decreases in violence-related injury. The Cardiff Model is a systematic and data-driven partnership between health, law enforcement agencies, and local government. Briefly, patients seen in the ED for a violent injury report the precise location of the incident to registration staff. This information is stripped of personal identifiers and shared by the hospital information technology systems on a monthly basis with a dedicated crime analyst. The aggregated data then are used to generate maps of hotspots that are used in turn to dictate intervention efforts, which are thus mapped to actual, rather than to perceived, risk. As more US communities partner with health systems to pilot the Cardiff Model and similar approaches, they create examples of equitable violence prevention that can help them recover from our epidemic of violence [61–63].

True, these structural issues seem enormous. But emergency clinicians can act on them by, for example, advocating within their institutions to evaluate the prevalence of interpersonal and/or community violence in their patient population. Emergency physicians can arrange discussions with administrators in their home institutions to ensure that leadership is aware of the issues their patient populations are experiencing. Trauma-informed care training should be instituted and IAT should be implemented for all health care providers, including physicians. Once biases are acknowledged, culturally competent trainings can be implemented and required for all providers; such measures are likely not one-off efforts and may need reinforcement. Emergency physicians can also ensure that leadership within their institution is making investments into hiring, training, and providing workforce development for their at-risk patient populations (e.g., mentoring, resume development, community health worker peer programs). Hospitals can also work with researchers to spatially observe trends and locations of violent crimes to ensure efficient and sufficient resources are allocated for HVIP's regionally [64]. Lastly, physicians have the privilege to have their voices heard by policymakers. Making time to meet with local, state, and federal level policymakers, such as the US Department of Justice and local District Attorneys, to advocate for funding of programs, including social justice policies that affect victims of violence, housing equity, justice reform, and others, can reduce community violence. As an example, New Jersey's Attorney General recently allocated $20 million in grants for up to nine HVIPs to be implemented throughout the state [65].

Recommendations for Emergency Medicine Practice

Basic

- Recognize the impact of trauma in the lives of individuals they are caring for and its impacts on ourselves.
- Use each interaction with victims of violence as a "teachable moment" in order to understand without judgment what they have experienced and advocate for behavior change.
- Uphold the humanity of each individual and resist the urge to legitimize our implicit and explicit biases.
- Have providers assess their implicit biases to know what they are [https://implicit.harvard.edu/implicit/takeatest.html].

Intermediate

- Make a commitment to shaping the care in your ED to adhere to trauma-informed principles.
- Lead, or advocate for, the development of a hospital-based violence intervention program (HVIP) or other resource tailored specifically to the needs of ED patients who are survivors of violence.
- Initiate Quality Assurance/Quality Improvement activities focused on improving the care of ED patients who are survivors of violence through increased trauma-informed training and care provided.

Advanced

- Create workforce development opportunities to create alternatives for community members. Hire, train, and advance the professional growth of local community members within institutions (e.g., community health worker peers).
- Advocate for funding for the Office of Victims of Crime, HVIPs, and mental health services.
- Advocate for more firearm and injury research funding for the National Institutes of Health and Centers for Disease Control and Prevention.
- Advocate for policies requiring that Level One trauma centers and hospitals that see a large number of victims of violence have an HVIP.

Teaching Case

Clinical Case

Marcus is a 21-year-old male who presents to the ED as a trauma activation, shot in the right chest while sitting outside of his home. Upon arrival, he is noted to have a GCS of 15 with a patent airway, decreased breath sounds on the right, and hypotensive to the 80s systolic. A tube thoracostomy returns 1.5 liters of blood immediately from his right chest. Massive transfusion protocol is initiated, and Marcus is taken emergently to the operating room wherein he undergoes a thoracotomy, repair of a pulmonary vein injury, and an exploratory laparotomy.

Marcus is stabilized operatively and transferred to the surgical intensive care unit. His notes reflect that he rarely engages in physical therapy and often seems remote and uncooperative, choosing instead to remain in bed. Nurses complain that he is quick to anger and occasionally curses at them when they push him to do things that he does not want to do, so they limit how frequently they go into his room and are often heard saying things like "I see why this happened to him." After a ten-day hospital course, Marcus is discharged with pain medications, wound care supplies, and outpatient follow-up with the trauma team.

Upon leaving the hospital, Marcus returns home where he lives with his mother, wife, and 2 children. He often avoids going outside and reports to his wife feeling anxious at times. He starts smoking marijuana more frequently and now drinks a half-pint of cognac each day, which causes tension with his mother, who is herself a recovering alcoholic.

Three weeks after returning home from the hospital, Marcus is shot in the head while walking home from the corner liquor store. Despite aggressive resuscitations attempts, he dies in the trauma bay.

Teaching Points

1. Violence is a recurrent issue with high mortality, with the highest risk of recurrent injury within 30 days of the initial ED visit.
2. Victims of violence will often manifest symptoms of trauma or PTSD which can include feelings of numbness, hyper-arousal, re-experiencing trauma, nightmares, avoidance behaviors, irritability, and emotional mood swings.
3. Victims of violence may use alcohol, marijuana, or other drugs to mask the negative emotional symptoms they are experiencing.
4. Care plans should be holistic and address not just the physical, but also the emotional wounds the patient sustains. HVIPs and other social resources are beneficial to victims of violence.

Discussion Questions

1. What symptoms of trauma or PTSD was Marcus manifesting during his time in the hospital and at home during recovery?
2. How would a trauma-informed care model have been beneficial to Marcus and the providers taking care of him?

3. What resources are available in your institution to help patients like Marcus? What is necessary to improve/supplement them? What stakeholders are needed to try to implement such a model, if one does not exist at your institution?

References

1. Wintemute GJ. The epidemiology of firearm violence in the twenty-first century United States. Annu Rev Public Health. 2015;36:5–19. https://doi.org/10.1146/annurev-publhealth-031914-122535.
2. Beard JH, Morrison CN, Jacoby SF, Dong B, Smith R, Sims CA, et al. Quantifying disparities in urban firearm violence by race and place in Philadelphia, Pennsylvania: a cartographic study. Am J Public Health. 2017;107(3):371–3. https://doi.org/10.2105/ajph.2016.303620.
3. Walker GN, McLone S, Mason M, Sheehan K. Rates of firearm homicide by Chicago region, age, sex, and race/ethnicity, 2005–2010. J Trauma Cute Care Surg. 2016;81(4 Suppl 1):S48–53. https://doi.org/10.1097/ta.0000000000001176.
4. Sered D. Young men of color and the other side of harm: Addressing disparities in our responses to violence. New York: Vera Institute of Justice; 2014. Retrieved from http://www.vera.org/sites/default/files/resources/downloads/men-of-color-as-victims-of-violence-v3.pdf.
5. Centers for Disease Control and Prevention. Web-based injury statistics query and reporting system (WISQARS) fatal injury data. 2019.
6. Centers for Disease Control and Prevention. Web-based injury statistics query and reporting system (WISQARS). 2019.
7. Sims DW, Bivins BA, Obeid FN, Horst HM, Sorensen VJ, Fath JJ. Urban trauma: a chronic recurrent disease. J Trauma. 1989;29(7):940–7.
8. Goins WA, Thompson J, Simpkins C. Recurrent intentional injury. J Natl Med Assoc. 1992;84(5):431–5.
9. Dowd MD, Langley J, Koepsell T, Soderberg R, Rivara FP. Hospitalizations for injury in New Zealand: prior injury as a risk factor for assaultive injury. Am J Public Health. 1996;86(7):929–34.
10. Corso PS, Mercy JA, Simon TR, Finkelstein EA, Miller TR. Medical costs and productivity losses due to interpersonal and self-directed violence in the United States. Am J Prev Med. 2007;32(6):474–82. https://doi.org/10.1016/j.amepre.2007.02.010.
11. Cooper R, David R. The biological concept of race and its application to public health and epidemiology. J Health Polit Policy Law. 1986;11(1):97–116.
12. Jones CP. Levels of racism: a theoretic framework and a gardener's tale. Am J Public Health. 2000;90(8):1212–5.
13. Kilpatrick DG, Acierno R. Mental health needs of crime victims: epidemiology and outcomes. J Trauma Stress. 2003;16(2):119–32.
14. Anixt JS, Copeland-Linder N, Haynie D, Cheng TL. Burden of unmet mental health needs in assault-injured youths presenting to the emergency department. Acad Pediatr. 2012;12(2):125–30. https://doi.org/10.1016/j.acap.2011.10.001.
15. Rich JA, Grey CM. Pathways to recurrent trauma among young black men: traumatic stress, substance use, and the "code of the street". Am J Public Health. 2005;95(5):816–24.
16. Breslau N, Davis GC, Andreski P, Peterson E. Traumatic events and posttraumatic stress disorder in an urban population of young adults. Arch Gen Psychiatry. 1991;48(3):216–22.
17. Breslau N, Kessler RC, Chilcoat HD, Schultz LR, Davis GC, Andreski P. Trauma and posttraumatic stress disorder in the community: the 1996 Detroit Area Survey of Trauma. Arch Gen Psychiatry. 1998;55(7):626–32.
18. Breslau N, Wilcox HC, Storr CL, Lucia VC, Anthony JC. Trauma exposure and posttraumatic stress disorder: a study of youths in urban America. J Urban Health. 2004;81(4):530–44.

19. Goldmann E, Aiello A, Uddin M, Delva J, Koenen K, Gant LM, et al. Pervasive exposure to violence and posttraumatic stress disorder in a predominantly African American Urban Community: the Detroit Neighborhood Health Study. J Trauma Stress. 2011;24(6):747–51.
20. Jaycox LH, Marshall GN, Schell T. Use of mental health services by men injured through community violence. Psychiatr Serv. 2004;55(4):415–20.
21. Reese C, Pederson T, Avila S, Joseph K, Nagy K, Dennis A, et al. Screening for traumatic stress among survivors of urban trauma. J Trauma Acute Care Surg. 2012;73(2):462–8.
22. Corbin TJ, Purtle J, Rich LJ, Rich JA, Adams EJ, Yee G, et al. The prevalence of trauma and childhood adversity in an urban, hospital-based violence intervention program. J Health Care Poor Underserved. 2013;24(3):1021–30.
23. Boccellari A, Alvidrez J, Shumway M, Kelly V, Merrill G, Gelb M, et al. Characteristics and psychosocial needs of victims of violent crime identified at a public-sector hospital: data from a large clinical trial. Gen Hosp Psychiatry. 2007;29(3):236–43.
24. Cunningham R, Knox L, Fein J, Harrison S, Frisch K, Walton M, et al. Before and after the trauma bay: the prevention of violent injury among youth. Ann Emerg Med. 2009;53(4):490–500.
25. Zun LS, Downey L, Rosen J. The effectiveness of an ED-based violence prevention program. Am J Emerg Med. 2006;24(1):8–13. https://doi.org/10.1016/j.ajem.2005.05.009.
26. Cheng TL, Haynie D, Brenner R, Wright JL, Chung SE, Simons-Morton B. Effectiveness of a mentor-implemented, violence prevention intervention for assault-injured youths presenting to the emergency department: results of a randomized trial. Pediatrics. 2008;122(5):938–46. https://doi.org/10.1542/peds.2007-2096.
27. Bonderman J. Working with victims of gun violence. Office for Victims of Crime Bulletin. Washington, DC: U.S. Department of Justice; 2001.
28. US Department of Justice Office of Victims of Crime. Transforming victim services Final report in vision 21. 2013. https://ovc.ojp.gov/library/publications/vision-21-transforming-victim-services-final-report.
29. Cooper LA, Roter DL, Carson KA, Beach MC, Sabin JA, Greenwald AG, et al. The associations of clinicians' implicit attitudes about race with medical visit communication and patient ratings of interpersonal care. Am J Public Health. 2012;102(5):979–87. https://doi.org/10.2105/AJPH.2011.300558.
30. Hall WJ, Chapman MV, Lee KM, Merino YM, Thomas TW, Payne BK, et al. Implicit racial/ethnic bias among health care professionals and its influence on health care outcomes: a systematic review. Am J Public Health. 2015;105(12):e60–76.
31. Dehon E, Weiss N, Jones J, Faulconer W, Hinton E, Sterling S. A systematic review of the impact of physician implicit racial bias on clinical decision making. Acad Emerg Med. 2017;24(8):895–904. https://doi.org/10.1111/acem.13214.
32. Van Ryn M, Burgess DJ, Dovidio JF, Phelan SM, Saha S, Malat J, et al. The impact of racism on clinician cognition, behavior, and clinical decision making. Du Bois Rev. 2011;8(1):199–218. https://doi.org/10.1017/S1742058X11000191.
33. Beach MC, Price EG, Gary TL, Robinson KA, Gozu A, Palacio A, et al. Cultural competence: a systematic review of health care provider educational interventions. Med Care. 2005;43(4):356–73. https://doi.org/10.1097/01.mlr.0000156861.58905.96.
34. Fischer KR, Bakes KM, Corbin TJ, Fein JA, Harris EJ, James TL, et al. Trauma-informed care for violently injured patients in the emergency department. Ann Emerg Med. 2019;73(2):193–202. https://doi.org/10.1016/j.annemergmed.2018.10.018.
35. Substance Abuse and Mental Health Services Administration (SAMHSA). Trauma and Violence. 2019.
36. Dominguez-Gomez E, Rutledge DN. Prevalence of secondary traumatic stress among emergency nurses. J Emerg Nurs. 2009;35(3):199–204; quiz 73–4. https://doi.org/10.1016/j.jen.2008.05.003.
37. Roden-Foreman JW, Bennett MM, Rainey EE, Garrett JS, Powers MB, Warren AM. Secondary traumatic stress in emergency medicine clinicians. Cogn Behav Ther. 2017;46(6):522–32. https://doi.org/10.1080/16506073.2017.1315612.

38. Kerig PK. Enhancing resilience among providers of trauma-informed care: a curriculum for protection against secondary traumatic stress among non-mental health professionals. J Aggress Maltreat Trauma. 2019;28(5):613–30. https://doi.org/10.1080/10926771.2018.1468373.
39. Health Affairs Blog. Implicit bias curricula in medical school: student and faculty perspectives. Health Affairs. 2020.
40. Schwartz S, Hoyte J, James T, Conoscenti L, Johnson R, Liebschutz J. Challenges to engaging black male victims of community violence in healthcare research: lessons learned from two studies. Psychol Trauma. 2010;2(1):54–62. https://doi.org/10.1037/a0019020.
41. Báez AA, Miller R, Giraldez E, Michaud Y, Rogers S. Victims of violence: a survey of emergency care providers attitudes and perceptions. Acad Emerg Med. 2007;14(Suppl 1):S67.
42. O'Doherty L, Hegarty K, Ramsay J, Davidson LL, Feder G, Taft A. Screening women for intimate partner violence in healthcare settings. Cochrane Database Syst Rev. 2015;(7):Cd007007. https://doi.org/10.1002/14651858.CD007007.pub3.
43. Butler B, Agubuzu O, Hansen L, Crandall M. Illinois trauma centers and community violence resources. J Emerg Trauma Shock. 2014;7(1):14–9. https://doi.org/10.4103/0974-2700.125633.
44. Koita K, Long D, Hessler D, Benson M, Daley K, Bucci M, et al. Development and implementation of a pediatric adverse childhood experiences (ACEs) and other determinants of health questionnaire in the pediatric medical home: a pilot study. PLoS One. 2018;13(12):e0208088. https://doi.org/10.1371/journal.pone.0208088.
45. Sumner SA, Mercy JA, Dahlberg LL, Hillis SD, Klevens J, Houry D. Violence in the United States: status, challenges, and opportunities. JAMA. 2015;314(5):478–88.
46. Rich JA. Wrong place, wrong time: trauma and violence in the lives of young black men. Baltimore: JHU Press; 2009.
47. Powell W, Adams LB, Cole-Lewis Y, Agyemang A, Upton RD. Masculinity and race-related factors as barriers to health help-seeking among African American men. Behav Med. 2016;42(3):150–63. https://doi.org/10.1080/08964289.2016.1165174.
48. American Psychological Association. APA guidelines for psychological practice with boys and men. 2018.
49. Sohn H. Racial and ethnic disparities in health insurance coverage: dynamics of gaining and losing coverage over the life-course. Popul Res Policy Rev. 2017;36(2):181–201. https://doi.org/10.1007/s11113-016-9416-y.
50. Purtle J, Rich LJ, Rich JA, Cooper J, Harris EJ, Corbin TJ. The youth nonfatal violent injury review panel: an innovative model to inform policy and systems change. Public Health Rep. 2015;130(6):610–5. https://doi.org/10.1177/003335491513000610.
51. National Academies of Sciences, Engineering, and Medicine. Communities in action: pathways to health equity. Washington, DC: The National Academies Press; 2017.
52. Jacoby SF, Dong B, Beard JH, Wiebe DJ, Morrison CN. The enduring impact of historical and structural racism on urban violence in Philadelphia. Soc Sci Med (1982). 2018;199:87–95. https://doi.org/10.1016/j.socscimed.2017.05.038.
53. Hillier AE. Redlining and the home Owners' loan corporation. J Urban Hist. 2003;29(4):394–420. https://doi.org/10.1177/0096144203029004002.
54. Hillier AE. Who received loans? Home owners' loan corporation lending and discrimination in Philadelphia in the 1930s. J Plan Hist. 2003;2(1):3–24. https://doi.org/10.1177/1538513202239694.
55. Pietila A. Not in my neighborhood: how bigotry shaped a great American City. Chicago: Ivan R. Dee; 2010.
56. Rothstein R. The color of law: a forgotten history of how our government segregated America. 1st ed. New York/London: Liveright Publishing Corporation, a division of W.W. Norton & Company; 2017.
57. McClure E, Feinstein L, Cordoba E, Douglas C, Emch M, Robinson W, et al. The legacy of redlining in the effect of foreclosures on Detroit residents' self-rated health. Health Place. 2019;55:9–19. https://doi.org/10.1016/j.healthplace.2018.10.004.
58. Gabrielson R, Jones RG, Sagara E. Deadly force, in black and white. ProPublica; 2014.
59. Richardson JB Jr, Vil CS, Cooper C. Who shot ya? How emergency departments can collect reliable police shooting data. J Urban Health. 2016;93(1):8–31.

60. Rich J. Moving toward healing: trauma and violence and boys and young men of color. Robert Wood Johnson Foundation (Issue Brief); 2016. https://bma.issuelab.org/resources/25846/25846.pdf.
61. Florence C, Shepherd J, Brennan I, Simon T. Effectiveness of anonymised information sharing and use in health service, police, and local government partnership for preventing violence related injury: experimental study and time series analysis. BMJ. 2011;342:d3313. https://doi.org/10.1136/bmj.d3313.
62. Mercer Kollar LM, Sumner SA, Bartholow B, Wu DT, Moore JC, Mays EW, et al. Building capacity for injury prevention: a process evaluation of a replication of the Cardiff Violence Prevention Programme in the Southeastern USA. Inj Prev. 2019; https://doi.org/10.1136/injuryprev-2018-043127.
63. National Academies of Sciences, Engineering, and Medicine. Community violence as a population health issue: proceedings of a workshop. Washington, DC: The National Academies Press; 2017.
64. Clery MJ, Dworkis DA, Sonuyi T, Khaldun JS, Abir M. Location of violent crime relative to trauma resources in Detroit: implications for community interventions. West J Emerg Med. 2020;21(2):291–4.
65. The State of New Jersey Office of the Attorney General. AG Grewal announces $20 million in available grants to establish nine hospital-based violence intervention programs across New Jersey. 2019. https://www.nj.gov/oag/newsreleases19/pr20190924a.html.

Firearm Injury Prevention in the Emergency Department

19

Megan L. Ranney, Patrick M. Carter, and Rebecca M. Cunningham

Key Points

- Rates and absolute numbers of firearm injuries are increasing in the US.
- Two-thirds of firearm deaths among adults are suicides, while among children and adolescents 60% of deaths are due to homicide.
- The most common reason for women to die of a firearm injury is intimate partner violence.
- We recommend a patient-centered approach to discussions about firearm injury (e.g., non-judgemental, respectful, empathetic).
- Screening and counseling for firearm injury prevention should be considered in high-risk patient groups especially those with depression or suicidal thoughts; those with a history of non-partner (i.e., peer) or partner (i.e., domestic) violence; children; and those with dementia.
- Validated interventions exist to reduce the risk of violence among at-risk teens. Other interventions have expert opinion and preliminary evidence.
- Emergency physicians are leading the charge to develop and disseminate best practices and evidence to reduce the risk of firearm injury.

M. L. Ranney (✉)
Department of Emergency Medicine, Alpert Medical School, Brown University, Rhode Island Hospital, Providence, RI, USA

American Foundation for Firearm Injury Reduction in Medicine (AFFIRM Research), Providence, RI, USA

Brown-Lifespan Center for Digital Health, Providence, RI, USA
e-mail: megan_ranney@brown.edu

P. M. Carter · R. M. Cunningham
University of Michigan Injury Prevention Center, Department of Emergency Medicine, School of Medicine, Department of Health Behavior and Health Education, School of Public Health, University of Michigan, Ann Arbor, MI, USA
e-mail: cartpatr@med.umich.edu; stroh@med.umich.edu

H. J. Alter et al. (eds.), *Social Emergency Medicine*,
https://doi.org/10.1007/978-3-030-65672-0_19

Foundations

Background

Firearm injuries are a significant and uniquely American social health problem. Although rates of suicide and non-firearm interpersonal violence are similar to other high-income countries, US firearm fatality rates remain the highest of all industrialized nations, with 83.7% of firearms deaths in industrialized countries occurring in the US [1]. In 2017, firearms resulted in nearly 40,000 deaths nationwide and were a leading cause of injury-related mortality, surpassing motor vehicle crash deaths for the first time and second only to opioid overdose mortality [2]. Of these deaths, nearly 60% resulted from firearm suicides, 38% were due to firearm homicides, and approximately 1% were the result of unintentional or "accidental" firearm injuries [2]. While active shooter incidents, or events where an individual is engaged in killing or attempting to kill multiple people in a populated area (e.g., mass and school shootings), garner significant media attention, they represent a small fraction of the overall fatality burden attributable to firearms. In 2016–2017, there were 50 active shooter incidents throughout the US, responsible for a total of 221 firearm-related deaths [3].

Examining trends over time, firearm fatalities have increased nearly 26% over the past decade, largely the result of a 31% increase in firearm suicides and a 16% increase in firearm homicides [2]. Estimates by the FBI indicate that active shooting incidents also appear to be increasing, with a 16% increase in annual numbers of events during the first decade of the twenty-first century [4]. Trends in non-fatal firearm injuries remain more difficult to catalog than fatalities due to current sampling methods, but best available estimates suggest that the total numbers of non-fatal firearm injuries among emergency department (ED) patients also have been increasing over time, with a 50% increase in the age-adjusted rate of assault-related non-fatal firearm injury from 2007 to 2017 [5].

Significant health and social disparities exist in the populations most affected by these firearm injuries. Examined across the lifespan, firearm fatalities disproportionately affect older youth (age 14–24), young adult (ages 25–34), and elderly (65+) populations, although underlying intent differs among these age groups [2]. Among youth and young adult populations (age 15–34), 59% of firearm deaths are due to homicide and 38% due to suicide, while firearm deaths among the elderly almost exclusively result from suicide (91%) [2]. By contrast, while unintentional firearm injuries represent the smallest fraction of firearm deaths overall, rates among children and adolescents under age 18 are twice as high as those for the remainder of the population [2]. Across all ages and injury intents, males are twice as likely to be injured or die from firearm injuries compared with females [2]. Despite this, it is important to note that over half of all female homicides result from intimate partner violence, with firearms the most frequent cause and responsible for more than half of female homicides [6]. Nationwide, absolute numbers of firearm fatalities are highest in urban settings; however, over the past 10 years rural communities have seen disproportionately higher rates of firearm fatalities (13.8 per 100,000 people) than suburban (11.5 per 100,000 people) or urban (9.7 per 100,000 people) communities. Substantive differences exist in the intents underlying firearm deaths in rural and urban communities: rural rates of firearm suicide and unintentional firearm injuries are twice and four times those observed in urban communities, respectively. By contrast, firearm homicide rates are 60% higher in urban than in rural communities. Firearm fatalities disproportionately impact Black populations, driven largely by firearm homicide [2]. In contrast, firearm suicide rates are highest among White and American Indian/Alaskan Native populations [2]. Many of these recognized disparities, especially racial and ethnic factors, likely reflect underlying socio-economic factors, including poverty, structural racism, differential environmental and geographic exposures, availability of firearms, and availability and access issues related to critical health and social services, especially mental health [7–10].

Evidence Basis

As the first and often the only healthcare providers to care for victims of firearm injury, emergency physicians are on the front line of the firearm injury epidemic. We also have a key role in prevention-based efforts, just as we do for other public and social health issues entering through our doors. Research focused on characterizing this public health problem and developing and instituting evidence-based prevention practices has been limited during the past two decades, as a result of both federal restrictions on research funding and the politicization of this injury prevention topic [11, 12]. In fact, one systematic review of the literature examining clinically based screening and intervention practices found that few firearm-specific interventions exist or have a strong evidence base [13]. Recent increases in public awareness have begun to renew interest in research, with emergency medicine as a leader in this field of science. The recent establishment of funded firearm research centers such as the Firearm Safety Among Children and Teens (FACTS) Consortium and the University of California Firearm Violence Research Center (UCFC), both led by emergency physicians, are stimulating novel evidence-based research for the field. Much of the initial work of these centers has been focused on cataloging the current state of the science [14–18] and developing consensus around the research needed for the field [19, 20]. While much work remains to be done, consensus and a preliminary evidence base exists on practical measures that emergency physicians can take today to enhance the safety of their patients and reduce the risk of firearm injury and deaths. The remainder of this chapter details current recommendations for preventing firearm injury and death, the evidence base behind them, and the role of emergency physicians in these efforts.

Emergency Department and Beyond

Bedside

As physicians, we are focused on the safety of our patients. Just as we talk about pool safety and car safety—and respect the rights of our patients to have pools and cars—our work in firearm injury prevention aims to reduce risks of harm and promote health. It is critical that we incorporate routine but culturally competent screening for firearm access among high-risk patients in the ED. In general, an ED-based conversation about firearm access should be just one part of a larger discussion about health risks. Given the current cultural narrative about firearm ownership, we must contextualize our questions about firearms.

To facilitate culturally competent questions about firearm injury risk, we suggest that physicians educate themselves on firearms and firearm safety. In one recent study, only 16% of EM physicians reported handling a firearm in the past year [21]. Although we are not encouraging physicians to change their firearm ownership practices, a basic familiarity with firearms and safe storage is encouraged, just as we have a basic familiarity with pools and car seats, whether or not we have used them.

Several excellent resources are available to fill this knowledge gap. For example, a brief overview of types of firearms, basic ammunition, firearm actions, parts of a firearm, types of gun locks, and ammunition storage options is available at the webpage for the FACTS consortium [22]. Pilot programs providing hands-on training for physicians are also in development and may provide future widespread education [23].

Inherent to all these potential interventions is the key concept that ED providers and staff are engaged in "Trauma-informed Care." This approach is based on four pillars: knowledge of the effect of trauma on the psyche, recognition of signs of such trauma, avoiding further repeat trauma, and the development of appropriate policies and procedures [24].

We recommend specifically engaging patients from 5 high-risk groups in discussions [25–27]. Specifically:

1. *Patients with suicidal ideation or who are otherwise at-risk for suicide*: Research shows that 90% of people who attempt suicide with a gun die, whereas about 10% of attempts by other methods result in death. Firearm access is one of the most well-established risk factors for death by suicide [28, 29]. Less than half of patients discharged from the ED with suicidal ideation are currently asked about firearm access, even when emergency physicians know the importance of such

questions [30, 31]. Current best practices for advising patients on access to firearms and other lethal means of suicide, known as lethal means counseling, are available from the Suicide Prevention Resource Center [32] and the American College of Emergency Physicians [33–35]. Researchers are also working on developing validated technology-based tools for emergency physicians to talk to firearm-owning patients and parents in a culturally competent way. Critical to these discussions is respect for patients' reasons for gun ownership, and an emphasis on the expected *temporary* nature of removing access to a firearm [36].

2. *Patients with a history of assault injury*: Firearm homicide is a leading cause of death for youth (age 14–20) regardless of race or ethnicity, and violent injuries are responsible for greater than 600,000 adolescent ED visits per year [37]. EDs are a key contact point for violently injured youth at risk for repeat violence [38]. Such youth have low rates of school attendance [39–41] and primary care access [42–44]. In a longitudinal study of violently injured youth recruited from an ED setting, investigators found that within 2 years, 37% returned for a repeat violent injury, and 59% experienced firearm violence [38, 45–48]. Thus, selective screening and risk stratification is indicated for this high-risk population. Early prevention of firearm violence includes de-escalation of fighting with peers/partners. *SafERteens 2.0* is an evidence-based brief intervention to prevent youth violence for up to 1 year following the ED visit. It has been translated into routine ED clinical practice. It has been demonstrated to be cost-effective and has been noted by the CDC as a best practice in the CDC technical package for violence prevention [49, 50]. The *SafERteens 2.0* website has resources for implementing this program in an ED setting, including: 1) training manuals/videos to screen patients for a history of recent violence behaviors and conduct the brief intervention to reduce youth violence, either remotely by establishing a call center or in-person by on-site staff; and 2) implementation support, including online screening questions and clinical therapy decision support for staff to use in real time [51]. The website includes resource brochures that can be adapted for varied clinical settings and includes instructions to create an optional text message booster post-discharge.

   ED providers may also consider implementing universal screening of high-risk youth for future firearm violence risk using the SaFETy score [52]. Although evidence is limited outside of urban youth in a single setting, this score is currently the only ED-based screening tool for future firearm violence risk [52]. This four-item questionnaire predicted firearm involvement over the subsequent 2 years in a sample of youth that reported past-month marijuana use. Of note, the screen more accurately predicted future firearm violence than did the ED presentation for an assault injury. This ED-based screen may be useful for stratifying hospital or community-based resources (Fig. 19.1).

3. *Patients reporting intimate partner violence (IPV)*: According to the well-validated Danger Assessment instrument, the abusive partner's access to a firearm is a leading predictor of risk of death [53–56]. All patients disclosing intimate partner violence should be screened for firearm access. Currently, best practice in response is to link a patient to standard domestic violence resources, such as social work, the National Domestic Violence Hotline, or a local shelter.

S (Serious Fighting): In the past 6 months, including today, how often did you get into a serious physical fight?
Never = 0 points
Once = 1 point
Twice = 1 point
3-5 Times = 1 point
6 or more times = 4 points

F (Friend Weapon Carrying): How many of your friends have carried a knife, razor, or gun?
None = 0 points
Some = 0 points
Many, Most or All = 1 point

E (Community Environment): In the past 6 months, how often have you heard guns being shot?
Never = 0 points
Once or Twice = 0 points
A Few Times = 0 points
Many Times = 1 point

T (Firearm Threats): How often, in the past 6 months, including today, has someone pulled a gun on you?
Never = 0 points
Once = 3 points
Twice or More = 4 points

Firearm SaFETy Score	Risk of Future Firearm Violence
0	18.2%
1-2	40.0%
3-5	55.8%
6-8	81.3%
9-10	100.0%

Fig. 19.1 Calculating the firearm injury SaFETy Score [52]

Leaving a relationship is also a strong predictor of homicide risk; presence of a firearm should also increase the importance of thoughtful, unhurried safety planning. In some states, the ED physician may be required to report the partner's firearm access to the police [57].

4. *Children*: The American Academy of Pediatrics recommends that firearm safe storage should be among standard anticipatory guidance topics. Given that firearm injury is the leading cause of death for teens, and that most teens who commit suicide do so with a family member's gun, these discussions should continue into adolescence. These discussions are not always appropriate in the ED setting, but should be raised in the appropriate clinical contexts, such as

among families with a depressed teen seeking ED treatment for suicidal ideation or attempt. More resources are available from the AAP's ASK campaign [58]. An example of how to manage pitfalls in discussions with teens and parents is also available from FACTS online training for physicians [59].

5. *Patients with dementia*: An increasing proportion of older Americans are suffering from cognitive decline and dementia. Data suggest that firearm access among these adults is similar to that among the general population (~30%) [60]. As dementia is known to be associated with a higher risk of both suicide and paranoia, discussions about safer storage of firearms with caregivers of demented patients may be in order [61]. This is an area of active research.

   A last note: limited resources or guidance is currently available for emergency physicians interacting with patients who are threatening or at risk of mass violence. In many states, "extreme risk protection orders" (ERPOs, otherwise known as "red flag laws") can be activated by law enforcement officials or family members. Best practices for when and how ERPOs should be used are still in development, but current expert guidance suggests that, if a clinician is concerned about a patient committing an act of mass violence, they should talk to local risk management and consider calling law enforcement or psychiatry for guidance.

Hospital/Healthcare System

Emergency physicians have long been leaders in the field of injury prevention. Our role in developing and validating firearm injury prevention strategies should be equally significant.

Beyond the ED as a point of contact for prevention, the current standard of care for violently injured youth (i.e., wound care/same-day discharge) is increasingly recognized as inadequate [62], with broad agreement that hospitals and EDs are key settings for violence prevention [62, 63]. In response, numerous hospital-based violence reduction programs [64–72] were developed for violently injured youth [67]. Such programs provide an array of services, including linkages to community services [73], peer mentorship by former gang members [64], and care management [65, 66]. Often these current programs focus on admitted youth, omitting the 84% of violently injured youth discharged directly from the ED [38]. Although promising for motivated patients, more study is needed to establish whether these programs are efficacious in reducing repeat violence, and to define key components, including optimal populations for intervention, the dose of intervention needed, and the effective elements of such programs [74]. Further resources are available through the Health Alliance for Violence Intervention (HAVI) [75].

Emergency physicians can also lead the implementation of best practice screening and intervention programs for at-risk groups and can develop or promote educational efforts and community collaboration.

Societal Level

Emergency physicians can take numerous actions to change the epidemic of firearm injury on a societal level. We can collaborate with victims and at-risk groups; we can reframe the lay narrative that firearm injury prevention is about "gun control"; and we can advocate for a stronger scientific evidence base. Most importantly, we can share stories of how this epidemic affects individual patients and society.

As emergency physicians, we have a privileged view on the reality of firearm injury. Until recently, our experiences caring for victims of violence have been hidden or sanitized in the public arena. In 2019, the "#ThisIsOurLane" movement, led primarily by emergency physicians, elevated physicians' voices and stories in shaping public debate [76].

We can advocate for research funding and collection of more accurate epidemiologic data [77]. Federal funding for firearm injury prevention research is less than 2% of what would be predicted based on the burden of mortality in the US [12], and the data collected by the CDC on non-fatal firearm injury lacks nuance and accuracy [78]. This shortfall directly led to the evidence gaps noted in the sections above, and leaves physicians creating best practices on suboptimal data that would not be accepted for the care of other diseases [79]. Thanks in part to physician advocacy, the CDC and NIH received $25 million in appropriations for firearm injury prevention research in 2019. This amount, however, remains insufficient and will need to be augmented in future budget years.

In addition to individual advocacy, engagement with professional societies is important, as is paying attention to how those societies are aligning their efforts with stated goals around firearm violence prevention. Although no professional society is or should be focused around a single issue [80], ED physicians and other healthcare professionals can help our membership organizations more accurately reflect the state of the evidence.

We can cooperate with local law enforcement and government entities to define and disseminate best practices for healthcare providers. A key example of this work were the efforts by pediatricians and others to oppose the so-called "Docs versus Glocks" law in Florida which restricted physicians' ability to discuss firearm access with patients [27, 81]. Another example is the collaboration between physicians and the Massachusetts Attorney General's office to develop and disseminate legal guidance on at-risk patients' storage of firearms [82].

Finally, emergency physicians have demonstrated a capacity bridge divides and should continue to do so. Some of the most transformative work to prevent firearm injury has been conducted by physicians who are working with the firearm community, including gun shop owners, to improve screening and interventions, and to disseminate collaborative approaches like Forefront and the Gun Shop Project (both of which work with gun shop owners to reduce suicide risk) [83–85]. Others are creating new organizations, such as the American Foundation for Firearm Injury Reduction in Medicine, which purposefully unite different perspectives (gun owners and non-gun-owners, rural and urban practitioners, nurses and physicians, etc.) to create innovative healthcare-based solutions to the firearm injury epidemic.

Within the field of social emergency medicine, we know the importance of social context, and we know that we have never fixed an epidemic by insisting on abstinence or purposefully diminishing the views of individuals most impacted by the public health problem. In the area of firearm injury prevention, we have an opportunity to create true change through collaboration.

Recommendations for Emergency Medicine Practice

Basic

- Understand the perspective of trauma-informed care and use patient-centered language with attention to implicit bias in caring for victims of violence [24].
- Understand the perspective of firearm owners and approach discussions with respect and a desire for shared goal-setting [36].
- Build knowledge and skills on firearm types and parts and safe storage options [22].
- Ask at-risk patients (i.e., suicidal ideation, community violence, partner violence, dementia, parents) about firearm access.
- Know local policy and resources for lethal means restriction [25, 26].
- Build skills in asking and counseling patients on lethal means restriction [22, 59].

Intermediate

- Work with local and national resources to develop programs for at-risk patients:
  - SafERteens 2.0 works to prevent youth from fighting with peers and dating partners and may aid in preventing escalation to firearm violence [51].
  - Consider implementation of national "Counseling on Access to Lethal Means" trainings [22, 59].
  - Consider SaFETy score for screening youth ages 14–24 for risk of peer or partner firearm violence in the next 2 years [52].
  - Consider wrap-around or hospital-based case management services, using Health Alliance for Violence Intervention resources as a model, recognizing the limited evidence-base for these programs and the need for further evaluation [75].
- Work with state and federal legal teams:
  - To develop guidance for colleagues on what is permissible, and what is required, when taking care of a patient at risk of firearm injury [82, 86]
  - To develop resources for victims of crime.
  - To develop guidance on the use of Extreme Risk Protection Orders (so-called "red flag laws"), if available in your state.
- Collaborate with local victim-services and social-service organizations to develop support services for at-risk patients (e.g., domestic violence and suicide prevention hotlines; community-based violence prevention programs).

Advanced

- Advocate for funding for evidence generation:
  - Research funding to inform evidence-based practices for the prevention of firearm injury and death.
  - Epidemiologic data on non-fatal firearm injury.
  - Evaluation of in-ED and community-based interventions.
- Be an active participant in professional societies' and other national organizations' firearm injury prevention efforts.
- Collaborate across the partisan divide to create a health-focused approach to firearm injury prevention.

Teaching Case

Clinical Case

It is a busy single-coverage shift in one of your partner emergency departments, located on the periphery of your city. Your next patient, a 65-year-old man, presents with the chief complaint "family forced him to come in." The triage nurse noted that "family says patient hasn't been himself since his wife died last month." Vital signs are all within normal limits. The patient is reportedly on few medications (metoprolol 25 mg daily, ASA 81 mg daily) and has no allergies. When you walk in the room, you note a man who is sitting dejectedly on the bed, with two younger family members next to him. You ask who they are, and they say that they are his daughter & son-in-law. When you ask the patient for history, he says "My family just won't leave me alone. I don't need to be here, I'm just sad. Wouldn't you be?" He answers all other questions about the physical review of systems with a terse "no." You examine him and find no acute abnormalities, other than his flat affect and slow movement. Before continuing, you get called overhead for a consultant returning your page. In your brain, you're thinking "Hmm. I have to ask a few more questions about mental health, but he's probably just grieving. I'll put in some orders to rule out the bad stuff and then head back in there in a little bit."

When you get off the phone with the consultant, you see the daughter standing nearby. She says "I'm really worried about my dad. Since my mom died, he's barely eating. He won't talk to us. The house is a mess. Can you please help him?" You ask if he's threatened to hurt himself or others, and she says "He says he wish he could just die." You explain that unless he meets the criteria for involuntary certification—or he decides he wants help, himself—you can't force care on him, and passive suicidal ideation alone is insufficient to support involuntary certification.

You go back into the room, and ask the family to step out. Bringing a stool in with you, you look the patient in the eye. "Listen, I know you're upset about your wife's death. That's normal. But your family is worried about you. I need to ask you a few more questions." He nods in assent. You start through your assessment: "How much have you been drinking? Have you used any drugs? Do you see things that aren't there, or hear voices that others don't hear?" He shakes his head no. You continue: "I've got to ask, have you had any thoughts about hurting yourself?" Here, he nods his head yes, and you gently say: "Tell me more. Have you ever tried anything? Do you have a plan for what you would do?" He shakes his head no. "OK, I have a few more questions just to help keep you safe: Do you have a firearm at home?" He nods yes, and starts to quietly cry. "What has stopped you?" you ask.

"My kids and grandkids," he says. "They wouldn't be able to handle it if they lost me, too."

Being at your community affiliate, you are able to use telehealth to complete the mental health assessment. You and the on-call mental health specialist are able to determine that the patient is the moderate risk: he has a number of strong protective factors, but also a number of strong risk factors. He does not want to be hospitalized. He doesn't meet the criteria for an involuntary hold, but he agrees to stay at his daughter's house for the next couple weeks, and to see a psychiatrist at your crisis stabilization unit the next day.

But you are still worried about his firearm access. You look up your state's laws and see that it is legal for his son-in-law to store your patient's gun at his gun range until the patient is doing better. You are relieved that you asked those few extra questions to get at your patient's true risk for gun injury.

Teaching Points

1. All suicidal patients should be asked about firearm access as part of the standard mental health assessment.
2. Just as with any health risk behavior, most patients will accept discussing firearm injury risk factors, but it is important to have this discussion in a respectful manner. This is best supported by having some familiarity with firearms and individuals' motivations for firearm ownership.

Discussion Questions

1. What barriers do you expect in putting these screening and counseling techniques into practice?
2. What resources would be helpful to you and your colleagues in reducing these barriers?
3. What needs to change in our society to facilitate attention to firearm injury prevention?

References

1. Grinshteyn E, Hemenway D. Violent death rates in the US compared to those of the other high-income countries, 2015. Preventive Medicine. 2019;123:20–6. https://doi.org/10.1016/j.ypmed.2019.02.026. https://www.sciencedirect.com/science/article/abs/pii/S0091743519300659?via%3Dihub
2. Centers for Disease Control and Prevention, National Center for Health Statistics [database on the Internet]. Compressed mortality file, 1999–2017. 2019. Available from: https://wonder.cdc.gov/cmf-icd10.html. Accessed 29 Apr 2019.

3. Active Shooter Incidents in the United States in 2016 and 2017. Washington, D.C.: Federal Bureau of Investigation, U.S. Department of Justice; 2018.
4. Blair JP, Schweit KW. A study of active shooter incidents in the United States between 2000 and 2013. Washington, D.C.: Texas State University and Federal Bureau of Investigation, U.S. Department of Justice; 2014.
5. Centers for Disease Control and Prevention, National Center for Injury Prevention and Control. Web-based Injury Statistics Query and Reporting System (WISQARS). 2016. http://www.cdc.gov/injury/wisqars/. Accessed 29 Apr 2019.
6. Petrosky E, Blair JM, Betz CJ, Fowler KA, Jack SP, Lyons BH. Racial and ethnic differences in homicides of adult women and the role of intimate partner violence—United States, 2003–2014. MMWR Morb Mortal Wkly Rep. 2017;66(28):741.
7. Sarche M, Spicer P. Poverty and health disparities for American Indian and Alaska Native children: current knowledge and future prospects. Ann N Y Acad Sci. 2008;1136:126–36. https://doi.org/10.1196/annals.1425.017.
8. Williams DR, Wyatt R. Racial bias in health care and health: challenges and opportunities. JAMA. 2015;314(6):555–6. https://doi.org/10.1001/jama.2015.9260.
9. Cubbin C, LeClere FB, Smith GS. Socioeconomic status and injury mortality: individual and neighbourhood determinants. J Epidemiol Community Health. 2000;54(7):517–24.
10. Cubbin C, LeClere FB, Smith GS. Socioeconomic status and the occurrence of fatal and non-fatal injury in the United States. Am J Public Health. 2000;90(1):70.
11. Carter PM, Cunningham RM. Adequate funding for injury prevention research is the next critical step to reduce morbidity and mortality from firearm injuries. Acad Emerg Med. 2016;23(8):952–5. https://doi.org/10.1111/acem.12982.
12. Stark DE, Shah NH. Funding and publication of research on gun violence and other leading causes of death. JAMA. 2017;317(1):84–5. https://doi.org/10.1001/jama.2016.16215.
13. Roszko PJ, Ameli J, Carter PM, Cunningham RM, Ranney ML. Clinician attitudes, screening practices, and interventions to reduce firearm-related injury. Epidemiol Rev. 2016;38(1):87–110. https://doi.org/10.1093/epirev/mxv005.
14. Oliphant S, Mouch C, Rowhani-Rahbar A, Hargarten S, Jay J, Hemenway D et al. A scoping review of patterns, motives, and risk and protective factors for adolescent firearm carriage. J Behav Med. 2019;42(4):763–10. https://doi.org/10.1007/s10865-019-00048-x.
15. Ngo Q, Sigel E, Moon A, Stein S, Massey L, Rivara F et al. State of the science: a scoping review of primary prevention of firearm injuries among children and adolescents. J Behav Med. 2019;42(4):811–29. https://doi.org/10.1007/s10865-019-00043-2.
16. Schmidt C, Rupp L, Pizarro J, Lee D, Branas C, Zimmerman M. Risk and protective factors related to youth firearm victimization, perpetration, and suicide: a scoping review and directions for future research. J Behav Med. 2019;42(4):706–23. https://doi.org/10.1007/s10865-019-00076-7.
17. Ranney M, Karb R, Ehrlich P, Bromwich K, Cunningham R, Beidas R. What are the long-term consequences of youth exposure to firearm injury, and how do we prevent them? A scoping review. J Behav Med. 2019;42(4):724–40. https://doi.org/10.1007/s10865-019-00035-2.
18. Zeoli A, Goldstick J, Mauri A, Wallin M, Goyal M, Cunningham R. The association of firearm laws with firearm outcomes among children and adolescents: a scoping review. J Behav Med. 2019;42(4):741–62. https://doi.org/10.1007/s10865-019-00063-y.
19. Ranney ML, Fletcher J, Alter H, Barsotti C, Bebarta VS, Betz ME, et al. A consensus-driven agenda for emergency medicine firearm injury prevention research. Ann Emerg Med. 2017;69(2):227–40. https://doi.org/10.1016/j.annemergmed.2016.08.454.
20. Cunningham RM, Carter PM, Ranney ML, Walton M, Zeoli AM, Alpern ER, et al. Prevention of Firearm Injuries Among Children and Adolescents: Consensus-Driven Research Agenda from the Firearm Safety Among Children and Teens (FACTS) Consortium. JAMA Pediatrics. 2019;173(8):780–9. https://doi.org/10.1001/jamapediatrics.2019.1494.
21. Ketterer AR, Ray K, Grossestreuer A, Dubosh N, Ullman E, Pirotte M. Emergency physicians' familiarity with the safe handling of firearms. West J Emerg Med. 2019;20(1):170–6. https://doi.org/10.5811/westjem.2018.11.39822.

22. Firearm Safety Among Children and Teens. Safe Storage Video. 2018. https://www.youtube.com/watch?v=tExYuQKaFec&feature=youtu.be. Accessed 29 Apr 2019.
23. Betz ME, Bebarta VS, DeWispelaere W, Barrett W, Victoroff M, Williamson K, et al. Emergency physicians and firearms: effects of hands-on training. Ann Emerg Med. 2019;73(2):210–1. https://doi.org/10.1016/j.annemergmed.2018.11.034.
24. Fischer KR, Bakes KM, Corbin TJ, Fein JA, Harris EJ, James TL, et al. Trauma-informed care for violently injured patients in the emergency department. Ann Emerg Med. 2019;73(2):193–202. https://doi.org/10.1016/j.annemergmed.2018.10.018.
25. The physician's role in promoting firearm safety. American Medical Association; 2018. https://edhub.ama-assn.org/provider-referrer/5823. Accessed 29 Apr 2019.
26. What You Can Do. UC Davis Health; 2019. https://health.ucdavis.edu/what-you-can-do/. Accessed 29 Apr 2019.
27. Wintemute GJ, Betz ME, Ranney ML. Yes, you can: physicians, patients, and firearms. Ann Intern Med. 2016;165(3):205–13. https://doi.org/10.7326/M15-2905.
28. Miller M, Barber C, White RA, Azrael D. Firearms and suicide in the United States: is risk independent of underlying suicidal behavior? Am J Epidemiol. 2013;178(6):946–55. https://doi.org/10.1093/aje/kwt197.
29. Miller M, Azrael D, Hemenway D. Household firearm ownership and suicide rates in the United States. Epidemiology. 2002;13(5):517–24. https://doi.org/10.1097/01.EDE.0000023967.88203.AE.
30. Betz ME, Miller M, Barber C, Miller I, Sullivan AF, Camargo CA Jr, et al. Lethal means restriction for suicide prevention: beliefs and behaviors of emergency department providers. Depress Anxiety. 2013;30(10):1013–20. https://doi.org/10.1002/da.22075.
31. Betz ME, Barber CW, Miller M. Firearm restriction as suicide prevention: variation in belief and practice among providers in an urban emergency department. Inj Prev. 2010;16(4):278–81. https://doi.org/10.1136/ip.2009.025296.
32. Suicide Prevention Resource Center. CALM: counseling on access to lethal means. 2018. https://www.sprc.org/resources-programs/calm-counseling-access-lethal-means. Accessed 29 Apr 2019.
33. American College of Emergency Physicians. iCAR2E: a tool for managing suicidal patients in the ED. 2018. https://www.acep.org/patient-care/iCar2e/. Accessed 29 Apr 2019.
34. Office of the Surgeon General, National Action Alliance for Suicide Prevention. 2012 National strategy for suicide prevention: goals and objectives for action: a report of the US Surgeon General and of the National Action Alliance for Suicide Prevention. 2012.
35. Capoccia L, Labre M. Caring for adult patients with suicide risk: a consensus based guide for emergency departments. Waltham: Education Development Center, Inc, Suicide Prevention Resource Center; 2015.
36. Betz ME, Wintemute GJ. Physician counseling on firearm safety: a new kind of cultural competence. JAMA. 2015;314(5):449–50. https://doi.org/10.1001/jama.2015.7055.
37. Centers for Disease Control and Prevention. Web-based injury statistics query and reporting system. Atlanta: Centers for Disease Control and Prevention; 2017. http://www.cdc.gov/injury. Accessed 10 Mar 2017.
38. Cunningham RM, Ranney M, Newton M, Woodhull W, Zimmerman M, Walton MA. Characteristics of youth seeking emergency care for assault injuries. Pediatrics. 2014;133(1):e96–e105. https://doi.org/10.1542/peds.2013-1864.
39. Chatterji P. Illicit drug use and educational attainment. Health Econ. 2006;15(5):489–511. https://doi.org/10.1002/hec.1085.
40. Ellickson P, Bui K, Bell R, McGuigan KA. Does early drug use increase the risk of dropping out of high school? J Drug Issues. 1998;28(2):357–80.
41. Bray JW, Zarkin GA, Ringwalt C, Qi JF. The relationship between marijuana initiation and dropping out of high school. Health Econ. 2000;9(1):9–18. https://doi.org/10.1002/(Sici)1099-1050(200001)9:13.0.Co;2-Z.
42. Pitts SR, Niska RW, Xu J. National Hospital Ambulatory Medical Care Survey: 2006 emergency department summary. Hyattsville: National Center for Health Statistics; 2008.

43. Grove DD, Lazebnik R, Petrack EM. Urban emergency department utilization by adolescents. Clin Pediatr (Phila). 2000;39(8):479–83.
44. McCormick MC, Stoto MA. HIV screening. Pediatrics. 2000;105(6):1375.
45. Carter PM, Walton MA, Roehler DR, Goldstick J, Zimmerman MA, Blow FC, et al. Firearm violence among high-risk emergency department youth after an assault injury. Pediatrics. 2015;135(5):805–15.
46. Cunningham RM, Carter PM, Ranney M, Zimmerman MA, Blow FC, Booth BM, et al. Violent reinjury and mortality among youth seeking emergency department care for assault-related injury. JAMA Pediatr. 2015;169(1):63. https://doi.org/10.1001/jamapediatrics.2014.1900.
47. Carter PM, Dora-Laskey A, Heinze J, Walton MA, Zimmerman MA, Cunningham RM. Longitudinal predictors of arrest among assault injured youth seeking ED care: results from the Flint youth injury study. Ann Emerg Med. Under Review.
48. Bohnert KM, Walton MA, Ranney M, Bonar EE, Blow FC, Zimmerman MA, et al. Understanding the service needs of assault-injured, drug-using youth presenting for care in an urban emergency department. Addict Behav. 2015;41:97–105.
49. Centers for Disease Control and Prevention, National Center for Injury Prevention and Control. Violence prevention in practice: hospital-community partnerships. Atlanta. https://vetoviolence.cdc.gov/apps/violence-prevention-practice/node/146#!/. Accessed 29 Apr 2019.
50. Sharp AL, Prosser LA, Walton M, Blow FC, Chermack ST, Zimmerman MA, et al. Cost analysis of youth violence prevention. Pediatrics. 2014;133(3):448–53. https://doi.org/10.1542/peds.2013-1615.
51. The University of Michigan. SafERteens 2.0. 2018. https://www.saferteens.org/home/. Accessed 29 Apr 2019.
52. Goldstick JE, Carter PM, Walton MA, Dahlberg LL, Sumner SA, Zimmerman MA, et al. Development of the SaFETy score: a clinical screening tool for predicting future firearm violence risk. Ann Intern Med. 2017; https://doi.org/10.7326/M16-1927.
53. Campbell JC, Webster D, Koziol-McLain J, Block C, Campbell D, Curry MA, et al. Risk factors for Femicide in abusive relationships: results from a multisite case control study. Am J Public Health. 2003;93(7):1089–97. https://doi.org/10.2105/ajph.93.7.1089.
54. Wintemute GJ, Wright MA, Drake CM. Increased risk of intimate partner homicide among California women who purchase handguns. Ann Emerg Med. 2003;41(2):281–3.
55. Snider C, Webster D, O'Sullivan CS, Campbell J. Intimate partner violence: development of a brief risk assessment for the emergency department. Acad Emerg Med. 2009;16(11):1208–16. https://doi.org/10.1111/j.1553-2712.2009.00457.x. ACEM457 [pii].
56. Campbell JC, Webster DW, Glass N. The danger assessment: validation of a lethality risk assessment instrument for intimate partner femicide. J Interpers Violence. 2009;24(4):653–74. https://doi.org/10.1177/0886260508317180.
57. Vigdor ER, Mercy JA. Do laws restricting access to firearms by domestic violence offenders prevent intimate partner homicide? Eval Rev. 2006;30(3):313–46. https://doi.org/10.1177/0193841X06287307.
58. Firearm Safety Among Children and Teens. Parent handout on safe storage. https://www.icpsr.umich.edu/icpsrweb/content/facts/factsheets.html. Accessed 30 Apr 2019.
59. Firearm Safety Among Children and Teens. Counseling video public. 2018. https://youtu.be/F_kFTsYu2pY. Accessed 29 Apr 2019.
60. Morgan ER, Gomez A, Rivara FP, Rowhani-Rahbar A. Household firearm ownership and storage, suicide risk factors, and memory loss among older adults: results from a statewide survey. 2019. 171(3):220–222. https://doi.org/10.7326/M18-3698.
61. Betz ME, McCourt AD, Vernick JS, Ranney ML, Maust DT, Wintemute GJ. Firearms and dementia: clinical considerations firearms and dementia. Ann Intern Med. 2018;169(1):47–9. https://doi.org/10.7326/M18-0140.

62. Cunningham R, Knox L, Fein J, Harrison S, Frisch K, Walton M, et al. Before and after the trauma bay: the prevention of violent injury among youth. Ann Emerg Med. 2009;53(4):490–500. https://doi.org/10.1016/j.annemergmed.2008.11.014. S0196-0644(08)02019-2 [pii]
63. National Institutes of Health. Statement: Preventing violence and related health-risking social behaviors in adolescents. An NIH state-of-the-science conference; 13–15 October 2004.
64. Becker MG, Hall JS, Ursic CM, Jain S, Calhoun D. Caught in the crossfire: the effects of a peer-based intervention program for violently injured youth. J Adolesc Health. 2004;34(3): 177–83.
65. Cooper C, Eslinger DM, Stolley PD. Hospital-based violence intervention programs work. J Trauma. 2006;61(3):534–7; discussion 7–40. 00005373-200609000-00002 [pii]. https://doi.org/10.1097/01.ta.0000236576.81860.8c.
66. Cheng TL, Wright JL, Markakis D, Copeland-Linder N, Menvielle E. Randomized trial of a case management program for assault-injured youth. Ped Emerg Care. 2008;24(3):130–6.
67. Karraker N, Cunningham RM, Becker MG, Fein JA, Knox LM. Violence is preventable: a best practices guide for launching & sustaining a hospital-based program to break the cycle of violence. 2011.
68. Dicker R, editor. Violence prevention for trauma centers: a feasible start [Poster 2901]. Denver: Injury and Violence in America; 2005.
69. De Vos E, Stone DA, Goetz MA, Dahlberg LL. Evaluation of a hospital-based youth violence intervention. Am J Prev Med. 1996;12(5 Suppl):101–8.
70. Zun LS, Downey L, Rosen J. The effectiveness of an ED-based violence prevention program. Am J Emerg Med. 2006;24(1):8–13.
71. Cheng TL, Haynie D, Brenner R, Wright JL, Chung SE, Simons-Morton B. Effectiveness of a mentor-implemented, violence prevention intervention for assault-injured youths presenting to the emergency department: results of a randomized trial. Pediatrics. 2008;122(5):938–46. https://doi.org/10.1542/peds.2007-2096.
72. Fein JA, Mollen CJ, Greene MB. The assault-injured youth and the emergency medical system: what can we do? Clin PediatrEmerg Med. 2013;14(1):47–55.
73. Zun LS, Downey LV, Rosen J. Violence prevention in the ED: linkage of the ED to a social service agency. Am J Emerg Med. 2003;21(6):454–7.
74. Affinati S, Patton D, Hansen L, Ranney M, Christmas AB, Violano P, et al. Hospital-based violence intervention programs targeting adult populations: an Eastern Association for the Surgery of Trauma evidence-based review. Trauma Surg Acute Care Open. 2016;1(1):e000024. https://doi.org/10.1136/tsaco-2016-000024.
75. National Network of Hospital-Based Violence Intervention Programs (NNHVIP). 2019. https://nnhvip.org/. Accessed 29 Apr 2019.
76. Ranney ML, Betz ME, Dark C. #ThisIsOurLane — firearm safety as health care's highway. N Engl J Med. 2018;380(5):405–7. https://doi.org/10.1056/NEJMp1815462.
77. Betz ME, Ranney ML, Wintemute GJ. Frozen funding on firearm research: "doing nothing is no longer an acceptable solution". Western J Emerg Med. 2016;17(1):91–3. https://doi.org/10.5811/westjem.2016.1.29767.
78. Campbell S, Nass D, Nguyen M. The CDC says gun injuries are on the rise. But there are big problems with its data. The trace 2018.
79. Taichman D, Bornstein SS, Laine C. Firearm injury prevention: AFFIRMing that doctors are in our lane. Ann Intern Med. 2018;169(12):885–6. https://doi.org/10.7326/M18-3207.
80. Cunningham RM, Zimmerman MA, Carter PM. Money, politics, and firearm safety: physician political action committees in the era of "This is Our Lane". JAMA Network Open. 2019;2(2):e187823. https://doi.org/10.1001/jamanetworkopen.2018.7823.
81. Betz ME, Ranney ML, Wintemute GJ. Physicians, patients, and firearms: the courts say "yes". Ann Intern Med. 2017;166(10):745–6. https://doi.org/10.7326/M17-0489.

82. Commonwealth of Massachusetts. Resources for talking to patients about gun safety. 2019. https://www.mass.gov/lists/resources-for-talking-to-patients-about-gun-safety.
83. University of Washington. Forefront suicide prevention. http://www.intheforefront.org/. Accessed 29 Apr 2019.
84. Suicide Prevention Resource Center. Common ground: reducing gun access. 2016. https://www.sprc.org/video/reducing-access-to-means. Accessed 29 Apr 2019.
85. Vriniotis M, Barber C, Frank E, Demicco R, the New Hampshire Firearm Safety C. A suicide prevention campaign for firearm dealers in New Hampshire. Suicide Life Threat Behav. 2015;45(2):157–63. https://doi.org/10.1111/sltb.12123.
86. Massachusetts Medical Society. Firearm violence resources: talking to patients about gun safety. 2018. http://www.massmed.org/Patient-Care/Health-Topics/Firearm-Violence-Resources/#.XMd8VOhKhdo. Accessed 30 Apr 2019.

Incarceration: The Intersection of Emergency Medicine and the Criminal Justice System

20

Susi Vassallo

Key Points
- The public is required to provide health care for prisoners who by reason of their loss of liberty cannot themselves access such care.
- The US Constitution makes health care a right for prisoners, although standards of care in prisons can be difficult to enforce.
- Emergency medicine providers are not agents of the state, corrections officials, or police. While respecting security concerns and safety, the emergency medicine provider's loyalty is to the patient.

Foundations

Background

Prisons, jails, and detention centers are examples of spaces designed to incarcerate or deprive individuals of liberty based on existing law. Detention centers are most often established in jails and prisons. The term "correctional" facilities stems from the idea that these facilities are not only for the purpose of punishment, but also for rehabilitation.

There are 1,465,200 inmates in the prison population in the US, a 1.6% decline from 2017 to the end of 2018. County and city jails held an additional 738,400 inmates nationwide at midyear in 2018, with an average inmate length of stay of 25 days [1]. The number of people passing through jails in 2018 was 10.7 million, a decline of 21% since 2008. However, this decrease is within the context of a 500%

S. Vassallo (✉)
Department of Emergency Medicine, Bellevue Hospital Center and NYU Langone Medical Center, New York City, NY, USA
e-mail: Susi.Vassallo@nyulangone.org

H. J. Alter et al. (eds.), *Social Emergency Medicine*,
https://doi.org/10.1007/978-3-030-65672-0_20

increase in incarceration over the last 40 years. The United States is the world leader of incarceration [2]. The Black imprisonment rate has fallen 21% since 2006, yet people of color remain disproportionately represented in the prison system [3]: Black and Latinx individuals comprise approximately 28% of the US population in 2017 but accounted for 56% of incarcerated people [2]. Recent criminal justice reforms have resulted in further decreases in the numbers of incarcerated individuals [3]. In addition, during the COVID-19 pandemic some nonviolent inmates have been released as a means to decrease jail crowding and lessen the danger of viral contagion to staff, inmates, and the community, although these policies are inconsistent and have not been adequate to prevent severe outbreaks and mortality in prisons [4].

Jails hold individuals pre-conviction and those unable to post bond, as well as individuals who have been sentenced to less than 1 year. In general, city or county governmental authorities administer jails. Prisons are administered by state or federal entities and hold individuals sentenced to more than 1 year. Crowding and transfers result in movement among correctional facilities. This matters to emergency medicine providers because often, diagnostic workups and treatment plans do not follow patients from one facility to another. This means that many diseases and conditions go undiagnosed, or un- or under-treated for long periods. In view of these systemic weaknesses in the criminal justice system, an emergency medicine provider will often choose to expedite a patient's care, admitting patients for needed workups that may not be completed even if recommended at the time of discharge. There is no outpatient safety net for prisoners or guarantee that the correctional facility personnel will have the resources or personnel for whatever aftercare is recommended by the emergency department.

Prisoners have a Constitutional right to health care. In 1976, in the case of Estelle v. Gamble in the US Supreme Court, a prisoner in a Texas state prison filed a civil rights action against prison officials and the chief medical officer [5]. In Estelle, the Supreme Court ruled that deliberate indifference to a prisoner's serious medical needs was cruel and unusual punishment under the Eighth Amendment of the Constitution. The Eighth Amendment provides "excessive bail shall not be required, nor excessive fines imposed, nor cruel and unusual punishments inflicted." This amendment has been interpreted by the courts to embody "broad and idealistic concepts of dignity, civilized standards, humanity, and decency" [6]. Unlike the common emergency medicine meaning of the word serious, a medical, psychiatric, or dental condition does not need to be life or limb threatening to be serious. Under the amendment, a serious medical condition is one diagnosed by a physician as mandating treatment or is so obvious that even a layperson would easily recognize the necessity for a doctor's attention. Conditions are serious if they cause pain, discomfort, or a threat to good health.

Denial or unreasonably delayed access to care, the failure to administer treatment prescribed by a physician, and the denial of a professional medical opinion are actionable under the Eighth Amendment and Fourteenth amendments of the US Constitution [7]. The patient has a right to professional, timely access to care ordered by a clinician. The Supreme Court, in *Farmer v Brennan* (Farmer v. Brennan 511 US 825 (1994)) clarified *Estelle* (Estelle v. Gamble 429 US 97 (1976)) [5, 8]. According to this decision, prison officials are liable when they know of a substantial risk of serious harm but fail to take reasonable steps to abate that risk. For example, if a prison doctor knows a patient has insulin-dependent diabetes and fails to prescribe insulin, this would be judged to put the patient at substantial risk of serious harm. Substantial means measurable or statistically significant. Holding prisoners in cells with a high heat index, for example, puts prisoners at substantial risk of serious harm [9] and violates the Constitution (Ball v. LeBlanc No 14-30067 (5th Cir. 2015)) [10].

Though prisoners have legal rights and protections, prisoners are not free to exercise these rights at will. For example, access to health care is dependent on the "sick call" system in place at a given facility. The prisoner must request health care either verbally or in writing. The written request is picked up from the cell side by either medical or security staff. The frequency of the gathering of these sick call requests (called "kites") differs in each correctional system but is usually once or twice daily. The level of training of the medical staff member who responds to the sick call request differs throughout the correctional system. Symptomatic treatment by licensed practical nurses is common, yet making a diagnosis is outside the scope of nursing practice. Treating only symptoms without a diagnosis can lead to serious complications for some patients. Sometimes staff may consider the patient's complaint a ruse to get a trip out of prison or believe the patient is complaining for other secondary gain [11]. Patients may present to prison sick call repeatedly for the same complaint and be treated only for their symptoms, without addressing their underlying disease process.

Another obstacle to care is the reporting structure. Rather than reporting to a health authority that is separate from security staff, medical staff is required to report to the warden on medical matters [12, 13]. The warden is almost never a medical professional and may have motivations or competing interests that result in failing to address gaps in medical care generally or for specific prisoners. There may be budgetary gaps or competing needs, such as the need to use prison transport vans for transportation needs other than outside appointments. The public is largely unaware of the everyday living conditions of confinement and the adequacy of medical care for prisoners [14]. Most recently, the COVID-19 pandemic has highlighted that correctional facilities are hot spots for the spread of the virus to correctional officers, inmates, and the surrounding communities [15].

Medical staffing of correctional facilities is difficult as the prestige and benefits are generally lower than medical jobs based in the community. Some staff are demoralized by staffing shortages and work conditions. There are few opportunities for continuing medical education or advancement. Staff may be disheartened by their inability to deliver a consistently high caliber of care due to resource limitations. In correctional facilities, security takes precedence over medical care, sometimes putting security personnel at odds with the mission of medical staff [12]. Prisoners may not appear for medical appointments if they are locked down or there is no one to escort the prisoner to the clinic. Emergency providers do not know and cannot anticipate how follow-up care at the facility will proceed [16].

Evidence Basis

Each year 11.5 million prisoners are released from prisons and jails in America. Ninety-five percent of prisoners in state prisons will eventually be released. Prisoners have disproportionally high rates of infectious diseases, cardiovascular disease, asthma, tuberculosis, and mental health diagnoses compared with the nation as a whole [7, 17–19]. The stark racial disparities in the current American system of mass incarceration exacerbate existing racial health disparities that are already present in the non-incarcerated population [3, 20]. When correctional facilities or emergency medicine providers deliver substandard care, both the prisoner and the public can be impacted by the consequences. As an example, approximately 1.3 to 1.4 million prisoners infected with hepatitis C were released from prison in 1996 [17]. In July of 2020, a massive COVID-19 outbreak in the San Quentin prison in California resulted in many hospitalizations and also a need to provide testing for correctional officers via the emergency department due to a lack of preparedness and testing capacity on behalf of the prison system [21]. In addition, gaps in care can result in prisoners presenting to the emergency department with advanced illnesses. For example, cardiac or gastrointestinal symptoms are sometimes treated symptomatically for months without diagnosis or necessary specialty referral. These cases come to the public's attention in class action civil actions, news reports, and in the experiences of emergency medicine providers receiving patients from correctional facilities [15, 22–24].

The burden of physical and mental illness is higher in the prison population than in the community [19]. Suicide rates in jails are high. The suicide rate in jail is higher than in the general population [25]. In nonurban jail settings, most suicide victims are White men, intoxicated, and non-violent. White men are six times more likely to die by suicide in jails than Black men and three times more likely to die by suicide than Hispanic prisoners [26]. The risk of suicide appears highest early after initial incarceration: a quarter of suicides occur in the first 24 hours after incarceration and an equal number of deaths by suicide occur in the next 2 to 14 days. The majority of suicide victims in prisons die by hanging [27]. Because emergency physicians often assess prisoners just after arrest and before arraignment, suicidal ideation must be assessed and taken with great seriousness. Cynicism regarding the young intoxicated "just arrested" patient's suicidal ideation may be deadly [22, 28, 29]. Self-injurious behavior is also disproportionately prevalent among individuals in prison [30] and includes acts such as the ingestion of foreign bodies, self-cutting, overdose, and hitting the head purposefully. Foreign bodies include paper clips, razor blades, toothbrushes, and eating utensils. Sharp objects may or may not be wrapped in tape to render them less dangerous. Many objects can be removed endoscopically. Worsening psychiatric illness is associated with increasing episodes of foreign body ingestion. The behavior of ingesting foreign objects often becomes more frequent and the number of objects ingested increases [30–32].

Excessively hot temperatures in correctional facilities are dangerous to the health of prisoners, and facility staff. By a series of class action suits, excessive temperatures in prisons have been found to present a substantial risk of serious harm to prisoners and to be unconstitutional. The increasing morbidity and mortality associated with heat is extensively documented in the scientific literature. Heat stress rises with increasing temperature and humidity, and a quick guide is the heat index as published by the National Weather Service[1]. Morbidity and mortality increase with increasing heat index. Heatstroke is not the only danger. Worsening of underlying conditions including respiratory and cardiovascular disease and mental illness is well documented [33–35].

Emergency Department and Beyond

Bedside

Communication between the emergency department (ED) and correctional facilities is limited in many systems. Because of these gaps, emergency medicine providers must take extra care to educate their patients about medical findings, and listen to the patient repeat the goals of care to assure understanding. Prisoners are not normally provided with visit results, diagnosis, and follow-up treatment plans due to security concerns that preclude divulging the exact date of a follow-up visit to an inmate. In some cases, the patient will not receive a copy of the discharge instructions and the follow-up plan, which means he/she will need to understand and remember even more than a patient who does receive a written plan. Emergency medicine providers can clearly communicate all pertinent diagnostic and follow-up information back to the facility housing the patient and thoroughly document in the medical chart, especially in cases of trauma [36]. Emphasize the follow-up care recommendations to the officers transporting the prisoner, and document all results, medical decision making, and detailed plans for follow-up care (e.g., needed laboratory or radiography studies, medication that must be obtained/administered, need for specialist consultations) on the transfer papers and clearly define the time frame within which this follow-up must occur. Legal action by prisoners against prison officials will seldom result in acceleration or timely completion of a patient's care.

The Standard of Care

Emergency medicine providers should strive to provide the same standard of care for prisoner patients as for patients who are free. For example, national guidelines such as those promulgated by the Centers for Disease Control, the American College of Emergency Physicians, the American Academy of Emergency Medicine, the

[1] http://www.noaa.gov/

American Public Health Association, the American Diabetes Association [9], the Infectious Disease Society of America, the American College of Cardiology, and the American Heart Association, apply to prisoner patients as they would apply to free patients.

Request for Body Cavity Searches

Police may come to the hospital with a prisoner and ask the emergency medicine provider to obtain blood or do a body cavity (mouth, rectum, or vagina) search for legal purposes. Emergency medicine providers are not agents of the state and should not use their skills for nonmedical purposes [37]. No provider in a therapeutic relationship with the patient should do a body cavity search for contraband or other forensic reasons. Finding contraband results in punishment for the patient and goes against the first do no harm doctrine of medicine. The patient may be placed in a cell and observed for the passage of contraband if that is a concern. Police may come to the hospital with a prisoner and ask the emergency medicine provider to obtain blood toxicology testing for legal purposes. Many patients simply consent. For those who do not, it is assault to proceed. Police may provide their own medical provider from the police department to meet the forensic needs.

It is common for emergency medicine providers to "medically clear" prisoners so that they can leave a medical ED and transfer to a psychiatric facility, go to court for arraignment, or to return to a correctional facility. Sometimes transportation to another facility may be take hours to days and once a patient is discharged, the emergency medicine provider cannot control what happens to the patient. Consider the following scenarios. A patient with Type 1 diabetes receives a dose of insulin in the ED and is returned to jail only to receive no additional insulin and to return the next day with diabetic ketoacidosis [29, 38]. A patient taking benzodiazepines presents to the ED with tachycardia and signs of withdrawal. The patient improves with the administration of benzodiazepines in the ED. The patient may return to jail and receive no more benzodiazepines and start to act strangely and be put into a padded cell, become unstable, and die due to unrecognized benzodiazepine withdrawal. A patient with uncontrolled hypertension is prescribed a medication in the emergency department yet the medication is not continued in the correctional facility. A patient with asthma improves after receiving treatment with albuterol and is medically cleared. He complains of shortness of breath in the facility but receives no care, and dies of asthma.

There is nothing in the scientific literature setting a standard for the emergency medicine provider with regard to medical clearance in the ED. Therefore, the threshold for holding an incarcerated patient in the protected environment of the hospital must be low.

The Right of Refusal of Care

Prisoners have the right to refuse medical care. This right is protected by the liberty interest of the Fourteenth Amendment and common law [39]. Refusal of care is accompanied by the clinician's assessment of the patient's mental capacity to refuse. As competence is a legal determination made by a judge in court, a clinician's assessment refers instead to "capacity" [40]. Medical treatment without the consent of the patient may constitute an assault and battery. Prisoners have the right to make "bad" or" wrong" decisions [41].

However, the right of a prisoner to refuse medical care is not absolute. When there is a strong public health reason to administer treatments, such as the treatment of active tuberculosis or other infectious diseases such as SARS-CoV-2, the right of refusal may be overridden [39]. In addition, the patient may have a severe medical condition and refuse emergency care, but the person may be hospitalized if the risk of death is considered too great to safely return to the general prison environment. An example is a patient with end-stage kidney disease who refuses dialysis; if the patient is deemed to have capacity to refuse care, he/she cannot be forced to receive dialysis but can be hospitalized. Additionally, the Supreme Court ruled that prisoners with serious mental illness may be involuntarily medicated with antipsychotic drugs if the inmate is dangerous to himself or others and the treatment is in the inmate's medical interest [42]. There are frequently extenuating circumstances that lead a prisoner to refuse care. An approaching court date or visitation by family may play into the prisoner's refusal. Emergency medicine providers must address these concerns for the patient to accept needed care.

Withholding a specific treatment in order to compel or coerce the prisoner to submit to a medically desired treatment plan (e.g., withholding pain medicine unless the prisoner agrees to a blood draw or wound closure) is unethical and may constitute deliberate indifference. The patient has a right to refuse. Fundamentally, the right of refusal is part of informed consent and the Constitution protects prisoners' right of informed consent. An informed and detailed review of the law surrounding the right of refusal is available elsewhere [39].

Privacy

Emergency medicine providers recognize that patient privacy is frequently violated in the ED [43]. This happens because of the organization of the ED, the nature of emergencies, crowding, and thoughtlessness. Information is power in correctional facilities. It may endanger the patient when another prisoner or a correctional staff member overhears private medical or other information about the prisoner during the ED stay. Such knowledge may be leveraged for the patient's disadvantage. For example, intelligence concerning sexual orientation, mental health, medical history, HIV status, or allegations of wrongdoing by police may subject the prisoner to discrimination or violence.

Indirect transmission of the type of sensitive information listed above from the provider to the correctional officer is unethical without a patient's consent or knowledge. Correctional officers do not have a right to know the patient's medical information [44]. However, with the patient's consent, in some circumstances, shared medical information provides an additional level of patient safety when correctional officials know to be alert for the symptoms of the patient's medical condition. For example, a patient's behavior that is due to an underlying medical condition can influence the responses of corrections officers and police and may be helpful information in their management of an inmate. For example, conditions that result in altered behavior and an inability to follow orders such as head injury, seizures, hypoglycemia, substance withdrawal, pain, fever, and mental illness are conditions that can result in altered behavior and an inability to follow orders, which can be dangerous in the correctional system environment.

Hospital/Healthcare System

When incarcerated individuals present to the ED repeatedly, it is usually a sign of an undiagnosed or missed condition. However, in prison, repeated presentation for the same complaint, particularly after a trip to the ED, may elicit disbelief [11]. This mindset means that an ED provider's common "come back if you get worse or develop additional symptoms" is not a realistic discharge instruction. Unlike an individual who is not incarcerated, there is no guarantee that a prisoner will be allowed to return, particularly for a subtle or ill-understood patient complaint.

Understanding the systemic limitations of access to health care must influence ED decision making for prisoner patients. It is a good rule of thumb that once the patient leaves the emergency department, the patient will be perceived by prison staff as "medically cleared." While medically clearance from the ED is a term that has no definition or standards, it is intuitively understood to mean that the patient has been evaluated and is "safe for incarceration." This highlights the responsibility emergency medicine providers have to assure that a thorough workup is completed and that specific discharge instructions are provided. As mentioned above, it is often safest for the patient to be admitted to the hospital if the natural history of the disease process is ill defined or has a significant risk of recrudescence [45]. For example, individuals withdrawing from alcohol will not have access to continued pharmaceutical treatment after being stabilized in the ED and their symptoms can recur upon discharge, so a lower threshold for admission may be appropriate in such cases.

Emergency medicine providers have an opportunity to effect improvement in the care of prisoners by reporting back to prison officials and health care providers when they identify lapses in quality of care. Most hospitals' electronic health records are not accessible by correctional facility personnel. Some correctional facilities have no electronic records for patients; others rely on a patchwork system. For example, the pharmacy medication administration record may be electronic but clinic notes may not be. In New York City's Rikers Island Jails [12], one of the largest jails in the US, the system of communication between the hospital and jail is a handwritten form that is filled out in pen, put in an envelope, handed to the corrections officer and carried back to the jail. These forms include notations such as "CT head and abdomen done" and "Medically cleared." The information contained on the form is invariably limited, with no medical decision making recorded. There is little space in which to write and most emergency medicine providers do not receive training on these forms or the limitations of the jail health care system. Hospital-based emergency medicine providers rarely have an opportunity to tour jail health care facilities, much less spend time shadowing a prison-based clinician.

Upon release from prison, individuals often face difficulty with continuity of care and obtaining necessary prescription medications. For example, they may be released with only a limited supply of medications and without a clear plan for obtaining more. This issue may be exacerbated if they lose their insurance coverage while in prison. Patients may therefore present to the emergency department, where they can be assisted in reestablishing health care in the community. In some communities, special Transitions Clinics exist to help people as they transition from incarceration to the community [46].

Societal Level

National health care guidelines apply to prisoner patients as they would apply to free patients, yet in practice, the aforementioned barriers faced by prisoners regarding timely and high-quality health care and preventive medicine make this difficult. Each year nearly 10 million prisoners are released from prisons and jails in America, and often the ED is their only clear path by which to receive follow-up care [47]. Reducing the epidemic of mass incarceration is a top priority for many policy makers, and there are also many other steps that society can take to improve care for individuals who are incarcerated. A number of cities have or are piloting innovative programs that offer access to medication-assisted treatment for prisoners with substance use disorders, social services in jail to help smooth re-entry via connection to ongoing social services, and relationships with local outpatient clinics that provide specialized care for those who have been recently released from jail or prison. In addition, most prisoners who are insured by Medicaid experience discontinuation of coverage if incarcerated for more than 90 days, and renewal can take 3 months, leaving already vulnerable individuals without insurance after release. Social services providers within prisons can restart benefits from within prison prior to release and avoid this issue.

The *Prison Litigation Reform Act* (PLRA) presents additional barriers for prisoners. PLRA is a federal law that makes it more difficult for prisoners to pursue legal claims in federal court. Before the claim will be heard, prisoners must exhaust requests to prison officials for administrative remedies. The exception occurs when the prisoner is in imminent danger such as if denied treatment for an ongoing serious medical problem or is subjected to environmental conditions that cause or aggravate such problems. In some cases, the risk of future injury may be sufficient [39]. Thus, even in a societal context, often the most expeditious and realistic means for an emergency medicine provider to improve medical care for a prisoner patient is to initiate the care of the patient [45].

Recommendations for Emergency Medicine Practice

Basic

- Similar to stroke care, sexual harassment prevention, privacy and security, and countless other required trainings, emergency medicine providers need formal training regarding key principles of the care of prisoners, including information on the system of incarceration, recidivism, and jail and prison-based health care.
- Emergency medicine providers should familiarize themselves with the standards of screening for and treatment of infectious agents that can present a substantial risk for correctional facility spread given the congregate living environments.
- Use the opportunity for one to one interaction with your patient to educate them about their diagnoses and advocate for them by assuring detailed and clear follow-up instructions are provided to their facility at discharge.

Intermediate

- Emergency medicine providers can educate themselves with books about health and incarceration. Examples include Clinical Practice in Correctional Medicine (Michael Puisis editor); Public Health Behind Bars: From Prison to Communities (Robert Greifinger editor); multiple publications from the National Commission on Correctional Health Care concerning Standards for Health Service in Prisons, Jails and Juvenile Detention Centers; the American Correctional Association (ACA) Performance-Based Standards for Correctional Health Care in Adult Correctional Institutions; and Life and Death in Rikers Island by Homer Venters, M.D.
- Interested clinicians and learners can find opportunities to learn about specialized care and models for prisoner health services provision including hospice and palliative care transitions clinics linking prisoners with health care upon release, care for and prevention of infectious diseases, mental health of juveniles, immigrant detention centers, prisons, federal, state and city authorities, civil rights organizations, security personnel and specialists, and the intersection of law and medicine.

Advanced

- The National Commission on Correctional Health offers certification in correctional health care. (CCHP, Correctional Care Health Professional).
  - Certification is a good introduction to correctional health care for emergency clinicians and can aid in understanding correctional health care standards.

  - Seek professional development in the specialty of correctional medicine: The American College of Correctional Physicians is dedicated to the professional development of physicians in the specialty (the American Board of Medical Specialties does not yet recognize correctional medicine).

    Join The Academy of Correctional Health Professionals (correctionalhealth.org), a membership partner of the NCCHC. Members receive the Journal of Correctional Health Care and the Academy Insider, a weekly e-news brief.
- Examine whether your ED or institution can develop contracts with health authorities in specific local correctional facilities that refer patients, so that a hospital-based emergency medicine provider can serve as a formal liaison.
  - The correctional health liaison must be a certified correctional health professional by the National Commission on Correctional Health Care. (NCCHC.org) and must actively practice emergency medicine.
  - The liaison physician must undergo clearance at the correctional facility, and can attend warden and health care authority meetings, facilitate communication between the correctional facility and the emergency department and create educational goals and curricula for emergency medicine clinicians.
  - The liaison has the authority to work with correctional staff to arrange for hospital-based emergency medicine providers to observe health care delivery monthly. This could include observing the chronic care clinics, sick call, pill call, pharmacy operations, and talking with prisoners.

Teaching Case

Clinical Case

A 55-year-old male prisoner presents to the hospital emergency department with a chief complaint of multiple seizures over the last 6 months, increasing in frequency. The past medical history is significant for a history of seizures secondary to a traumatic brain injury, and diabetes. He normally has only one seizure every 6 months. The patient has been compliant with his levetiracetam and metformin medication regimen. He has occasional headaches that are thought to be residual from the traumatic brain injury. In the emergency department, the patient's vital signs are normal and the neurological exam is normal. A CT scan of the brain is normal. Laboratory testing is unremarkable. The neurology service is consulted and recommends an increase in the levetiracetam dose and clinic follow-up in 1 week.

The recommendation for increasing the levetiracetam reached the prison medical staff the day after the emergency department visit. A physician assistant ordered the increased dosage from the prison pharmacy. The pharmacy technician entered the new order into the records such that 3 days later the patient would have received the increased dose at pill call. However, a weekend intervened, causing delay. The patient did not receive the new dosage. (Pill call is the distribution of medicine to prisoners at their cell or dormitory. In some instances, prisoners walk to the window of the pharmacy to receive medications directly at the window.)

Two days after the emergency department visit, the patient had another seizure in the prison. The patient was awake but nonverbal after the seizure. Officers believed the prisoner was malingering in order to get a "trip out of jail." Officers told the prisoner he was "just at the hospital" and did not need to go back. Officers called the medical team and the nurse practitioner determined the patient was fine as he was talking and interacting. She called the doctor at home and the doctor said there is no indication for the patient to go back to the hospital as his neurology appointment is rescheduled for the following month, he has already been to the emergency department at the hospital, the levetiracetam dose was increased, and the patient has had seizures and headaches for years. The doctor has no way of knowing over the phone, and does not ask for the nurse to confirm, whether the prisoner is receiving the increased dose of levetiracetam.

The next morning the patient submits a sick call request stating that he is having more seizures, and he needs a refill for Tylenol. The sick call request is reviewed and the patient is assigned to see the physician assistant holding clinic two mornings later. Two mornings later there is a brief scuffle in the dorm and the prisoners are locked down. No patient gets to sick call in the morning. Every patient is reassigned to the afternoon sick call clinic. In the afternoon, the patient refuses to come out of his cell for sick call. He tells officers that he is scheduled for family visitation. He refuses to sign a refusal form that he is refusing sick call clinic evaluation. He states he is not refusing, he just needs to make his family visit. He explains to officers that his wife has come to see him on the bus. She comes once every 2 weeks. It a four-hour bus ride each direction and they have three children at home.

The patient does not get up for breakfast in the morning. Officers bang on the bars and there is no response. A call for back up goes out over the radio and 10 minutes later several officers have arrived and enter the cell. The patient is warm to the touch and 911 is called. Officers start chest compressions. No officer has brought the automated external defibrillator to the scene. The ambulance arrives at the front gate of the prison grounds. Because it is not possible to simply open all the gates and doors, it takes the ambulance another 10 minutes to get from the front gate of the prison to the patient. Paramedics find that the patient is in asystolic cardiac arrest and begin advanced cardiac life support and transport the patient to the emergency department of the hospital. Resuscitative efforts are unsuccessful.

Teaching Points

1. Although it is the prisoner's constitutional right to care that is ordered, there are many barriers to receiving this care. The non-medical functions of prison life

influence the nature of the medical care. The delivery of medical services in the nation's prisons and jails is beset with problems and conflicts which are virtually unknown to other health care services [46].
2. Emergency medicine offers an opportunity to provide high quality, expeditious care for prisoners.
3. Do not be reassured that "someone is watching the patient." The teaching case illustrates that both a lack of both physical supervision and an assumption that a prisoner is malingering can lead to ruinous outcomes.

Discussion Questions
1. What might the emergency providers have done differently at the initial visit that could have potentially resulted in a different outcome for the patient?
2. Emergency medicine providers may voice that the patient was "just there for a trip out of jail" or is malingering for secondary gain. What techniques can help guard against this bias?
3. What are the opportunity costs to engaging in health care from the patient's perspective?

References

1. Institute for Crime and Justice Policy Research. World prison population list. 12th ed. https://www.prisonstudies.org/news/icpr-launches-12th-edition-world-prison-population-list. Accessed January 9, 2021.
2. Criminal Justice Facts | The Sentencing Project, 2020. https://www.sentencingproject.org/criminal-justice-facts/. Accessed August 16, 2021.
3. NAACP | Criminal Justice Fact Sheet, 2020. https://naacp.org/resources/criminal-justice-fact-sheet. Accessed August 16, 2021.
4. Widra E. and Wagner P. 2021. Jails and prisons have reduced their populations in the face of the pandemic, But not enough to save lives. Prison Policy Initiative. https://www.prisonpolicy.org/blog/2020/08/05/jails-vs-prisons-update-2/ accessed Aug 13, 2021.
5. Estelle v. Gamble 490 US 97 (1976).
6. The Legal Information Institute of Cornell Law School. https://www.law.cornell.edu/supremecourt/text/429/97. Accessed August 13, 2021.
7. Wilper AP, Woolhandler S, Boyd JW, Lasser KE, McCormick D, Bor DH, et al. The health and health care of US prisoners: results of a nationwide survey. Am J Public Health. 2009;99(4):666–72. https://doi.org/10.2105/AJPH.2008.144279.
8. Farmer v. Brennan 511 USA 825 (1994). *Oyez*, www.oyez.org/cases/1993/92-7247. Accessed August 16, 2021.
9. Diabetes management in correctional institutions: position statement. Diabetes Care. 2007;30 Suppl 1:S77–84. https://doi.org/10.2337/dc07-S077. Accessed August 14, 2021.
10. Ball v. LeBlanc No 14-30067. (5th Cir. 2015).
11. Greifinger RB. The acid bath of Cynicism. Correct Care. 2015;29(2).
12. Venters H. Life and death in Rikers Island. Baltimore: John Hopkins University Press; 2019.
13. US Immigration and Customs Enforcement: Performance-based national detention standards, 2011. https://www.ice.gov/doclib/detention-standards/2011/pbnds2011r2016.pdf.

14. MacDonald R PA, Venters H. The triple aims of correctional health: Patient safety, population health and human rights. J Health Care Poor Underserved. 2013;24(3):1226–34.
15. Goldensohn R. 2014. City renewed contract with Rikers health provider despite 4 inmate deaths. DNA info. https://www.dnainfo.com/new-york/20140924/east-elmhurst/state-renewed-contracts-with-rikers-health-provider-despite-4-inmate-deaths/ Accessed August 14, 2021.
16. Lewis v. Cain, CIVIL DOCKET NO.: 3:15-CV-318 (M.D. La. Mar. 31, 2021).
17. National Commission on Correctional Health Care. The Health Status of Soon to Be Released Inmates: A Report to Congress. Volume 2; Document No. 189736. September 2004. https://www.ojp.gov/pdffiles1/nij/grants/189736.pdf.
18. National Commission on Correctional Health Care. The Health Status of Soon-To-Be-Released Inmates: A Report to Congress. Volume 1; Document No. 189735, September 2004. https://www.ojp.gov/pdffiles1/nij/grants/189735.pdf.
19. Binswanger IA, Krueger PM, Steiner JF. Prevalence of chronic medical conditions among jail and prison inmates in the USA compared with the general population. J Epidemiol Community Health. 2009;63:912–19.
20. Gramlich J. Black imprisonment rate in the US has fallen by a third since 2006. Pew Research Fact Tank, May 6, 2020. https://www.pewresearch.org/fact-tank/2020/05/06/share-of-black-white-hispanic-americans-in-prison-2018-vs-2006/. Access date 1/9/21. 2020.
21. Cassidy M. San Quentin officials ignored coronavirus guidance from top Marin County Health Officer, Letter Says. San Francisco Chronicle. 2020. https://www.sfchronicle.com/crime/article/San-Quentin-officials-ignored-coronavirus-15476647.php. Accessed August 14, 2021.
22. Rohde D. City Board Members Fault St. Barnabas Hospital over prison care. New York Times. 17 September 1998.
23. Novick LF RE. A study of 128 deaths in New York City Correctional facilities 1971–1976; implications for prisoner health care. Med Care. 1978.
24. White K. Thirty -two short stories about death in prison. 2019.
25. Noonan ME. Mortality in Local Jails 2000–2014 [database on the Internet]. Bureau of justice statistics. Washington, DC: US Department of Justice 2016. Available from: BJS.gov. Accessed August 14, 2021.
26. Pope L. Delany-Brumsey A. Creating a Culture of Safety. Vera. 2016. https://www.vera.org/publications/culture-of-safety-sentinel-event-suicide-self-harm-correctional-facilities/culture-of-safety/overview. Accessed August 14, 2021.
27. Hayes LM. National study of jail suicide: 20 years later. J Correct Health Care. 2012;18(3):233–45. https://doi.org/10.1177/1078345812445457.
28. Greifinger RB. Public health behind bars from prisons to communities. https://www.springer.com/gp/book/9780387716947. Springer-Verlag New York, 2007.
29. Gill GV, MacFarlane IA. Problems of diabetics in prison. BMJ. 1989;298(6668):221–3. https://doi.org/10.1136/bmj.298.6668.221.
30. Dalal PP, Otey AJ, McGonagle EA, Whitmill ML, Levine EJ, McKimmie RL, et al. Intentional foreign object ingestions: need for endoscopy and surgery. J Surg Res. 2013;184(1):145–9. https://doi.org/10.1016/j.jss.2013.04.078.
31. Knoll IV, JL. Suicide in correctional settings: assessment, prevention, and professional liability. J Correct Health Care. 2010;16(3):188–204. https://doi.org/10.1177/1078345810366457.
32. Evans DC, Wojda TR, Jones CD, Otey AJ, Stawicki SP. Intentional ingestions of foreign objects among prisoners: A review. World journal of gastrointestinal endoscopy. 2015;7(3):162–8. https://doi.org/10.4253/wjge.v7.i3.162.
33. Cole V. Livingston. CIVIL ACTION NO. 4:14-CV-1698, S.D. Tex. Judgment Law, 2020.
34. Gates v. Cook, 376 F.3d 323 (5th Cir. 2004).
35. Roth A. Do Heat Sensitive Inmates Have a Right to Air Conditioning? NPR. July 24, 2014. https://www.npr.org/2014/07/24/334049647/do-heat-sensitive-inmates-have-a-right-to-air-conditioning. Accessed August 14, 2021.
36. Strote J, Verzemnieks E, Walsh M. Emergency department documentation of alleged excessive use of force. Am J Forensic Med Pathol. 2013;34(4):363–5.

37. Physicians for Human Rights. Dual loyalty and human rights in health professional practice. 2003. Accessed Feb 15, 2021: https://phr.org/wp-content/uploads/2003/03/dualloyalties-2002-report.pdf.
38. Keller AS, Link RN, Bickell NA, Charap MH, Kalet AL, Schwartz MD. Diabetic ketoacidosis in prisoners without access to insulin. JAMA. 1993;269(5):619–21.
39. Boston J. Prisoners' self-help litigation manual. 4th ed. Cary: Oxford University Press; 2010.
40. Appelbaum PS, Grisso T. Assessing patients' capacities to consent to treatment. N Engl J Med. 1988;319(25):1635–8. https://doi.org/10.1056/NEJM198812223192504.
41. King, K. Martin Lee, LM. Goldstein S. EMS, Capacity and Competence. Stat Pearls Publishing. 2019. https://www.ncbi.nlm.nih.gov/books/NBK470178/ Accessed August 15, 2021.
42. Washington v. Harper, 494 U.S. 210, (1990).
43. Lin YK, Lin C-J. Factors predicting patients' perception of privacy and satisfaction for emergency care. Emerg Med J. 2011;28:604–8.
44. Baker EF, Moskop JC, Geiderman JM, et al. Law enforcement and emergency medicine; an ethical analysis. Ann Emerg Med. 2016;68:599–607.
45. Hurst A, Castaneda B, Ramsdale E. Deliberate Indifference: Inadequate Health Care in U.S. Prisons. Annals of internal medicine. 2019. https://doi.org/10.7326/m17-3154.
46. Transitions Clinic. San Francisco, 2020. Transitionsclinic.org. Accessed on August 15, 2021.
47. Incarceration and Reentry. Office of the Assistant Secretary for Planning and Evaluation. https://aspe.hhs.gov/topics/human-services/incarceration-reentry-0. Accessed August 16, 2021.

Human Trafficking

21

Bryn Mumma, Wendy Macias-Konstantopoulos,
and Hanni Stoklosa

Key Points

- ED visits provide a key opportunity for recognizing and assisting trafficked individuals.
- Emergency clinicians need to be educated regarding human trafficking, recognize risk factors and indicators in their patients, and assess for human trafficking.
- Healthcare systems can and should develop guidelines, programs, and collaborations to improve the care of trafficked patients.
- Societal support systems and legal protections should be bolstered to address human trafficking.

Foundations

Background

Human trafficking is defined by the United Nations as the recruitment, transportation, transfer, harboring or receipt of persons using force, threat, or coercion for the purpose of exploiting someone for commercial sex acts or labor services [1]. Human

B. Mumma (✉)
Department of Emergency Medicine, University of California, Davis, Sacramento, CA, USA
e-mail: bemumma@ucdavis.edu

W. Macias-Konstantopoulos
Department of Emergency Medicine, Massachusetts General Hospital, Boston, MA, USA
e-mail: wmacias@mgh.harvard.edu

H. Stoklosa
Department of Emergency Medicine, Brigham and Women's Hospital, Boston, MA, USA
e-mail: hstoklosa@bwh.harvard.edu

H. J. Alter et al. (eds.), *Social Emergency Medicine*,
https://doi.org/10.1007/978-3-030-65672-0_21

trafficking involves three components: action, means, and purpose. The action may be recruitment, transportation, transfer, harboring, or receipt of persons [2]. The means used may be force, fraud, and/or coercion; they include but are not limited to isolation, emotional abuse, economic abuse, and physical violence [3, 4]. The purpose may be for either labor or commercial sex acts to benefit the trafficker. Notably, individuals younger than 18 years of age engaging in commercial sex acts are automatically considered human trafficking victims under US federal law without the burden of proving the use of force, fraud, or coercion.

Human trafficking must be distinguished from smuggling and commercial sex. Human smuggling is the illegal transport of persons across borders. Smuggled individuals voluntarily enter into an agreement with the smuggler to facilitate the illegal crossing of an international border. In contrast, human trafficking is involuntary, and it does not require that victims be transported across city, state, or international borders [5]. Individuals can be trafficked in their own homes, and they can be trafficked by someone for whom they previously consented to work. Similarly, those who initially consented to be smuggled across a border may subsequently be exploited by use of force, fraud, or coercion on arrival at their destination. In this case, what may have started as human smuggling, a movement-based crime against the state, has also become a case of human trafficking, an exploitation-based crime against another person [6]. Commercial sex is the exchange of sexual services for things of value [7]. Commercial sex may be voluntary. In contrast, individuals who are sex trafficked are either under the age of 18 or coerced, forced, or defrauded into performing these services to benefit the trafficker.

Due to the clandestine nature of trafficking, estimating the scope of human trafficking is extremely challenging. Limited data are available, and statistics are often unreliable. In a 2017 report, the International Labor Office reported that 40 million people are victims of modern slavery worldwide, of whom 25% are children and 71% are women. Women and girls are thought to comprise 58% of labor and approximately 99% of sex trafficking victims [4], although the sex trafficking of boys and men is a less recognized event. Individuals of all sexes, genders, races, and ages can be trafficked. Worldwide, the regions with the highest reported rates of trafficking are Africa, Asia, and the Pacific. According to one estimate, over 400,000 trafficked persons are estimated to live within the US [8], with trafficking reported in all 50 states [9]. Over 40,000 cases of suspected human trafficking were reported to the National Human Trafficking Hotline from 2007 to 2017, with the number increasing annually [3].

Human trafficking is a public health problem. It affects not only the health of individuals who are trafficked but also the health of communities in which it takes place. Just as other forms of violence – such as child abuse, intimate partner violence, and firearm violence – have been approached as public health issues, so should human trafficking. A public health approach examines societal, community, and individual levels of vulnerability and designs approaches to prevent trafficking from occurring and respond effectively to mitigate harm when it occurs. This approach requires an understanding of the different forms of trafficking, the root causes of trafficking, and the ways in which individuals enter trafficking [4, 10].

While individuals of all backgrounds can be trafficked, certain populations are at greater risk for human trafficking [3, 11–14]. Risk factors for human trafficking are conditions that marginalize and isolate individuals, and they overlap substantially with social determinants of health in other conditions. Information about risk for trafficking may be gathered during a standard medical and social history. Healthcare providers should recognize potential indicators of human trafficking and perform additional inquiry when they are present [15–17]. Some currently understood risk factors and indicators for human trafficking are listed in Tables 21.1a and 21.1b.

Little data exist on protective factors in human trafficking [22]. Given the overlap between human trafficking and other forms of violence, protective factors for interpersonal violence may extend to human trafficking, though likely not in all cases. These include stable and nurturing family relationships, strong friendships, caring adult role models, tangible social support, high neighborhood collective efficacy, and collaboration among community agencies [23–25].

Table 21.1a Risk factors for human trafficking [12, 15–21]

Risk factors
Recent relocation
Immigration
Substance use
Runaway youth
Homelessness
Mental health disorder
History of childhood physical or sexual abuse
LGBTQI identity
Poverty
Violence or conflict in home/community
Involvement in child protective services
Involvement in juvenile detention or delinquency

Table 21.1b Red flags/indicators for human trafficking [15–19, 21]

Indicators
Works long and/or unusual hours
Required to "check in" frequently
Owes a debt to someone that is difficult to pay off
Occupational injury due to lack of protective equipment
Appears fatigued or malnourished
Appears fearful, anxious, or submissive
Lacks identification documents
Lacks awareness of their location or date
Inconsistent or scripted history
Not allowed to speak for themselves or be alone
Repeated traumatic injuries
Tattoos that are sexually explicit or suggest ownership

Evidence Basis

Studies show 50–88% of trafficked persons in the US access medical care during their exploitation; therefore, healthcare settings provide a critical opportunity for recognizing and intervening in cases of human trafficking [26–28]. With 56–80% of trafficked individuals accessing care in the emergency department (ED), the ED is the most common source of healthcare for trafficked individuals, and identification of trafficked individuals is feasible in the emergency setting [26, 29, 30]. However, substantial knowledge gaps exist around the care of human trafficking victims among healthcare providers [29, 31–33]. While some health systems have implemented programs to address this knowledge gap and screen patients for trafficking, widespread education is necessary.

Trafficked individuals present to the ED and other healthcare settings for a variety of reasons. They have significant and complex physical and mental health needs [10, 18, 27, 34–36]. Common physical health conditions result from poor working conditions, physical violence, toxic or environmental exposures, and deprivation of basic needs such as sleep and nutrition [18, 27, 34–36]. Conditions include traumatic injuries, headaches, abdominal pain, back pain, pregnancy, and sexually transmitted infections [18, 27]. Trafficked individuals use substances, either as a coping mechanism for their situation or because their trafficker forces them to use substances [18, 27]. Depression, anxiety, and post-traumatic stress disorder are common among trafficked individuals [18, 27, 35, 36].

Emergency Department and Beyond

Bedside

Trauma-informed care is an essential part of responding to and caring for human trafficking victims. In brief, trauma-informed care is an approach to patient care that recognizes how traditional medical interviews, examinations, and settings may exacerbate the vulnerabilities of trauma survivors. It focuses on safety, trust, collaboration, and empowerment [21, 37, 38]. A trauma-informed approach is preferred for the care for all patients, especially those who have been trafficked.

When risk factors or indicators of human trafficking are present, emergency providers (EPs) should assess for human trafficking. This conversation should take place in a private space with the patient alone. If needed, a professional interpreter should be involved rather than a friend or family member, as these individuals may be a trafficker or agent of the trafficker [17, 21, 38]. Limits of confidentiality in relation to state mandatory reporting requirements for domestic violence [39], in states where these exist, and child abuse [40], including human trafficking in patients under 18 years old, should be discussed [21, 38]. The patient should be empowered to decline to answer any questions they wish not to discuss [17]. The conversation should be patient-centered, focused on identifying and meeting the patient's needs and providing education on resources rather than identifying the patient as a victim of human trafficking. Providers should use normalizing and non-judgmental language [17]. Several screening questions have been developed for both adult and pediatric patients. While these questions require validation prior to widespread implementation, they may be useful to help guide screening in the ED [17, 29, 41]. Questions should be limited to those necessary to identify the patient's medical and psychosocial needs [38]. Regardless of whether a patient discloses that they are being trafficked, the EP may always offer resources to meet the patient's needs. In the US, the National Human Trafficking Hotline (1-888-373-7888) can assist providers and patients with referrals to local resources [42]. Some patients find it easier to remember 888-ER-ER-888, since the leading "1" is unnecessary on most mobile phones. Finally, EPs should involve law enforcement only if the patient desires and with their explicit consent, as law enforcement involvement may result in the patient being arrested or deported [43]. Involving law enforcement without explicit consent can undermine trust and erode victims' perceived options for accessing future assistance and support [21].

Hospital/Healthcare System

EDs are the "safety net" for society's most vulnerable marginalized members, including those who are being trafficked. Because EDs provide care 24/7, making them accessible whenever a trafficked individual is able to seek care, and because of EMTALA obligations and the freedom from the risks of a continuity relationship, trafficked individuals may choose the ED for medical care. Helping them safely obtain medical services need not fall on clinicians alone; many EDs also have social workers or discharge planners who can assist trafficked individuals during their visit.

However, some characteristics of EDs make them less effective at identifying and caring for trafficked individuals, inadvertently aiding traffickers who prefer that the victimization remain undetected. As EDs become more crowded, care is provided in hallways and other non-private locations. Visitors are generally allowed to accompany patients during their ED stay. Both of these factors may lead physicians to modify their history and physical exam such that they fail to detect or address features relevant to human trafficking and may make it more challenging to question an at-risk individual privately [44]. EPs often face time pressures, making it difficult for them to find time to engage in a conversation about abuse, including human trafficking, with a patient. Patients frequently see a different provider during each ED visit, preventing the development of rapport and trust [17] that are critical to trafficked individuals disclosing and discussing their situations.

Despite these challenges, several EDs and hospital systems have implemented programs to identify and assist trafficked individuals [45, 46]. While these programs vary among hospitals, reputable programs share some characteristics. Programs are either led or informed by the experiences of human trafficking survivors [47]. Education and training on human trafficking is provided to all members of the healthcare team to ensure universal awareness of the indicators of and response to human trafficking. A protocol or guideline with key steps in assessment as well as reporting requirements and referral information is created for providers to use as a reference [17, 48]. Some hospital systems have incorporated trafficking assessment into their general assessments for abuse [49]. One such system incorporated a silent notification method, such as placing a designated sticker on their urine specimen cup, that allows patients to indicate they would like to speak with a provider alone about confidential concerns [50].

Lack of coordinated medical care between the ED and outpatient setting and between physical and psychiatric healthcare providers can make it challenging for trafficked individuals to obtain care for all of their healthcare needs, which include physical health, mental health, and social needs [51]. While integrated clinics that address the unique needs of trafficked individuals exist, most trafficked individuals in other health systems need to visit several providers and agencies to obtain these services [52, 53].

The addition of diagnostic codes specific to human trafficking in the International Classification of Diseases, Tenth Revision, Clinical Modification (ICD-10-CM) [54] presents both opportunities and challenges for EPs and healthcare systems. ICD codes are widely used to identify patients with specific conditions for research studies and epidemiologic surveillance, and inclusion of codes specific for human trafficking allows this population to be identified and studied [55]. EDs and hospitals may use these codes to track the impact of programs to identify and assist patients who are being trafficked and to uncover patterns in the presentations of trafficked individuals. Public health officials may use these codes to monitor human trafficking at a local or regional level and target prevention efforts. At the level of the healthcare provider, ICD-10-CM codes for human trafficking may facilitate communication among providers and be another reminder for providers to use trauma-informed principles when approaching these patients. However, as has been the case for survivors of sexual assault or intimate partner violence, the codes must be used with caution, to avoid stigmatizing patients with a diagnosis of human trafficking and to avoid risking the patient's safety if a trafficker views the diagnosis of human trafficking [55]. When applied carefully and appropriately, ICD-10-CM codes for human trafficking have the potential to provide valuable data to improve care for trafficked individuals.

Societal Level

Social, cultural, and economic factors facilitate human trafficking, and addressing these factors is an important step in preventing human trafficking. At-risk children and young adults should be supported and educated regarding trafficking and traffickers' tactics, using evidence-based strategies rooted in public health [16]. Poverty, homelessness, unemployment, and natural disasters create circumstances that can lead to human trafficking or hinder exiting a trafficking situation [4, 52]. Social protection systems that provide food, housing, and basic needs should be developed and expanded to prevent vulnerable individuals from being trafficked or retrafficked if attempting to exit [4, 52, 56]. Reducing the social stigma and improving access to healthcare and social services for individuals with mental health disorders, substance use disorders, and LGBTQI identities may reduce the risk of these individuals being trafficked and facilitate exit and recovery from trafficking [52]. Judicial efforts to expunge the criminal records of trafficking victims forced to engage in illegal activities while being trafficked are an important intervention in ensuring the employability and well-being of survivors [52]. Based on existing data, women are disproportionately affected by trafficking, and fostering gender equality in society is another strategy to reduce trafficking in women.

Legal protections can also be used to prevent human trafficking and protect those who are trafficked. Many individuals are trafficked in an informal economy that is not regulated by the government and thus are not subject to minimum wage laws, mandated breaks, workplace safety regulations, and other worker protections. Extending labor and social rights into the informal labor sector is another strategy to mitigate human trafficking [4]. In the US, the Victims of Trafficking and Violence Protection Act of 2000 established human trafficking as a federal crime and mandated that restitution be paid to trafficking victims [2]. It created pathways for prosecuting traffickers and protecting victims. In some situations, a pathway to legal residence is available for undocumented immigrants who are trafficking victims. This pathway is particularly important because concerns regarding immigration status and fear of deportation are key reasons that trafficked individuals do not disclose their trafficking to a healthcare provider [57].

Recommendations for Emergency Medicine Practice

Basic

- EPs should educate themselves about human trafficking so that they recognize risk factors and indicators for human trafficking in their patients. In some states, this education is a requirement for licensure [58].
- When risk factors and indicators are present, EPs may be prompted to assess for human trafficking using a trauma-informed approach.
- When trafficking is suspected or confirmed, EPs can refer adults to the National Human Trafficking Hotline or appropriate local resources. They should follow state reporting requirements, including federally mandated reporting for minors [Justice for Victims of Trafficking Act of 2015].
- EDs should display information on human trafficking and related resources in language accessible to their patient population. In some states, EDs are required by law to display posters containing this information [MO § HB1246; CA § AB2034].

Intermediate

- EDs can and should develop a program for educating and training all staff – including but not limited to physicians, nurses, social workers, security officers, registration staff, and technicians – to recognize indicators of human trafficking. Staff at any level can be empowered then to assess for human trafficking themselves or to share their concerns with another EP.
- Medical schools and teaching hospitals should include education on human trafficking for their students and trainees [59, 60].
- EDs should have policies and procedures to ensure that all patients receive appropriate care and resources and that EPs comply with mandatory reporting requirements [61].
- In areas where human trafficking is known to be prevalent, EDs and health systems should establish a multidisciplinary work group to evaluate and improve the department and health system response to human trafficking [62].

Advanced

- EDs and health systems have worked with local medical and community organizations to coordinate referrals to medical, psychiatric, and social services for trafficked individuals. They can also work with local partners to understand the epidemiology and characteristics of trafficking in their community.
- EDs and health systems should support research and quality improvement efforts to improve the healthcare and outcomes of trafficked individuals.
- Emergency medicine professional organizations should take a stance against human trafficking, such as the policy statement by the American College of Emergency Physicians [63], and include programming related to human trafficking.
- EPs can advocate for increased resources to prevent trafficking and meet the myriad complex social and legal needs of trafficking survivors.
- EPs can join professional groups focused on addressing human trafficking, such as HEAL Trafficking and the Social Emergency Medicine Section of the American College of Emergency Physicians.

Teaching Case

Clinical Case

A 23-year-old transgender male with past medical history significant for bipolar disorder, migraine headaches, gastroesophageal reflux, and methamphetamine use presents to the ED with a chief complaint of suicidal ideation and anxiety. He endorses feeling very sad and anxious, with these feelings worsening over the past week. He endorses difficulty sleeping. He has been living in a house with a family and has limited access to food and money. He has also been forced to use methamphetamine and cocaine. He feels that he has nothing to live for, and he plans to take all of his medication to end his pain. He overdosed once before in a suicide attempt. He denies current access to firearms.

He also endorses right knee pain from an assault several weeks ago. Following the assault, he was evaluated in the ED. At that time, he was given a knee immobilizer that he no longer has. He describes the pain as a dull ache that is non-radiating, worse with ambulation and better with rest.

His current medications include bupropion, omeprazole, quetiapine, and sumatriptan. He has not been taking any of his medications recently because he has been unable to afford them.

His social history is remarkable for a history of domestic violence, for which he has previously received services through a local organization.

On physical examination, his vital signs are within normal limits. He appears his stated age and is sitting comfortably on the gurney. His musculoskeletal exam is remarkable for right knee tenderness to palpation with no effusion or ecchymosis, no instability, and normal range of motion. His psychiatric exam is remarkable for poor eye contact. His thought processes are linear. He endorses suicidal ideation, primarily related to getting away from his living situation in which he feels trapped. He exhibits some paranoia and impulsivity but has adequate insight and judgment. He exhibits splitting behavior between members of the healthcare team.

Recognizing his risk factors and indicators of human trafficking, you ask additional questions to explore the possibility that he is being trafficked. On further conversation, he states that he moved in with this family a year ago, when they "adopted" him. He initially trusted them but now feels betrayed. He states this family has forced him to have sex with other men.

Both the psychiatry and social work services are consulted. The patient feels that he would no longer be suicidal if he had an alternate, safe place to live. The psychiatrist recommends close outpatient psychiatric care, and the social worker is able to secure placement in a local residential program for human trafficking victims. He declines to file a police report or involve law enforcement at this time.

Teaching Points

1. This patient has risk factors for human trafficking including LGBTQI status, mental health disorder, and history of domestic violence. Indicators of human trafficking include multiple ED visits related to physical violence and mental health conditions, limited access to food and money, being forced to use drugs, and being coerced by his adoptive "family" to have sex with other men.
2. Like many trafficked individuals, this patient had multiple prior presentations to the ED during which his immediate needs were addressed but the underlying issue of human trafficking was not recognized. He also exhibits both physical and mental health manifestations of human trafficking, and he has co-existing psychiatric disease and substance use that likely both contributed to and resulted from his trafficking.

3. This patient's splitting behavior between members of the healthcare team could lead to him being considered a "difficult" patient. EPs should remember to use trauma-informed principles and consider the experiences that may have led this patient to interact with them in this way. Providers should allow time for the patient to develop trust and create an environment for the patient to disclose information when the patient feels comfortable doing so.
4. When trafficked adults do not want to leave their trafficker or seek help immediately, EPs should respect this decision. When trafficked adults do want assistance, such as in this case, EPs should engage local resources to meet patients' needs. In the US, EPs can give patients the National Human Trafficking Hotline information or assist them in memorizing the phone number if they feel it would be unsafe to have on their person.

Discussion Questions

1. How were the three key elements that comprise human trafficking manifest in this case?
2. What risk factors and indicators for human trafficking are present in this patient's case?
3. What resources did the EP employ to address the patient's needs in this case? If psychiatry and social work had not been immediately available, what resources could the EP have offered to the patient?

References

1. United Nations Office on Drugs and Crime. United Nations convention against transnational organized crime and the protocols thereto. New York: United Nations; 2004. https://www.unodc.org/documents/treaties/UNTOC/Publications/TOC%20Convention/TOCebook-e.pdf. Accessed 6 Feb 2019.
2. Victims of Trafficking and Violence Protection Act of 2000, H.R. 3244. 2000.
3. Polaris Project. Growing awareness. Growing impact. 2017 Statistics from the national human trafficking hotline and beFree textline 2017. http://polarisproject.org/sites/default/files/2017NHTHStats%20%281%29.pdf. Accessed 27 July 2018.
4. International Labour Organization and Walk Free Foundation. Global estimates of modern slavery: forced labour and forced marriage. Geneva; 2017.
5. United States Department of State. Human trafficking and migrant smuggling: Understanding the difference. 2017. https://www.state.gov/documents/organization/272325.pdf. Accessed 28 Mar 2019.
6. United States Department of State. Trafficking in persons report. 2015.
7. Overs C. Sex workers: part of the solution. 2002. https://www.who.int/hiv/topics/vct/sw_toolkit/115solution.pdf. Accessed 27 Mar 2019.
8. Global Slavery Index. Country data: United States. https://www.globalslaveryindex.org/2018/data/country-data/united-states/. Accessed 28 Mar 2019.
9. UNICEF. End trafficking https://www.unicefusa.org/sites/default/files/assets/pdf/End-Child-Trafficking-One-Pager.pdf. Accessed 27 July 2018.
10. Chisolm-Straker M, Stoklosa H, editors. Human trafficking is a public health issue: a paradigm shift in the United States. Cham: Springer International Publishing; 2017. p. 185–210.
11. Choi KR. Risk factors for domestic minor sex trafficking in the United States: a literature review. J Forensic Nurs. 2015;11(2):66–76.

12. Franchino-Olsen H. Vulnerabilities relevant for commercial sexual exploitation of children/domestic minor sex trafficking: a systematic review of risk factors. Trauma Violence Abus 2019:1524838018821956.
13. O'Brien JE, White K, Rizo CF. Domestic minor sex trafficking among child welfare-involved youth: an exploratory study of correlates. Child Maltreat. 2017;22(3):265–74.
14. Estes RJ, Weiner NA. The commercial sexual exploitation of children in the U. S., Canada/Mexico. 2002. https://abolitionistmom.org/wp-content/uploads/2014/05/Complete_CSEC_0estes-weiner.pdf. Accessed 28 Mar 2019.
15. Polaris Project. Recognize the signs. https://polarisproject.org/human-trafficking/recognize-signs. Accessed 28 Mar 2019.
16. UNITAS. Human trafficking 101. https://www.unitas.ngo/human-trafficking-101. Accessed 28 Mar 2019.
17. United States Department of Health and Human Services. Adult human trafficking screening toolkit and guide. 2018. https://www.acf.hhs.gov/otip/resource/nhhtacadultscreening. Accessed 8 Mar 2019.
18. Macias-Konstantopoulos W. Human Trafficking: the role of medicine in interrupting the cycle of abuse and violence. Ann Intern Med. 2016;165(8):582–8.
19. Shandro J, Chisolm-Straker M, Duber HC, Findlay SL, Munoz J, Schmitz G, et al. Human trafficking: a guide to identification and approach for the emergency physician. Ann Emerg Med. 2016;68(4):501–8.e1.
20. Institute of Medicine and National Research Council. Confronting commercial sexual exploitation and sex trafficking of minors in the United States: a guide for providers of victim and support services. Washington (DC): National Academies Press; 2014.
21. Alpert EJ, Ahn R, Albright E, Purcell G, Burke TF, Macias-Konstantopoulos WL. Human trafficking: guidebook on identification, assessment, and response in the health care setting. Waltham: Massachusetts General Hospital and Massachusetts Medical Society; 2014.
22. Rothman EF, Stoklosa H, Baldwin SB, Chisolm-Straker M, Kato Price R, Atkinson HG. Public health research priorities to address US human trafficking. Am J Public Health. 2017;107(7):1045–7.
23. Centers for Disease Control and Prevention. Intimate partner violence: risk and protective factors for perpetration 2018. https://www.cdc.gov/violenceprevention/intimatepartnerviolence/riskprotectivefactors.html. Accessed 29 Mar 2019.
24. Centers for Disease Control and Prevention. Child abuse & neglect: risk and protective factors 2019. https://www.cdc.gov/violenceprevention/childabuseandneglect/riskprotectivefactors.html. Accessed 29 Mar 2019.
25. Chisolm-Straker M, Sze JK, Einbond J, White J, Stoklosa H. A supportive adult may be the difference in homeless youth not being trafficked. Child Youth Serv Rev. 2018;91:115–20.
26. Chisolm-Straker M, Baldwin S, Gaigbe-Togbe B, Ndukwe N, Johnson PN, Richardson LD. Health care and human trafficking: we are seeing the unseen. J Health Care Poor U. 2016;27(3):1220–33.
27. Lederer LJ, Wetzel CA. The health consequences of sex trafficking and their implications for identifying victims in healthcare facilities. Ann Health Law. 2014;23(1):61–91.
28. Baldwin SB, Eisenman DP, Sayles JN, Ryan G, Chuang KS. Identification of human trafficking victims in health care settings. Health Hum Rights. 2011;13(1):E36–49.
29. Mumma BE, Scofield ME, Mendoza LP, Toofan Y, Youngyunpipatkul J, Hernandez B. Screening for victims of sex trafficking in the emergency department: a pilot program. West J Emerg Med. 2017;18(4):616–20.
30. Greenbaum VJ, Dodd M, McCracken C. A short screening tool to identify victims of child sex trafficking in the health care setting. Pediatr Emerg Care. 2018;34(1):33–7.
31. Beck ME, Lineer MM, Melzer-Lange M, Simpson P, Nugent M, Rabbitt A. Medical providers' understanding of sex trafficking and their experience with at-risk patients. Pediatrics. 2015;135(4):e895–902.
32. Chisolm-Straker M, Richardson LD, Cossio T. Combating slavery in the 21st century: the role of emergency medicine. J Health Care Poor U. 2012;23(3):980–7.

33. Titchen KE, Loo D, Berdan E, Rysavy MB, Ng JJ, Sharif I. Domestic sex trafficking of minors: medical student and physician awareness. J Pediatr Adol Gynec. 2017;30(1):102–8.
34. Pocock NS, Tadee R, Tharawan K, Rongrongmuang W, Dickson B, Suos S, et al. "Because if we talk about health issues first, it is easier to talk about human trafficking"; findings from a mixed methods study on health needs and service provision among migrant and trafficked fishermen in the Mekong. Globalization Health. 2018;14(1):45.
35. Oram S, Stockl H, Busza J, Howard LM, Zimmerman C. Prevalence and risk of violence and the physical, mental, and sexual health problems associated with human trafficking: systematic review. PLoS Med. 2012;9(5):e1001224.
36. Kiss L, Pocock NS, Naisanguansri V, Suos S, Dickson B, Thuy D, et al. Health of men, women, and children in post-trafficking services in Cambodia, Thailand, and Vietnam: an observational cross-sectional study. Lancet Glob Health. 2015;3(3):e154–61.
37. Wilson C, Pence DM, Conradi L. Trauma-informed care. 2013. http://socialwork.oxfordre.com/view/10.1093/acrefore/9780199975839.001.0001/acrefore-9780199975839-e-1063. Accessed 27 July 2018.
38. Macias-Konstantopoulos WL. Caring for the trafficked patient: ethical challenges and recommendations for health care professionals. AMA J Ethics. 2017;19(1):80–90.
39. Durborow N, Lizdas KC, O'Flaherty A. Compendium of state statutes and policies on domestic violence and health care. Family Violence Prevention Fund; 2010. https://www.acf.hhs.gov/sites/default/files/fysb/state_compendium.pdf. Accessed 29 Mar 2019.
40. Child Welfare Information Gateway. Mandatory reporters of child abuse and neglect. https://www.childwelfare.gov/pubPDFs/manda.pdf. 29 Mar 2019.
41. Armstrong S. Instruments to identify commercially sexually exploited children: feasibility of use in an emergency department setting. Pediatr Emerg Care. 2017;33(12):794–9.
42. National human trafficking hotline. https://humantraffickinghotline.org/. Accessed 29 Mar 2019.
43. Stoklosa H, Baldwin SB. Health professional education on trafficking: the facts matter. Lancet. 2017;390(10103):1641–2.
44. Stoklosa H, Scannell M, Ma Z, Rosner B, Hughes A, Bohan JS. Do EPs change their clinical behaviour in the hallway or when a companion is present? A cross-sectional survey. Emerg Med J. 2018;35(7):406–11.
45. Stoklosa H, Dawson MB, Williams-Oni F, Rothman EF. A review of U.S. health care institution protocols for the identification and treatment of victims of human trafficking. J Hum Trafficking. 2017;3(2):116–24.
46. Stoklosa H, Showalter E, Melnick A, Rothman EF. Health care providers' experience with a protocol for the identification, treatment, and referral of human-trafficking victims. J Hum Trafficking. 2017;3(3):182–92.
47. Dignity Health. Human trafficking response: survivor-led and survivor-informed. https://www.dignityhealth.org/hello-humankindness/human-trafficking/survivor-led-and-survivor-informed. Accessed 29 Mar 2019.
48. HEAL Trafficking. Protocols. 2019. https://healtrafficking.org/protocols-committee/. Accessed 29 Mar 2019.
49. Dignity Health. Human trafficking response: victim-centered and Trauma-informed. https://www.dignityhealth.org/hello-humankindness/human-trafficking/victim-centered-and-trauma-informed. Accessed 15 May 2020.
50. Egyud A, Stephens K, Swanson-Bierman B, DiCuccio M, Whiteman K. Implementation of human trafficking education and treatment algorithm in the emergency department. J Emerg Nurs. 2017;43(6):526–31.
51. Macias Konstantopoulos W, Ahn R, Alpert EJ, Cafferty E, McGahan A, Williams TP, et al. An international comparative public health analysis of sex trafficking of women and girls in eight cities: achieving a more effective health sector response. Journal Urban Health. 2013;90(6):1194–204.

52. Judge AM, Murphy JA, Hidalgo J, Macias-Konstantopoulos W. Engaging survivors of human trafficking: complex health care needs and scarce resources. Ann Intern Med. 2018;168(9):658–63.
53. Powell C, Asbill M, Louis E, Stoklosa H. Identifying gaps in human trafficking mental health service provision. J Hum Trafficking. 2018;4(3):256–69.
54. National Center for Health Statistics. International classification of diseases, 10th revision, clinical modification (ICD-10-CM). https://www.cdc.gov/nchs/icd/icd10cm.htm. Accessed 29 Mar 2019.
55. Macias-Konstantopoulos WL. Diagnosis codes for human trafficking can help assess incidence, risk factors, and comorbid illness and injury. AMA J Ethics. 2018;20(12):E1143–51.
56. Long E, Reid J, McLeigh J, Stoklosa H, Felix E, Scott T. Preventing human trafficking using data-driven, community-based strategies. https://www.communitypsychology.com/preventing-human-trafficking/. Accessed 26 Apr 2019.
57. Restore NYC. Healthcare access for foreign-national survivors of trafficking. 2019. https://static1.squarespace.com/static/59d51bdb6f4ca3f65e5a8d07/t/5c705af74e17b658d074c7fc/1550867206256/Healthcare+Access_Restore+2019.pdf. Accessed 26 Apr 2019.
58. Atkinson HG, Curnin KJ, Hanson NC. U.S. State laws addressing human trafficking: education of and mandatory reporting by health care providers and other professionals. J Hum Trafficking. 2016;2(2):111–38.
59. Stoklosa H, Lyman M, Bohnert C, Mittel O. Medical education and human trafficking: using simulation. Med Educ Online. 2017;22(1):1412746.
60. Cole MA, Daniel M, Chisolm-Straker M, Macias-Konstantopoulos W, Alter H, Stoklosa H. A theory-based didactic offering physicians a method for learning and teaching others about human trafficking. Acad Emerg Med Educ Tr. 2018;2(Suppl 1):S25–s30.
61. HEAL Trafficking. HEAL trafficking and hope for justice's protocol toolkit. 2019. https://healtrafficking.org/2017/06/new-heal-trafficking-and-hope-for-justices-protocol-toolkit-for-developing-a-response-to-victims-of-human-trafficking-in-health-care-settings/. Accessed 29 Apr 2019.
62. Powell C, Dickins K, Stoklosa H. Training US health care professionals on human trafficking: where do we go from here? Med Educ Online. 2017;22(1):1267980.
63. American College of Emergency Physicians. Policy statement: human trafficking. 2016. https://www.acep.org/patient-care/policy-statements/human-trafficking/. Accessed 29 Mar 2019.

Index

[1] Page numbers followed by “t” denote tables.

H. J. Alter et al. (eds.), *Social Emergency Medicine*,
https://doi.org/10.1007/978-3-030-65672-0

T